# INTELLIGENT ROBOTIC SYSTEMS FOR SPACE EXPLORATION

## THE KLUWER INTERNATIONAL SERIES IN ENGINEERING AND COMPUTER SCIENCE

ROBOTICS: VISION, MANIPULATION AND SENSORS

*Consulting Editor:* **Takeo Kanade**

***PERCEPTUAL ORGANIZATION AND VISUAL RECOGNITION,*** D. Lowe
ISBN: 0-89838-172-X
***ROBOT DYNAMICS ALGORITHMS,*** F. Featherstone
ISBN: 0-89838-230-0
***THREE- DIMENSIONAL MACHINE VISION,*** T. Kanade (editor)
ISBN: 0-89838-188-6
***KINEMATIC MODELING, IDENTIFICATION AND CONTROL OF ROBOT MANIPULATORS,*** H.W. Stone
ISBN: 0-89838-237-8
***OBJECT RECOGNITION USING VISION AND TOUCH,*** P. Allen
ISBN: 0-89838-245-9
***INTEGRATION, COORDINATION AND CONTROL OF MULTI-SENSOR ROBOT SYSTEMS,*** H.F. Durrant-Whyte
ISBN: 0-89838-247-5
***MOTION UNDERSTANDING: Robot and Human Vision,*** W.N. Martin and J. K. Aggrawal (editors)
ISBN: 0-89838-258-0
***BAYESIAN MODELING OF UNCERTAINTY IN LOW-LEVEL VISION,*** R. Szeliski
ISBN 0-7923-9039-3
***VISION AND NAVIGATION: THE CMU NAVLAB,*** C. Thorpe (editor)
ISBN 0-7923-9068-7
***TASK-DIRECTED SENSOR FUSION AND PLANNING: A Computational Approach,*** G. D. Hager
ISBN: 0-7923-9108-X
***COMPUTER ANALYSIS OF VISUAL TEXTURES,*** F. Tomita and S. Tsuji
ISBN: 0-7923-9114-4
***DATA FUSION FOR SENSORY INFORMATION PROCESSING SYSTEMS,*** J. Clark and A. Yuille
ISBN: 0-7923-9120-9
***PARALLEL ARCHITECTURES AND PARALLEL ALGORITHMS FOR INTEGRATED VISION SYSTEMS,*** A.N. Choudhary, J. H. Patel
ISBN: 0-7923-9078-4
***ROBOT MOTION PLANNING,*** J. Latombe
ISBN: 0-7923-9129-2
***DYNAMIC ANALYSIS OF ROBOT MANIPULATORS: A Cartesian Tensor Approach,*** C.A Balafoutis, R.V. Patel
ISBN: 07923-9145-4
***PERTURBATION TECHNIQUES FOR FLEXIBLE MANIPULATORS:*** A. Fraser and R. W. Daniel
ISBN: 0-7923-9162-4
***COMPUTER AIDED MECHANICAL ASSEMBLY PLANNING:*** L. Homen de Mello and S. Lee
ISBN: 0-7923-9205-1

# INTELLIGENT ROBOTIC SYSTEMS FOR SPACE EXPLORATION

*Edited by*


**Alan A. Desrochers**
*Center for Intelligent Robotic Systems for Space Exploration*
Rensselaer Polytechnic Institute



KLUWER ACADEMIC PUBLISHERS
Boston/London/Dorcrecht

**Distributors for North America:**
Kluwer Academic Publishers
101 Philip Drive
Assinippi Park
Norwell, Massachusetts 02061 USA

**Distributors for all other countries:**
Kluwer Academic Publishers Group
Distribution Centre
Post Office Box 322
3300 AH Dordrecht, THE NETHERLANDS

**Library of Congress Cataloging-in-Publication Data**

Intelligent robotic systems for space exploration / edited by Alan A.
Desrochers.
p. cm. -- (The Kluwer international series in engineering and
computer science ; SECS 168)
Includes Bibliographical references and index.
ISBN 0-7923-9197-7
1. Large space structures (Astronautics) -- Design and construction.
2. Robots--Control Systems. 3. Assembling machines.
I. Desrochers, A.A. (Alan A.) II. Series
TL875.I58 1992
629.47'3--dc20 91-44493
CIP



*Printed on acid-free paper.*

Printed in the United States of America

# Contents


**List of Figures** xi

**List of Tables** xv

**Contributors** xvii

**Preface** xix

**Acknowledgments** xxi

**1 Testbed for Cooperative Robotic Manipulators** 1

1.1 Introduction . . . . . 1
1.2 Purposes of the CIRSSE Testbed . . . . . 4
1.3 Space-Based Assembly . . . . . 4
1.4 Issues and Results for Space-Based Robotic Applications . . . . . 5
1.4.1 Lighting Characteristics of Space . . . . . 5
1.4.2 Zero-Gravity . . . . . 6
1.4.3 Integration . . . . . 6
1.4.4 Theory of Intelligent Machines . . . . . 8
1.4.5 Task Planning, Path Planning, and Collision Avoidance . 9
1.4.6 Control . . . . . 10
1.4.7 Remote Control . . . . . 11
1.4.8 Dual Arm Control . . . . . 11
1.4.9 Sensing . . . . . 12
1.4.10 Error Recovery . . . . . 12
1.4.11 Visual and Force Servoing . . . . . 13
1.5 Physical Description of the CIRSSE Testbed . . . . . 13
1.5.1 Manipulator Subsystems . . . . . 14
1.5.2 Sensor Subsystems . . . . . 16
1.5.3 Processor Subsystems . . . . . 17
1.5.4 Mobile Robot . . . . . 19
1.6 Design of the MCS for the CIRSSE Testbed . . . . . 20
1.6.1 Purpose of the MCS . . . . . 20

1.6.2 Functional Requirements . . . . . 22
1.6.3 Manpower Requirements . . . . . 27
1.7 High-Level Control of the CIRSSE Testbed . . . . . 27
1.7.1 Coordination Level of Intelligent Machines . . . . . 28
1.7.2 Petri Net Implementation . . . . . 29
1.7.3 Coordination Level Research . . . . . 31
1.8 Summary . . . . . 31
1.9 References . . . . . 32

**2 Automated Assembly** **39**
2.1 Introduction . . . . . 40
2.2 Facility and Hardware Description . . . . . 43
2.2.1 Robot and Motion Bases . . . . . 46
2.2.2 Truss and Joining Mechanism . . . . . 46
2.2.3 End-Effector . . . . . 49
2.2.4 Strut Storage . . . . . 51
2.2.5 System Control and Communication . . . . . 57
2.2.6 Operator Control . . . . . 57
2.3 Assembly Operations . . . . . 60
2.3.1 Strut Identification Convention . . . . . 60
2.3.2 Strut Pick-up From the Canister . . . . . 61
2.3.3 Motion Base Moves . . . . . 63
2.3.4 Robot Paths and Capture Sequence . . . . . 64
2.3.5 End-Effector Installation Operations . . . . . 66
2.3.6 Operator Pause and Reverse . . . . . 67
2.3.7 Error Recovery . . . . . 67
2.3.8 Tray Transfer Operations . . . . . 68
2.3.9 Truss Disassembly . . . . . 68
2.4 Software Structure . . . . . 69
2.4.1 Executive Program . . . . . 69
2.4.2 Robot Program . . . . . 76
2.4.3 System Data Structures . . . . . 76
2.4.4 End-Effector Software . . . . . 79
2.5 Tests . . . . . 80
2.6 Current Test Observations and Results . . . . . 82
2.6.1 Preliminary Assembly Results . . . . . 82
2.6.2 Test Observation Results . . . . . 86
2.7 Future Research Opportunities . . . . . 90
2.7.1 Vision System . . . . . 90
2.7.2 Installation of Panels . . . . . 95
2.7.3 Sequence Planning . . . . . 96
2.7.4 Path Planning . . . . . 96
2.7.5 Assembly of a Linear Truss Beam . . . . . 97
2.7.6 Second Generation End-Effector . . . . . 98

2.7.7 Graphics Simulation . . . . . 99
2.7.8 Operator Training and Human Factors Evaluation . . . . 99
2.8 Conclusions . . . . . 100
2.9 Appendix A . . . . . 101
2.10 APPENDIX B . . . . . 104
2.11 References . . . . . 109

**3 A Truss Joint for Robotic Assembly 111**
3.1 Space-Based Assembly . . . . . 112
3.1.1 EVA Structural Assembly . . . . . 112
3.1.2 Robotic Structural Assembly . . . . . 113
3.1.3 The ARASS Demonstration . . . . . 113
3.2 The Robot-Friendly Structural Joint Study . . . . . 115
3.2.1 Designing for Robotic Assembly . . . . . 116
3.2.2 Design Requirements . . . . . 117
3.2.3 Design Descriptions . . . . . 117
3.3 Test Descriptions . . . . . 126
3.3.1 Robot-Friendliness Tests . . . . . 126
3.3.2 Structural Tests . . . . . 130
3.4 Test Results and Evaluation . . . . . 131
3.4.1 Robotic Testing Results . . . . . 131
3.4.2 Structural Testing Results . . . . . 133
3.4.3 Evaluation . . . . . 134
3.5 Conclusions . . . . . 136
3.5.1 DFA Observations . . . . . 136
3.5.2 Real World Robotic Observations . . . . . 137
3.5.3 Concluding Remarks . . . . . 138
3.6 References . . . . . 139

**4 Hierarchical Planning for Space-Truss Assembly 141**
4.1 Introduction . . . . . 141
4.2 Part I: Assembly Sequence Planning . . . . . 143
4.2.1 Introduction . . . . . 143
4.2.2 Background . . . . . 143
4.2.3 Assembly Sequence Planning in the Space-Truss Domain . 145
4.2.4 Case Studies . . . . . 151
4.2.5 Discussion on Assembly Sequence Planning . . . . . 152
4.3 Part II: Assembly Path Planning . . . . . 155
4.3.1 Introduction . . . . . 155
4.3.2 Background . . . . . 155
4.3.3 Geometric Modeling . . . . . 157
4.3.4 Robot Path Planning . . . . . 159
4.3.5 Combining Global And Local Path Planning Methods . . 169
4.3.6 Results . . . . . 173

4.3.7 Discussion on Path Planning . . . 177
4.3.8 Configurations and Singularities . . . 180
4.4 References . . . 180

5 **Three-Dimensional Vision** **185**
5.1 Passive Techniques . . . 185
5.2 Active Techniques . . . 186
5.3 Calibration of Fixed Cameras . . . 187
5.3.1 Summary of Tsai's Calibration Technique for Fixed Cameras . . . 188
5.3.2 Refining Stage 1 of Tsai's Calibration Technique . . . 193
5.3.3 Methods for Determining the Stage 2 Parameters . . . 201
5.3.4 Sources of Camera Calibration Error . . . 207
5.3.5 Summary of Refinements to Calibration for Fixed Cameras 210
5.4 Calibration of a Laser Scanner . . . 211
5.4.1 The Mathematical Model of a Laser Scanner . . . 212
5.4.2 Calibration of the Intrinsic Parameters . . . 213
5.4.3 Calibration of the Extrinsic Parameters Using an LSE Method . . . 214
5.4.4 Direct Geometric Method for Calibrating a Laser Scanner 216
5.4.5 Locating a Laser Spot in a Camera Image . . . 221
5.5 Three-Dimensional Point Estimation Methods . . . 223
5.5.1 Overdetermined System of Linear Equations Method . . . 223
5.5.2 Midpoint of the Common Normal Method . . . 226
5.5.3 Comparison of Point Estimation Methods . . . 232
5.5.4 Point Estimation Using More than Two Calibrated Cameras 233
5.6 Description of the CIRSSE 3-D Vision System . . . 234
5.6.1 Five Camera Vision System . . . 235
5.6.2 Structured Light Source . . . 235
5.6.3 Calibration of the CIRSSE 3-D Vision System . . . 235
5.6.4 Demonstration of the CIRSSE 3-D Vision System . . . 237
5.7 References . . . 239

6 **CIRSSE Gripper and Controller System** **241**
6.1 Introduction . . . 242
6.2 Gripper Design Specifications . . . 242
6.3 Mechanical Design . . . 242
6.3.1 Sensors . . . 243
6.4 Controller Electronic Design . . . 243
6.4.1 Regulated Voltage Supplies . . . 245
6.4.2 Pneumatic Valve Drive . . . 245
6.4.3 Light Beam Sensor . . . 247
6.4.4 Force Sensors . . . 247
6.4.5 Position Sensor . . . 248

6.5 Controller Software Design . . . . 248
6.5.1 Power-on Hardware Initialization . . . . 248
6.5.2 Base Level Execution . . . . 248
6.5.3 Commands . . . . 250
6.5.4 Interrupt Level . . . . 250
6.5.5 Controller States . . . . 250
6.5.6 Design Issues . . . . 253
6.6 Summary . . . . 253
6.7 References . . . . 255

**7 Simulation of Space Manipulators 257**
7.1 Introduction . . . . 257
7.2 Notation . . . . 259
7.3 Single Manipulator Dynamics . . . . 263
7.3.1 Velocity and Acceleration Propagation . . . . 264
7.3.2 Simulation . . . . 268
7.3.3 A Simulation Case Study . . . . 269
7.3.4 Simulation Parameters . . . . 269
7.3.5 Simulation Experiments . . . . 271
7.3.6 Discussion . . . . 273
7.4 Simulation of Cooperating Manipulators . . . . 276
7.4.1 Recursive Inverse Dynamics . . . . 276
7.4.2 Closed-Form Inverse Dynamics . . . . 279
7.4.3 Interaction with a Common Load . . . . 282
7.4.4 Summary . . . . 282
7.4.5 Motion on a Mobile Base . . . . 283
7.4.6 Case Study . . . . 285
7.4.7 Control . . . . 285
7.5 Conclusion . . . . 286
7.6 References . . . . 290

**8 JPL Telerobot Testbed 297**
8.1 Background of the NASA/JPL Telerobot Testbed . . . . 298
8.2 System Description . . . . 300
8.2.1 System Architecture . . . . 300
8.2.2 Assembly Example . . . . 302
8.2.3 System Communication . . . . 303
8.2.4 Manipulator Control and Mechanization . . . . 303
8.2.5 Sensing and Perception . . . . 304
8.2.6 Run-Time Control . . . . 304
8.2.7 Task Planning and Reasoning . . . . 305
8.2.8 Operator Control Station . . . . 305
8.2.9 Monitoring and Diagnosis . . . . 306
8.3 Force and Motion Control Capability . . . . 306

8.3.1 Guarded Motion Primitive . . . 306
8.3.2 Contact Motion Primitive . . . 308
8.3.3 Compliant Motion Primitive . . . 308
8.4 World Model System . . . 310
8.4.1 Database Trees . . . 310
8.4.2 World Model Database Fields . . . 310
8.4.3 Object Geometry and Properties . . . 311
8.4.4 World Model Database Update . . . 312
8.4.5 World Model and Calibration . . . 313
8.4.6 Database Coordination . . . 315
8.4.7 Database Implementation . . . 315
8.5 Collision Detector . . . 316
8.5.1 Overview . . . 316
8.5.2 Entities and Object Modeling . . . 316
8.5.3 Collision Detection Algorithm — Low-Level . . . 317
8.6 Path Planner . . . 318
8.6.1 Cost Function . . . 319
8.6.2 Convergence of the Cost Function Minimization . . . 325
8.7 Task Primitives . . . 326
8.7.1 Establish_Gripper_Approach Routine . . . 332
8.7.2 Establish_Approach_Coarse Routine . . . 332
8.7.3 Establish_Approach_Fine Routine . . . 333
8.7.4 Establish_Grasp_Contact Routine . . . 333
8.7.5 Establish_Grasp_Level Routine . . . 335
8.7.6 Establish_Grasp_Close Routine . . . 337
8.8 Conclusions . . . 339
8.9 References . . . 339

# List of Figures


1.1 The CIRSSE Testbed . . . 3
1.2 Hierarchy of Intelligent Machines . . . 9
1.3 Distributed Petri Net . . . 30

2.1 Space-Based Antenna . . . 41
2.2 Truss-Supported Aerobrake Configuration . . . 42
2.3 Schematic of Automated Structures Assembly Laboratory . . . 44
2.4 Photograph of Automated Structures Assembly Laboratory . . . 45
2.5 Top View of Two-Ring Tetrahedral Truss Structure . . . 47
2.6 Tetrahedral Truss Structure Showing Strut Insertion Planes . . . 48
2.7 Strut/Node Joint Connector Hardware . . . 50
2.8 Schematic of End-Effector and Strut Insertion Concept . . . 52
2.9 End-Effector and Actuator Mechanisms . . . 53
2.10 Strut Storage Tray Pallet Details . . . 55
2.11 Strut Storage Tray Stacking Arrangement . . . 56
2.12 Schematic of Facility Computer Control System . . . 58
2.13 Operator's Workstation . . . 59
2.14 Schematic Illustrating Strut Naming Convention . . . 62
2.15 Automated Assembly System Software Hierarchy . . . 70
2.16 Operator's Menu Display . . . 72
2.17 NASREM Standard Reference Model Architecture . . . 73
2.18 Automated Assembly Program Architecture . . . 74
2.19 Robot Arm Path Segments . . . 82
2.20 Average Time Breakdown for Complete Truss Assembly and Disassembly . . . 83
2.21 Targets for Machine Vision System . . . 92
2.22 Machine Vision System Camera and Lighting . . . 94
2.23 Tetrahedral Beam Configuration . . . 97
2.24 Tetrahedral Truss with Cells and Sequence Numbers . . . 107
2.25 Facility Layout Diagram . . . 108

3.1 ARASS Workcell Configuration . . . 114
3.2 Threaded Collar Joint . . . 119

3.3 Collet/Flex Drive Joint . . . . . 120
3.4 Bolted/Flex Drive Joint . . . . . 121
3.5 Hammer-Head Joint . . . . . 122
3.6 Robot Friendly Structural Fasteners Evaluated by the Robotics and Mechanical Systems Laboratory . . . . . 123
3.7 Robot-Friendly Joint Selection Process . . . . . 127
3.8 Joint Coordinate System . . . . . 128
3.9 View of general test set-up. Visible are the robot, force/torque sensor, vision system camera, truss joint, scar, and node ball atop the flexible node simulator (see text). . . . . . 129
3.10 Location of Vision System Test Points (as seen from the end-effector camera) . . . . . 130
3.11 Pin-in-Slot Clocking Device . . . . . 137

4.1 Examples of Structural Volume Elements . . . . . 148
4.2 Examples of Non-Structural Volume Elements . . . . . 148
4.3 Examples of Structured Volume Element Connections . . . . . 148
4.4 Case Study with 10 Nodes and 31 Struts . . . . . 152
4.5 Disassembly Sequence for Structure with 10 Nodes and 24 Struts 153
4.6 Intermediate Disassembly States for Structure with 31 Nodes and 102 Struts . . . . . 154
4.7 Basic Geometric Model . . . . . 158
4.8 Shortest Distance Between Two Swept Sphere Cylinders . . . . . 158
4.9 Overview . . . . . 161
4.10 Sense of Rotation . . . . . 162
4.11 Basic Collision Avoidance Procedure for a Revolute Joint . . . . . 163
4.12 Basic Collision Avoidance Procedure for a Prismatic Joint . . . . . 165
4.13 Numbering Convention for the Joints and Links of the Robot . . 166
4.14 $\delta q_r(q)$ . . . . . 167
4.15 Two Ways of Inserting a Strut . . . . . 170
4.16 Choosing the Correct Sense of Rotation . . . . . 170
4.17 Splitting a Task into Two Subtasks . . . . . 171
4.18 Intermediate Steps Placed Around a Tetrahedron . . . . . 173
4.19 A Robot Arm of the CIRSSE Testbed . . . . . 174
4.20 A Simple Motion Task . . . . . 175
4.21 Completing a Tetrahedron . . . . . 176
4.22 Initial State of Obstacle Avoidance Demonstration . . . . . 176
4.23 Obstacle Avoidance Demonstration . . . . . 177
4.24 Step Size Selection . . . . . 179

5.1 Camera Calibration Coordinate Frame Definitions . . . . . 189
5.2 Internal Arrangement of a Laser Scanner . . . . . 212
5.3 Effect of Pincushion Distortion . . . . . 213
5.4 Determination of a Laser Scanner's Euler Angles . . . . . 217

5.5 Transformation Parameters from Frame $F$ to Frame $W$ . . . . . 217
5.6 Relationship Between the Floor and $F$'s $xy$ Plane . . . . . . . . . . 219
5.7 Single Image Plane . . . . . . . . . . . . . . . . . . . . . . . . . 224
5.8 Two Front Image Planes . . . . . . . . . . . . . . . . . . . . . . 227
5.9 Midpoint of the Common Normal . . . . . . . . . . . . . . . . . . 228
5.10 Triangle Completion Demonstration . . . . . . . . . . . . . . . . 238

6.1 The CIRSSE Gripper Shown in Reduced Scale . . . . . . . . . . . 244
6.2 Strain Gage and Amplifier on the Finger Mounting Details. Light Beam Sensor Element Internal Fingertip Detail. . . . . . . . . . . . 245
6.3 RPI Pneumatic Hand Control Electronics Board . . . . . . . . . . 246
6.4 CIRSSE Gripper Controller Software Architecture . . . . . . . . . 249
6.5 Top-Level Controller State Diagram . . . . . . . . . . . . . . . . 252
6.6 Grab Servo Mode State Transition Diagram . . . . . . . . . . . . 254

7.1 Overall Robot Link Structure . . . . . . . . . . . . . . . . . . . . 260
7.2 Geared Joint Model . . . . . . . . . . . . . . . . . . . . . . . . . 264
7.3 Flexible Joint Model . . . . . . . . . . . . . . . . . . . . . . . . 265
7.4 Link 2, PUMA 560 . . . . . . . . . . . . . . . . . . . . . . . . . 270
7.5 Initial Position and First Pick Location . . . . . . . . . . . . . . 272
7.6 Place Location and Final Position . . . . . . . . . . . . . . . . . 272
7.7 Trace of Tip Position During Motions . . . . . . . . . . . . . . . 273
7.8 Difference in Tip Position During Motions . . . . . . . . . . . . . 274
7.9 Degrees of Rotation Between Orientations . . . . . . . . . . . . . 274
7.10 Impact of Longer Trajectory Times on Difference in Tip Position, 10x Gear Ratio . . . . . . . . . . . . . . . . . . . . . . . . . . 275
7.11 Two Arm Mobile Platform Link Model . . . . . . . . . . . . . . . 277
7.12 Initial Configuration . . . . . . . . . . . . . . . . . . . . . . . . 287
7.13 First Two Motion Sequences . . . . . . . . . . . . . . . . . . . . 287
7.14 Last Two Motion Sequences . . . . . . . . . . . . . . . . . . . . 288
7.15 Cartesian Error of Load . . . . . . . . . . . . . . . . . . . . . . 288
7.16 Differences in Y Tip Forces . . . . . . . . . . . . . . . . . . . . 289

8.1 View of the JPL Telerobot Testbed . . . . . . . . . . . . . . . . 299
8.2 Architecture of the Telerobot Testbed . . . . . . . . . . . . . . . 301
8.3 Example Path Iterations for 2-Dimensional Case . . . . . . . . . 320
8.4 A Sigmoid Function Plotted as a Function of the Independent Scalar Variable $x$ . . . . . . . . . . . . . . . . . . . . . . . . . 322
8.5 ORU Grapple Lug and Corresponding Gripper Fingers . . . . . . 323
8.6 State Transition Graph for the "Grasp" Task Level Primitive . . 328

# List of Tables


2.1 Automated Assembly Data Structure . . . . . . . . . . . . . . . . . . 78
2.2 Assembly Sequence . . . . . . . . . . . . . . . . . . . . . . . . . . 105

3.1 Robot-Friendly Structural Joint Design Requirements for ARASS 118
3.2 Robot-Friendly Structural Joint Evaluation Matrix . . . . . . . . 135

6.1 Commands: RS-232 Serial Port . . . . . . . . . . . . . . . . . . . . 250
6.2 Parameter Registers . . . . . . . . . . . . . . . . . . . . . . . . . 251
6.3 Error Codes . . . . . . . . . . . . . . . . . . . . . . . . . . . . . 252

7.1 Mass and Inertia Parameters for Platform . . . . . . . . . . . . . . 271

8.1 Possible State Vector Element Values for Grasp Primitive State Transition Graph . . . . . . . . . . . . . . . . . . . . . . . . . . . . 329
8.2 State Transitions and State Transition Routine Names . . . . . . 330

# Contributors

**Jeff Bethel**
**Alan A. Desrochers**
**Keith R. Fieldhouse**
**Lester A. Gerhardt**
**Robert B. Kelley**
**Donald R. Lefebvre**
**Stephen H. Murphy**
**J. Russell Noseworthy**
**John Peiffer**
**Arthur M. Ryan**
**Arthur C. Sanderson**
**Jodi Tsai**
**James F. Watson, III**
Electrical, Computer, and Systems Engineering Department
Center for Intelligent Robotic Systems for Space Exploration
Rensselaer Polytechnic Institute
Troy, New York 12180-3590

**George F. Parma**
Robotics and Mechanical Systems Laboratory
Structures and Mechanics Division
NASA - Lyndon B. Johnson Space Center
Houston, TX 77058

**Rolf Münger**
Buchholzstrasse 1551
8053 Zurich, Switzerland

**Ralph Will**
**Marvin Rhodes**
**William R. Doggett**
**Catherine L. Herstrom**
**Carolyn Grantham**
**Cheryl L. Allen**
**P. Daniel Sydow**
**Eric G. Cooper**
NASA - Langley Research Center
Hampton, VA 23665

**Cuang C. Quach**
**Marion A. Wise**
Lockheed Engineering and Sciences Company
Langley Program Office
Hampton, VA 23665

**Rajive K. Mathur**
GTE Spacenet Corporation
McLean, VA 22102

**J. (Bob) Balaram**
**Henry W. Stone**
Jet Propulsion Laboratory
California Institute of Technology
Pasadena, CA 91109

# Preface

This volume represents the contributions of numerous researchers in the field of intelligent robotic systems for space exploration. It is based on work that has been done at the NASA Langley Research Center, the NASA Lyndon B. Johnson Space Center, the Jet Propulsion Laboratory, and the Center for Intelligent Robotic Systems for Space Exploration at Rensselaer Polytechnic Institute.

Chapter 1 describes basic robotic systems engineering problems and emphasizes those issues that are unique to the space environment. Some of these problems are addressed in this book and so an overview of this volume is intertwined with this description. This chapter also discusses some specific research and implementation work at Rensselaer for a dual arm testbed. A unified architecture is described that uses a common representation of all three levels in the hierarchical structure. This results in a mathematical approach to integration.

Much of the work in this book is related to assembly in space, and especially in-orbit assembly of large truss structures. Such structures have been proposed for constructing the space station, large antennas, and aerobrakes. A testbed has been developed at the NASA Langley Research Center for studying the problems that arise during earth-based robotic assembly of complex truss structures. This experience is reported in Chapter 2. This chapter helps to set the stage for several of the remaining chapters. Specifically, it addresses problems such as truss joint design, assembly sequence planning, trajectory planning for strut insertion, vision and sensing, end-effector design, supervised autonomy, and integration. Several of these problems are discussed in detail in the remaining chapters.

A key element in the assembly of a large truss structure is the truss joint that is used to hold the struts together. Chapter 3 documents the truss joint development work that has transpired at the Johnson Space Center during the past couple of years. Numerous "robot friendly" joints have been designed, built, and tested under several realistic scenarios. Several designs are described and a comparison of experimental results is presented. This performance evaluation of the joints leads to a design that is promising for robotic assembly, disassembly, and maintenance of truss structures.

Assembly of a structure composed of struts and nodes, requires a detailed plan. In particular, the sequence in which the struts are inserted is important because all partially completed structures must be physically stable. It is also necessary to maintain geometric access during the assembly so as not to preclude the insertion of future struts. This assembly sequence planning problem is the first part of Chapter 4. The second part is concerned with the actual path that is used by the robot to insert the strut. Trajectory planning algorithms are discussed for inserting the strut into a partially completed structure. Strategies for collision avoidance are included.

Strut insertion in these structures needs to be sensor-based. Vision is particularly useful in the supervised autonomy mode of operation. In this situation, strategically placed cameras can be used to guide a human operator through the assembly process. Such multi-camera systems operating in a three-dimensional environment is the subject of Chapter 5. Camera calibration methods for multi-camera systems is discussed in detail along with the use of laser pointing systems.

Chapter 6 describes the design of a robot gripper for grasping struts during insertion. The design of this gripper is motivated by the need for sensor-based control during the assembly of the struts and nodes. The complete hardware and software design is presented.

The control of robots during in-orbit operations gives rise to new problems. Chapter 7 shows how the zero-gravity environment affects the robot modeling and control problem. Of particular significance, are the induced forces that occur as a result of gyroscopic effects. These are important for heavily geared robotic arms such as the space shuttle Remote Manipulator System.

Another dual arm testbed has been developed at the Jet Propulsion Laboratory. The purpose of this testbed is to investigate the solutions to problems such as in-orbit satellite maintenance, orbital replacement unit change-outs, sensor-based control, and integration of the various subsystems. The implementation details and the experience gained over this five year development are reported in Chapter 8.

# Acknowledgments

Although I am listed as the editor of this volume, its compilation, editing, and assembly is the result of the work of many people. I would especially like to thank Betty Lawson for typing every word and formatting every paragraph and reference in this book. Additional thanks are in order for Jim Watson who helped with the editing and generously served as the captain of the LaTeX SWAT team. Other members of that team include Russ Noseworthy and Steve Murphy. Keith Fieldhouse also provided an invaluable service by converting some authors' original text into LaTeX .

I also appreciate the assistance, cooperation and patience of Tom Stone, publisher at Kluwer Academic Publishers.

Finally, the resources and the opportunity to organize and edit this volume are largely due to NASA grant NAGW-1333 in support of the work at Rensselaer's Center for Intelligent Robotic Systems for Space Exploration.

*Alan A. Desrochers*
*Rensselaer Polytechnic Institute*

# Chapter 1

# Testbed for Cooperative Robotic Manipulators

by: James F. Watson, III
Donald R. Lefebvre
Alan A. Desrochers
Stephen H. Murphy
Keith R. Fieldhouse
Electrical, Computer, and Systems Engineering Department
Center for Intelligent Robotic Systems for Space Exploration
Rensselaer Polytechnic Institute
Troy, New York 12180-3590

This chapter serves three purposes. Firstly, it is intended to give an overview of many of the robotic engineering problems encountered when considering space-based applications. Secondly, during this overview, the approaches to some of these problems, as currently addressed in the robotic testbed facility at Rensselaer's Center for Intelligent Robotic Systems for Space Exploration, are presented. Thirdly, throughout the survey of these space-based robotic engineering problems, the reader will be directed to the particular chapters in this book that treat these problems in more detail. It is intended that this chapter will motivate the contents of the book.

## 1.1 Introduction

Research in space-based robotic applications is being conducted by many investigators. The relevant research areas include machine intelligence, manipulator control, computer vision, sensing, and integration of subsystems. The Center for Intelligent Robotic Systems for Space Exploration (CIRSSE) includes a testbed facility for experimentation of robotic applications. This facility, located at Rensselaer Polytechnic Institute, Troy, New York, contains several robotic manipulators, sensors, and processors, and is utilized for a wide range of research interests. Figure 1.1 illustrates the CIRSSE Testbed.

The CIRSSE Testbed contains two PUMA robots, each with wrist-mounted force/torque sensors, a custom-built Robotic Transporter, consisting of two 3 degree-of-freedom (DOF) carts, and custom-built grippers. A variety of vision capabilities are provided by fixed stereo cameras, arm-mounted stereo cameras, and a laser scanner. Each PUMA is mounted on top of a cart, forming two independent 9-DOF manipulators. The purpose of this testbed is to provide a facility for research on space-based robotic applications. Specifically, the Theory of Intelligent Machines, dual arm control, path and task planning, and 3-D vision are being investigated. The development of several experiments involving the assembly of a strut-based truss structure have been focal points for the research activities. The testbed is also useful for investigating the integration problems that occur in an environment that is heavily dependent upon the acquisition, transmission, and processing of data in the form of information, knowledge, and control commands.

Robotic applications are characterized by control problems for the robotic manipulators, image processing tasks from the vision sensors, on-line task planning for assembly, and trajectory planning calculations. Testbeds designed for experiments in the field of robotics require many hardware and software subsystems to implement the required functionality. Design of such a testbed is a non-trivial task. Additional complications arise from the distributed nature of the hardware and software subsystems, the communication of data and commands between these subsystems, and the lack of well-defined functional and operational boundaries for the subsystems. This chapter includes a description of the CIRSSE Testbed. Specifically, the hardware and software systems that make this facility unique are discussed.

The Jet Propulsion Laboratory (JPL), Pasadena, California, has recently completed a five year program for the development of a dual arm robotic testbed. The purpose of the JPL testbed was to investigate the potential use of robotic technology for in-orbit satellite maintenance and repair. Problems in grasping, dual arm coordination, collision avoidance, path planning, and sensing were considered. Details of the JPL experience are reported in Chapter 8.

The remainder of this chapter is organized in the following manner. Purposes of the CIRSSE Testbed are presented in the following section. This is followed by a discussion on the robotic engineering problems that are unique to the space environment. Throughout this discussion, references are made to the activities described in this book to point out their relevance and contribution to robotic space engineering problems. Subsequently, descriptions of the CIRSSE

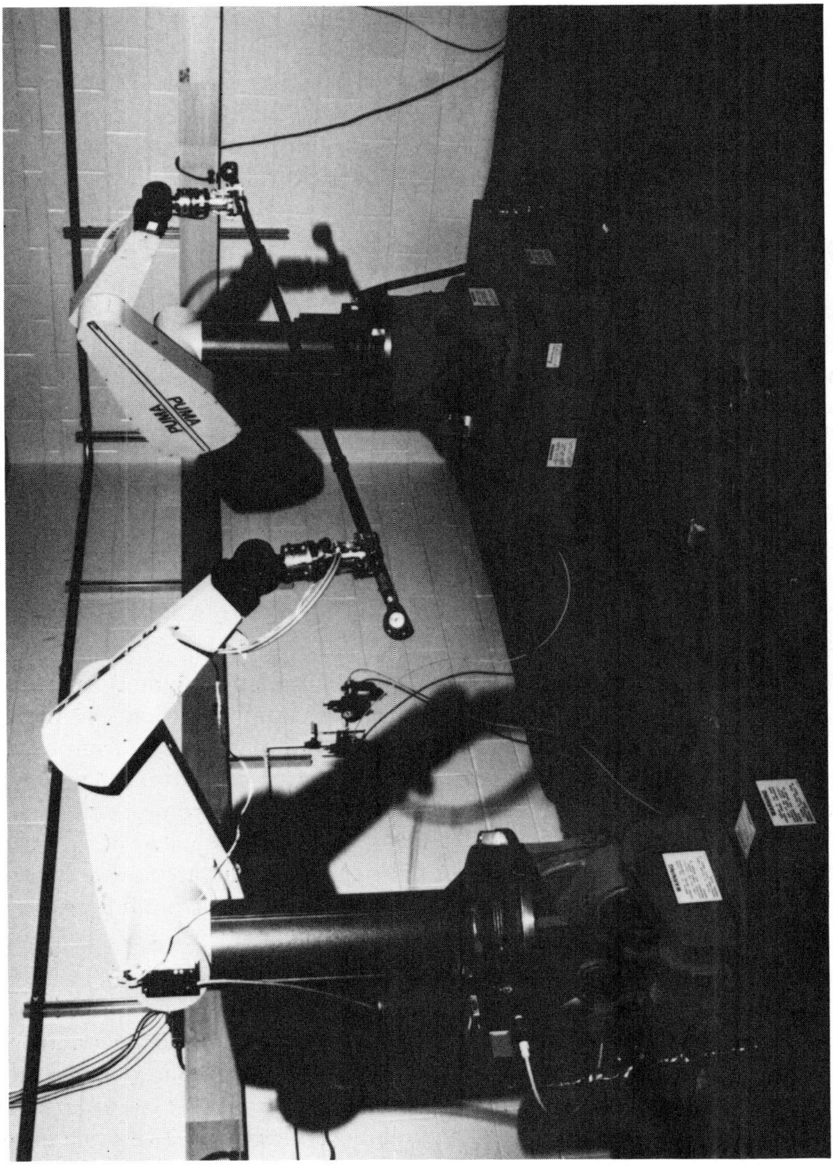

Figure 1.1: The CIRSSE Testbed

Testbed hardware components are given. The next section describes the need for and implementation of a Motion Control System for the testbed. Lastly, the aspects of the high-level control of the testbed are discussed.

## 1.2 Purposes of the CIRSSE Testbed

Automation of in-orbit activities is of primary interest to NASA. However, as simple as the word "automation" is, it encompasses many diverse and challenging research areas, including, planning and reasoning, manipulation and control, and vision and sensing. The CIRSSE Testbed provides a facility for experimentation in robotic automation. The testbed includes a variety of hardware subsystems for manipulation, sensing, and computation. Section 1.5 describes the testbed hardware in detail.

Automation of in-orbit activities is complicated by the environmental characteristics of space, e.g., there is a lack of gravity, lighting is difficult to control, and multiple manipulators may have to be used for a given task. The overall purpose of the CIRSSE Testbed is to promote research and experimentation in robotic automation, with emphasis on the complications of in-orbit applications. Section 1.4 elaborates on this purpose by discussing some of the engineering problems associated with space-based robotic applications. Most of these problems occur during automated assembly of structures in space.

## 1.3 Space-Based Assembly

A perennial experiment within CIRSSE is the automatic assembly of a strut-based truss structure. This experiment has a natural evolution: inserting a single strut to complete a triangle of struts; building this triangle from scratch using a single robot; building the triangle from scratch using two robots; building an entire tetrahedron of struts using two robots. Other complexities can be introduced, such as, using vision to determine the initial locations of the struts, and disturbing the position of the robot's base. Truss structure assembly serves as an excellent case study for integration, control, sensing, planning, and error detection and recovery.

First, object identification is required and range information is needed to accurately locate the struts and nodes. Second, an assembly sequence needs to be planned, [1], [2]. The sequence is important for maintaining stability of the partially completed structure and geometric access to the next nodes to be used in the assembly. The assembly sequence has to result in a set of plans for robot control and a strategy for determining which of the five cameras and other sensors will be used.

Depending on the complexity of the assembly subtask, either one or two arms are required. The two arm case involves issues of manipulator coordination. In either case, path planning and collision avoidance are required for task execution and safety. Sensory information acquired during the task requires processing. Controls may be affected by on-line algorithms using this sensor data.

The truss structure assembly scenario is centered around in-orbit assembly strategies for autonomous and telerobotic systems. Thus, the tilt and rotate axes of the Robotic Transporter can be used in a variety of ways: simulation of the relative motion between two arms that is likely to occur during docking, simulation of disturbances that would occur during the assembly of large truss structures, and simulation of the lack of reactive forces in space by providing motions that maintain the center of mass of the PUMA and payload in a given location.

Chapter 2 presents a case-study of strut insertion experiments. Issues in automation, task planning, path planning, etc., are presented in the context of NASA Langley's Automated Structures Assembly Laboratory.

The automated assembly of a truss structure is as dependent on the design of the parts to be assembled as it is on accurate control, sensing, and planning. Chapter 3 discusses the design and experimental evaluation of truss joints, undertaken by the Structures and Mechanics Division at the Johnson Space Center. The main objective was to ensure "robot-friendliness." A thorough analysis of several designs is presented.

Assembly sequence planning and path planning are the major topics of Chapter 4. Vision algorithms to help carry out the assembly are presented in Chapter 5. Chapter 6 presents the design and implementation of a sensor-based gripper for manipulating struts. More references to all of the chapters will be made in the next section.

## 1.4 Issues and Results for Space-Based Robotic Applications

This section presents several topics associated with space-based robotic applications. Each topic deserves a full discussion of the related engineering issues and research results, however, due to space considerations, this type of detail is presented only occasionally.

### 1.4.1 Lighting Characteristics of Space

Application of robotic vision to in-orbit tasks is complicated by the extreme lighting characteristics of space. Direct sunlight causes saturation of the camera sensing elements, creates glares on shiny, metallic surfaces, and casts shadows on objects of interest. Using vision for identification of objects, verification of part matings, or inspection of structures, is hampered by the inability to accurately detect features (e.g., edges, centroids) in the scene.

Strut insertion experiments conducted at CIRSSE have relied on the ability to move lighting sources in the lab to advantageous locations and paint parts in a manner that minimizes glare and maximizes the contrast of relevant features. Making vision algorithms more robust to uncontrollable lighting characteristics is an area of potential research.

### 1.4.2 Zero-Gravity

While accurately producing zero-gravity conditions within an earth-based testbed is difficult, experimentation in this type of environment is possible via simulation. Software simulation is possible, but it requires accurate models of manipulator dynamics and extensive programming of the simulation environment. Fortunately, however, for this case it is also possible to implement a "hardware" simulation—a *gravity compensation* controller. A gravity compensation controller for the PUMA has been developed using the detailed mechanical information provided in [3].

A gravity compensation controller has been implemented for the testbed's PUMA arm. Based on arm configuration, weight of the arm links, gear ratios at the joint motors, and the payload, this controller computes the motor torques needed to prevent the manipulator links from falling. Displacement of the arm by an external force is possible by simply overcoming the joint friction. Efforts to algorithmically compensate for this friction are on-going. Furthermore, gravity compensation for the 9-DOF redundant arms in the CIRSSE Testbed is being investigated. Considering the gravity compensation control as an additive component of the applied control signal, controllers and other subsystems can be tested in a simulated zero-gravity environment.

Chapter 7 examines the effects of zero-gravity on control algorithm implementation. Gyroscopic effects are considered and it is shown that significant forces are generated, especially in heavily geared arms like the space shuttle Remote Manipulator System (RMS).

### 1.4.3 Integration

Integration is a key issue for the design and operation of any complex system. The interconnection of various subsystems introduces concerns of communication mechanisms, data and command flows, and functional compatibility. System performance is affected by processor speeds, the quantity of data to be transmitted, communications delays, and algorithm efficiency. There is no clear cut technique for designing these systems to achieve a desired performance—integration, at best, is a trial and error process that may be speeded by the designer's experience and/or development tools.

Development of the CIRSSE Testbed as a useful facility for robotic experimentation is an on-going integration process. The testbed is characterized by control problems for the robotic manipulators, image processing, on-line task planning and trajectory planning, as well as communication between subsystems. Complex, integrated systems, such as the testbed, can be classified as Discrete Event Dynamic Systems (DEDS). Models for these systems must be able to capture the following characteristics:

> **Concurrency or parallelism:** The system architecture may permit many processes to execute simultaneously.
>
> **Asynchronous operations:** The evolution of system events is aperiodic. This may be due to non-constant process completion times (e.g., the time to perform a scene analysis depends on the complexity of the scene). Furthermore, two or more processes may have large differences in their nominal completion times, and the system architecture requires the completion of all these processes prior to continuing execution.
>
> **Deadlock:** It may be possible to reach a state in which no process within the system can continue execution. Obviously, this situation is undesirable, and is usually the result of the system design. An important feature of a good model is that it can detect deadlock, permitting time for correction and re-design prior to system implementation.
>
> **Conflict:** Two or more processes may require a common resource at the same time. Examples of this type of contention include two manipulators requesting the same camera for sensor information, or two tasks requesting the same CPU. Unlike the deadlock situation, it is possible for one of the processes to continue, provided the conflict can be resolved.
>
> **Event driven:** The system can be viewed as a sequence of discrete events. For example, a typical sequence might include collecting environmental data via automated vision functions, followed by on-line planning, manipulator control, and finally, force sensing. Since there are other processes in the system (e.g., another manipulator, or automatic safety monitors), the order of occurrence of events is not necessarily unique; the order is one of many allowed by the system architecture. In addition, one event may trigger several other events that may execute serially, in parallel, or in some combination of the two.

DEDS have been modeled and analyzed by Markov chains [4], queueing networks [5], perturbation analysis [6], Petri nets [7], and formal language theory [8]. For several reasons, the majority of the systems analysis research conducted within CIRSSE utilizes Petri net (PN) models. Petri nets were developed by Carl Petri, [9], and are a compact graphical representation that permits visualization of the system and the evolution of its states. PN models are derived in a relatively straightforward manner based on a knowledge of how the system works. More importantly, PN models have a well-developed mathematical foundation, whereby structural analysis (e.g., detection of deadlock and conflict) of the system is possible, as well as performance analysis, [7], [10], [11], [12], [13]. Software packages are available for PN structural and performance analyses, [14], [15].

In [16], the performance of the proposed CIRSSE Testbed architecture was investigated. The performance analysis criterion was taken as the response time of the system, (i.e., the time it takes to complete a task after a command is initiated). This response time includes image processing time, trajectory planning time, control execution time, etc. Performance analysis, in general, permits the system designer to answer numerous "what if" questions, (e.g., what if process speed is increased, more computers are added, communication protocols are changed?), and PN modeling, in particular, provides a methodology for handling such interrelated functions as control, communication, computation, and database access.

A major integration effort was undertaken at JPL during the past five years. This is also a dual arm testbed that uses vision and other sensory information. The implementation details and experimental results are the subject of Chapter 8.

The assembly of large truss structures at the Langley Research Center is also a major integration effort. A detailed history of that testbed and its accomplishments are provided in Chapter 2.

### 1.4.4 Theory of Intelligent Machines

Designing autonomous robotic systems to perform complex tasks is a formidable challenge. Issues of learning and reasoning, distribution of intelligence, robustness with respect to uncertainty, and error recovery must be addressed. The Theory of Intelligent Machines, [17], is an analytical formulation for structuring the execution of automated operations. Based on the *Principle of Increasing Precision With Decreasing Intelligence*, and the use of entropy as a unifying measure of uncertainty, a hierarchy for the Intelligent Machine has been proposed, [18], [19], [20], [21]. This hierarchy is shown in Figure 1.2.

As one of its primary purposes, the CIRSSE Testbed serves as a vehicle for the development and implementation of the Theory of Intelligent Machines. The *Organization Level*, the highest level in the hierarchy, performs the representation and task planning functions. As demonstrated in [22], [23], it is possible to implement this level as a Neural Network. The Organization Level maintains the world database of objects in the environment, the status and model of capabilities for the manipulators and sensors, and motion and sensing plans for individual tasks. Specific research issues and results for the area of task planning are presented in Chapters 2 and 4.

The second level in Figure 1.2, i.e., the *Coordination Level*, provides an interface between the Organization Level and the *Execution Level*. Specifically,

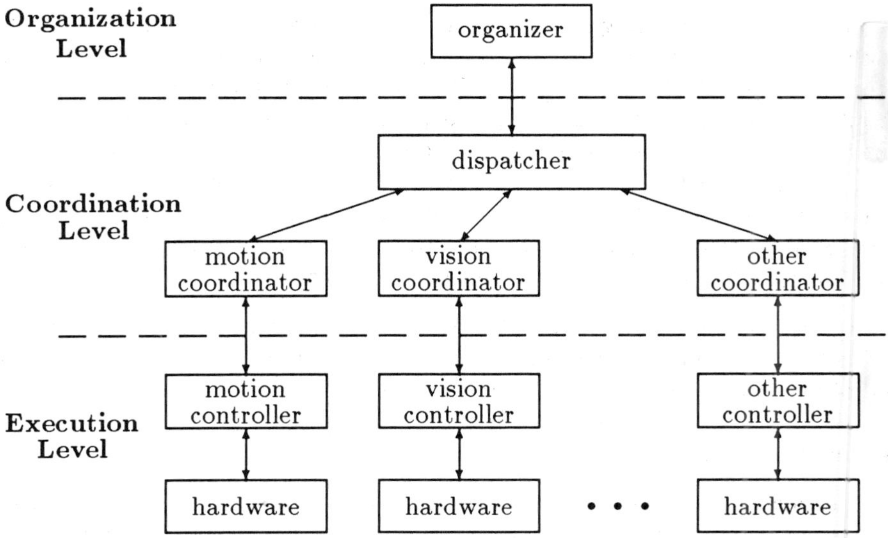


Figure 1.2: Hierarchy of Intelligent Machines

this intermediate level performs communication functions and generates tapes of primitive commands, which are implemented in the Execution Level. As proposed in [24], [25], individual Coordinators contained within this level can be implemented by Petri Net Transducers. A discussion of the research issues and results for the Coordination Level are given in Section 1.7.

The lowest level in the Hierarchy of Intelligent Machines is called the Execution Level. This level performs the actual control algorithms and/or sensing functions, and interfaces directly with the appropriate hardware. Consideration of the Execution Level within the context of the CIRSSE Testbed has led to the design and implementation of the Motion Control System (MCS) and the Vision Services System (VSS). These two software systems represent an interesting solution to the problem of effectively integrating several of the hardware components found in such a complex robotic testbed. Due to space considerations, only the MCS will be discussed in more detail, (in Section 1.6). The hardware for the VSS is presented as part of Chapter 5.

### 1.4.5 Task Planning, Path Planning, and Collision Avoidance

Automated systems are most useful when they can correctly respond to high-level commands, such as `build_truss`. Responses of this type require reasoning about the task and subtasks. For example, to build a large truss structure in space, a detailed assembly plan is required to ensure the mechanical accessibility of nodes and the integrity of the partial structure. The assembly sequence implies a partial motion plan, i.e., the starting and goal positions of manipulators are prescribed for each step. The motion plan, however, requires more detail for successful execution. Typically, the manipulator must move through a region of obstacles before arriving at the place of interest. Obstacle avoidance is complicated by the possibility of using many manipulators to build the truss—thus, some obstacles are non-stationary. Additionally, while carrying truss parts, the manipulator must also move in such a manner as not to cause a collision of this part with itself or other obstacles.

Active research areas within CIRSSE include task planning, path planning, and collision avoidance. Assembly of a truss structure from struts and nodes, and other types of assembly tasks, are considered, [26], [27]. Chapter 4 includes a discussion of task planning issues and solution techniques in the context of assembly problems.

Automatic path planning falls into one of two categories: static environments and dynamic environments. In either case, reaching an appropriate destination is desired while maximizing some performance criteria and avoiding collisions with obstacles. Research and simulation has been conducted for static environments, considering dual arms, [28], and a single arm carrying a strut, [29]. Chapter 4 also presents results in path planning for a single arm carrying a strut. Path planning for mobile robots in dynamic environments has been investigated, [30], [31], [32].

While planning for obstacle avoidance and lack of self-collisions may be possible, collisions may still occur because of inaccurate models of the environment, poor control of the joints, and/or lack of coordination between manipulators. Thus, it is also desirable to detect potential collisions on-line, [33], [34]. Additionally, the research issues concerning robot safety in general have been explored, [35], [36].

### 1.4.6 Control

Control problems are basic to every robotic application in space. Success of a particular task is dependent on quality trajectory generation for the manipulator joints, responsive servo loops, and in some cases, accurate control of internal and external forces. Research in the areas of adaptive control, [37], [38], [39], and force control, [40], [41], of robotic manipulators is being conducted at CIRSSE.

Space-based manipulators, such as the space shuttle RMS, often possess a mechanical flexibility. Thus, modeling and control of robotic manipulators with flexible joints and/or links is being researched, [42], [43], [44]. Additionally, there is on-going research concerning the control of redundant manipulators.

### 1.4.7 Remote Control

The remote control of manipulators encompasses many specific research topics having space-based applications to mobile robots, sensor data display methodologies, and operator-controlled obstacle avoidance.

The concept of remote operation of manipulators also includes telerobotics, teleoperation, and telepresence. Research issues in these areas include modeling the effects of time delay in the system, modeling human operators in the control loop, and the effective division of fully autonomous, semi-autonomous, supervised autonomous and operator-controlled tasks. The work reported by JPL in Chapter 8 is a good example of cutting-edge research in this field

Similarly, the NASA Langley Research Center has experimented with remote control of manipulators in the Automated Structures Assembly Laboratory. Specifically, they have used visual information obtained from cameras located in the workspace to aid a human operator during the assembly of large truss structures. A very detailed account of their supervised autonomous experiments during the last several years is reported in Chapter 2. In fact, this chapter addresses many of the space-related engineering problems that are important to future successful robotic applications in space. Each of the remaining chapters examines one or more of the problems raised by the experimental work at Langley.

### 1.4.8 Dual Arm Control

Dual arm manipulation significantly improves the operational flexibility over that of a single arm. Specifically, payload capacity is increased, workspace is enlarged, and more configurations of the manipulators are available for task and path planning. Each of these aspects provides a useful enhancement for space-based assembly operations. However, because of the additional DOFs of the system and the increased number of potential collisions, the issues of path planning and control become more complicated.

Dual arm control is important to such space-based applications as docking, manipulation of large payloads, and in-orbit assembly of large structures (e.g., the space station, antennas, solar arrays, aerobrakes), and satellite maintenance and repair. The last topic has been a focal point of the JPL Testbed, described in Chapter 8.

The dynamic equations of motion for the two 9-DOF CIRSSE Testbed manipulators holding a common object have been derived, [45], [46]. The full dynamic interactions from arms-to-carts and arm tip-to-arm tip are included in the formulation. The closed-form inverse dynamics are obtained by first developing the equations of motion for the open kinematic chain on a mobile platform, and then closing the chain through the forces on the load. The closed-chain forward dynamics is obtained from a recursive solution, and verified by simulation.

Dual arm control in a zero-gravity environment is a significant problem unique to space-based robotic engineering. There are gyroscopic effects due to motor rotation that can drastically affect the position of the manipulator's end-effector. This topic is studied in detail in Chapter 7.

### 1.4.9 Sensing

Sensor data is critical to using robots to perform fully autonomous, semi-autonomous, supervised autonomous, and operator-controlled tasks in space. Issues include providing adequate coverage for large work volumes and the applicability of sensors to the space environment. Tasks that involve only partial automation require that the sensor data be useful to both automated systems and the human operator. Integration of different sensors, and the data generated by them, is a current research issue.

The CIRSSE Testbed houses several cameras and force/torque sensors, as well as the necessary position sensors for the manipulator actuators. Visual and force sensing provide individual research areas in themselves. Issues include camera calibration (see Chapter 5), 3-D data acquisition, decomposing internal and external forces, and manipulator cooperation.

The work reported in Chapters 2, 6, and 8 also relies heavily on sensor information.

Research on the specific issues of using sensor information for control has been conducted. In [47], the reliability of combined sensor and control systems is investigated. Additionally, as a case study of using multiple sensors for control applications, experiments have been conducted to automatically insert printed circuit boards into card cages, [48].

### 1.4.10 Error Recovery

The overall performance of automated systems is enhanced by the ability to automatically recognize errors and recover from them. For example, assembly of truss structures by fully autonomous robotic systems requires the detection of failed part insertions, determination of causes of the failure, and planning for the recovery tasks. Even when a human operator is involved, error recovery procedures reduce the operator's workload by providing detection and analysis functions.

Error recovery processes would reside in the higher levels of the Intelligent Machine Hierarchy. Outside of CIRSSE, work has been done on modeling error recovery procedures with Petri nets, [49]. Results from this work could potentially be incorporated into the Petri net models for the Intelligent Machine Coordinators.

### 1.4.11 Visual and Force Servoing

Space-based robotic applications benefit from visual and force servoing capabilities, whereby the pre-planned manipulator trajectory is modified by sensor information so that some additional motion constraint is satisfied. For example, force servoing is used for compliant insertion, providing compensation for inaccuracies in the end-effector and/or the goal positions. Considering the large structures, manipulator links, and work volumes associated with space-based applications, assembly in the presence of these positional errors is expected. Payload constraints may require the use of more than one manipulator to perform a task, and therefore, internal forces of the manipulators must be controlled. Additionally, tasks that involve maintaining a fixed distance from or converging to a moving object, require visual information for trajectory adjustment.

Preliminary results indicate that the CIRSSE Testbed possesses enough sensor and computational resources to make possible rudimentary visual servoing. Specifically, experiments in automated strut following have been performed, [50]. The algorithm underlying these experiments uses arm-mounted cameras to identify and track a strut that is being held by a human operator. The setpoints for the joint controllers are updated at camera frame rate.

Stereo cameras can be used for object velocity estimation. From a sequence of stereo images, a dense map of the object displacement is generated. With this map, and knowledge of the camera orientations, velocity estimates are obtained from the frame sequence. These estimates can be used for control or verification of the velocity of a manipulator and/or the object it is carrying.

Force servoing encompasses the specific research areas of hybrid force, direct force, impedance, and position accommodation control. Experiments with strut insertion using impedance control have been performed. These experiments are concerned with controlling the forces and torques between the manipulator (i.e., the end-effector) and the environment while accomplishing the insertion. The force and torque data are collected by wrist-mounted force/torque sensors, and can be used to correct for small positional errors.

Using the 9-DOF redundant arms in the CIRSSE Testbed, experiments to control the forces internal to a manipulator are possible. Additionally, simulations have been performed and experiments proposed for controlling forces and torques exerted by the dual-arm system onto a commonly held object, [45], [46].

## 1.5 Physical Description of the CIRSSE Testbed

The CIRSSE Testbed consists of several hardware subsystems. Collectively, these subsystems provide a powerful facility for complex robotic experiments, including automated assembly of truss structures. Individually, the subsystems are also tools for research in specific areas of control, vision, sensing, and computing. The testbed subsystems can be grouped into the following three categories:

1. *manipulators*—those subsystems that move and/or change the state of the environment as a result of being controlled.

2. *sensors*—those subsystems that gather data on the environment in either an automatic and/or per request manner.

3. *processors*—those subsystems that collect, disseminate, and/or store data, and/or perform computations.

While some subsystems may overlap two or more of the categories, it is often the case that a most significant category can be identified for each subsystem.

In an effort to provide the reader with an understanding of how a robotic testbed like the one at CIRSSE might be configured, the remainder of this section gives a description of the hardware subsystems within the CIRSSE Testbed.

## 1.5.1 Manipulator Subsystems

Robotic experiments can be performed using any combination of the various manipulators in the testbed. The larger subsystems are either commercially available products or were custom designed outside of CIRSSE. However, as the results of individual research projects, some of the hardware has been designed within CIRSSE.

### PUMA Robots

As a primary source of manipulation, two 6-DOF PUMA robots are contained in the testbed: a PUMA 560 and a PUMA 600. For engineering purposes, the two robots are identical. These robots are commercially available from Unimation (Westinghouse, Corp.), Pittsburgh, Pennsylvania.

Each robot is connected to a controller that provides servoing logic and a communication interface, and also houses the joint motor power amplifiers. Either a teach pendant or terminal can be interfaced to the controller. For more flexibility with respect to designing hierarchical controllers for the PUMAs, these Unimation controllers are bypassed. This is accomplished by connecting the Q-Bus in the Unimation controller to another computer system. In the CIRSSE Testbed, the PUMAs are controlled by the Motion Control Cage, which is a multiprocessor, custom-designed, real-time computer system discussed in more detail below.

Joint encoder resolution is about 1/5000 of a degree. This resolution, combined with the controller performance and the joint friction, results in a Cartesian space repeatability error specification of $\pm 0.1$ millimeters. This specification reflects the amount of error that can be expected in the end-effector position under normal applications. A force/torque sensor and/or gripper can be mounted onto the wrist flange. The maximum payload is 5 pounds, including the weight of the sensor and/or gripper. Thus, lightweight gripper design is important.

The work volume of each PUMA is contained within a sphere of approximate radius of 34 inches. Regions within this sphere that cannot be reached by the wrist include a wedge-shaped volume at the back of the robot, and a tube-shaped volume around the robot's body. Maximum end-effector velocity is 1 meter/second.

### Robotic Transporter

This subsystem, custom-built for CIRSSE by K. N. Aronson, Inc., Arcade, New York, consists of two 3-DOF carts. Each cart has a large flange and is used to carry a PUMA robot. The carts can travel along a 12 foot linear track at 24 inches/second with a positional resolution of 0.03mm, and can tilt $\pm 45^o$ and rotate $\pm 150^o$ at 3 RPM with a resolution of 6 arc seconds.

The configuration of the PUMAs on the carts results in an 18-DOF system. The carts can be used to position the PUMAs for a particular experiment. While this is useful, the more dramatic uses of the transporter are realized when the system is considered as two 9-DOF manipulators working either independently or in cooperation. Compared to the PUMA by itself, each 9-DOF cart/PUMA manipulator now has a much greater work volume. Furthermore, each cart/PUMA manipulator has three redundant DOFs, thereby permitting many configurations of the manipulator for each Cartesian space specification of position and orientation.

The robotic transporter operates in one of two modes. In the first mode, the two carts can be powered and positioned independently of one another. Limit switches located on either end of the track and between the carts prevent excessive travel and collisions of the carts, respectively. The second mode permits the carts to be connected together, with one cart driving the other. Thus, the rotate and tilt axes on each cart remain independent, however, the linear travel of the two carts is constrained by a rigid link.

All axes on the transporter have twin pinion, backlash-adjustable drivers, allowing no lost motion between the motor shafts and the axes outputs. Each axis is driven by a DC-PWM servo motor/amplifier with a current-loop controller. A tachometer generator and encoder (800 lines/revolution) are mounted on each motor.

Integration of the transporter into the testbed became a research project area in itself. Specifically, a computer interface to the transport hardware was designed and implemented, as was a PID servo loop for the joints, [51]. Additionally, a technique for the global calibration of the transporter joints was developed, [52].

### Grippers

Two grippers were custom-designed at Rensselaer for use on the PUMAs, [53], [54], [55]. Each gripper has two fingers that can be actuated open and shut by a pneumatic cylinder. Each finger has a range of $35^o$, ($30^o$ open from the nominal position, and $5^o$ closed). A positional accuracy of approximately $0.15^o$ is achieved by using linear potentiometers and 8-bit analog-to-digital converters (ADCs).

The grippers provide two sensing functions in addition to grasping. Gripping force is measured by strain gauges located in either finger. A force range of 100 pounds per square inch (PSI), 80 PSI compression and 20 PSI extension, can be measured. Additional sensing is provided by an infrared crossfire sensor located near the end of the gripper fingers. Proximity to objects during grasping operations can be detected with this sensor. A detailed description of the CIRSSE grippers is given in Chapter 6.

## 1.5.2 Sensor Subsystems

### Vision

The vision system consists of five cameras and a programmable laser scanner. The cameras are used individually or pairwise to collect 2-D or 3-D scene data, respectively. The laser scanner is used as a pointing device or to provide structured light patterns.

The laser scanner, (commercially available from General Scanning, Inc., Watertown, Massachusetts), and two of the cameras (from Javelin Electronics, Inc., Torrance, California) are mounted on the ceiling above the Robot Transporter subsystem. These cameras are not mounted on servo-tripods and focusing is done manually. However, the cameras are arranged to provide a view of a large portion of collective workspace associated with the two cart/PUMA manipulators.

The laser scanner contains a 5 milliwatt laser. Experimentally, the frequency response of the mirror positioning servos in the scanner permits approximately 3500 points to be specified each second. In a commonly used region of the workspace, (i.e., in the plane of the flanges of the Robotic Transporter carts), the laser scanner can illuminate any point within an approximate area of 8 feet square. Each of the overhead cameras provides a 60 Hz interlaced black-and-white video signal, with a pixel array size of 576 (H) by 485 (V). Stereo vision by the overhead cameras is limited to the intersection of the cameras' fields of view. This region is considerably smaller than the field of view of an individual camera, but provides adequate coverage for one of the cart/PUMA manipulators.

Two other cameras are mounted on opposite sides of one of the PUMAs' forearms. These camera mountings are fixed relative to the PUMA forearm, but via motion of the PUMA and/or cart, the cameras can be positioned within the Testbed workspace. These cameras are commercially available from Toshiba American Consumer Products, Inc., Wayne, New Jersey, and are small (17.5 by 52.9 millimeters) and light (25 grams) devices suitable for mounting on the PUMA robots. Support hardware for each camera is remotely located. Each camera contains a 0.5 inch CCD sensor, resulting in an effective pixel grid of 570 (H) by 485 (V). Focusing is done manually.

The fifth camera, which is of the same type as the forearm cameras, is not permanently mounted. Possible experiments utilizing the fifth camera include having it mounted on a fixed tripod, on a servo-tripod, on a mobile robot, or on the forearm of the second PUMA robot.

Any two cameras can be accessed to provide range information about objects in the field of view, [56]. Accurate range data, i.e., 3-D location, for an object requires calibration of the dual camera system. Calibration is the process by which parameters of the camera(s) are determined so the mathematical relationships between the actual 3-D location of a point and its representation in the image plane(s) of the camera(s) is known. Chapter 5 discusses single- and multiple-camera calibration issues and techniques for performing accurate calibrations.

#### Force/Torque

A force/torque sensor is mounted on the wrist of each PUMA. These devices are commercially available from Lord (Assurance Technologies, Inc.), Garner, North Carolina. This type of sensor data is necessary for experiments involving compliant insertion tasks and the carrying of a common load between two or more manipulators. Remotely located support hardware samples the eight strain gauges located on each sensor, and provides a serial and/or parallel communications interface to other processor subsystems. The strain gauge values can be mathematically related to the 3-D force and 3-D torque vectors that the manipulator is experiencing at its gripper.

### 1.5.3 Processor Subsystems

The various sensor and manipulator subsystems within the testbed require processing—data must be collected, stored and disseminated, computations performed, and command signals generated. Differing requirements for the amount of data, speed and complexity of computations, and available communication interfaces make it necessary to have several processor subsystems.

### SUN Workstations

Within CIRSSE, several SUN Workstations are housed to provide general computing services. These computers are commercially available from Sun Microsystems, Inc., Mountain View, California, and are configured with the SunOS and X Windows operating systems. X Windows, while not public domain software, is available free from the Massachusetts Institute of Technology.

As part of the testbed operations, the SUNs are used for software design, compilation, and maintenance. The backbone of the CIRSSE simulation software consists of PRO-MATLAB, commercially available from The MathWorks, Inc., South Natick, Massachusetts, and CimStation, commercially available from SILMA, Inc., Cupertino, California. Additionally, SPNP [14], and GreatSPN [15], are used for Petri net modeling and simulation.

Network connections between the SUNs, the Motion Control Cage, and the Datacube are used to download software and upload experimental data. As the complete functionality of the Theory of Intelligent Machines is developed and comes on-line, the SUNs will serve an increasing role as the higher levels of the Intelligent Machine Hierarchy.

### Motion Control Cage

The primary processing subsystem for manipulators within the testbed is the Motion Control Cage. This subsystem is a VME-Cage configured with five MC68020 CPU boards (operating at 20 MHz), a network interface to the SUNs, and a 2-channel digital I/O board. Additionally, 5 MBytes of shared memory is available to complement the 1 MByte of local memory on each CPU board. The backplane of the VME-Cage, namely the VME-Bus, and the CPU boards mentioned above are commercially available from Motorola, Inc., Tempe, Arizona.

Each CPU board runs the VxWorks operating system, commercially available from Wind River Systems, Inc., Alameda, California. VxWorks is a single-CPU, single-user, multi-tasking operating system, (thereby, in essence, making the CIRSSE VME-Cage a five user system, with each user "owning" one CPU). Thus, to make the entire collection of CPUs and shared memory within the VME-Cage available to a single user, a set of services had to be developed on top of VxWorks. These services are the heart of the Motion Control System, which is described in Section 1.6.

The Unimation Controller for each PUMA is used to provide an interface to the robot's hardware and the current loops for the joint motors. The standard Unimation programming language (i.e., VAL) is not used, since all control computations are performed on the Motion Control Cage or other CIRSSE processors. Using a VME-Bus to Q-Bus memory mapper, the Motion Control Cage interfaces to the Unimation Controller for each PUMA. The company that produced the memory mapper used in the CIRSSE Testbed, no longer supports this product. However, several other companies produce these devices. The digital I/O board is used to interface the VME-Cage with the Robotic Transporter.

**Datacube**

The Datacube is a VME-based image processing system, commercially available from Datacube, Inc., Peabody, Massachusetts. The CIRSSE Testbed Datacube is configured with the ImageFlow software package. Image processing pipelines are created and controlled using an object-oriented approach.

Many of the primitive vision functions are implemented in hardware. For example, image thresholding, specification of a Region Of Interest (ROI), and convolution, are performed by specialized hardware. Software modules control these hardware functions by setting registers.

**M68HC11EVB**

The M68HC11EVB is a single-board computer commercially available from Motorola, Inc., Tempe, Arizona. It is used to perform the communication and control functions for the CIRSSE gripper. Other processor subsystems communicate directly with this single-board computer to access gripper functions. The board includes a watch-dog timer, an EEPROM, an analog-to-digital converter (ADC), and a serial interface for a remote terminal.

### 1.5.4 Mobile Robot

In a parallel effort to the CIRSSE Testbed, research on issues concerning mobile robots is being conducted in the Mobile Robot Laboratory (MRL). This laboratory also includes several manipulator, sensing, and processor subsystems.

The primary manipulator in the MRL is a TRC Labmate, commercially available from Transitions Research Corporation, Danbury, Connecticut. This mobile base can carry up to 200 pounds at maximum speeds of 1 meter/second. The MRL robot is configured with a 15–20 hour battery that powers the motor amplifiers, the on-board electronics, and a VME-Cage that is carried on top of the robot. The system is completely self-contained except for a network connection to a host processor. It is planned to replace this connection with a wireless link.

Eight pairs of ultrasonic and infrared sensors have been mounted on the Labmate robot. The ultrasonic sensors provide the primary source of range information, and are accurate between 0.2 to 5 meters. Detection of the edges on the floor is accomplished by the infrared sensors. These sensors are also used for gross position detection, as they are sensitive to a grid pattern painted on the laboratory floor. Additionally, a camera can be mounted on top of the robot.

Obstacles in the environment are simulated with boxes. These obstacles are used for experiments involving path planning in static environments. A pneumatic bumper installed on the Labmate is used as a safety device (i.e., it causes the motors on the robot to be disabled when it is deflected).

Inclusion of a mobile robot as another manipulator in the CIRSSE Testbed would demonstrate the integration of dramatically different systems. The mobile robot would provide increased workspace, and additional degrees-of-freedom for manipulation and/or camera positioning.

## 1.6 Design of the MCS for the CIRSSE Testbed

The previous sections described the equipment in the CIRSSE Testbed and its intended purposes. This section discusses the software system designed and implemented at CIRSSE to affect the commands to the testbed manipulators—succinctly, this software is called the Motion Control System (MCS). The need for such a system is motivated below by comparing two possible experiments for the testbed. This is followed by discussions of the functional requirements of the MCS and some of the design issues. Finally, an estimate of the manpower requirements for the design and implementation of the MCS is given.

### 1.6.1 Purpose of the MCS

The operational purpose of the testbed is to perform experiments. A great percentage, if not all, of these experiments will make use of the manipulators, and therefore require the generation of control signals and communication with hardware. While part of the problem is computing trajectories, writing torques, reading joint encoders, monitoring safety, closing the grippers, etc., the system design is complicated by differences in the experience levels of its intended users, the types of experiments that will be performed, and the capabilities of the hardware. Because of this complexity, development of the MCS is non-trivial. Evaluating the trade-off between designing and implementing such a system in-house, contracting for a custom-built system, and configuring an already existing system to a particular testbed environment, is a difficult and situation-dependent topic not covered here. The intention of this section is to provide the reader with an understanding of the design considerations made at CIRSSE. This should be an aid regardless of which method is chosen to bring a robotic testbed on-line.

To motivate the need and usefulness of a well-designed system, consider two possible experiments. The first experiment is designed to test a dual arm force control algorithm with the PUMAs carrying a common load. The Robotic Transporter is used to inject disturbances to the PUMA positions. The second experiment is designed to perform camera calibration by moving a reference object to various places in the workspace so the camera parameters can be computed.

While each experiment involves motion of the manipulators, there are several significant differences in the requirements made on the MCS:

- **Number of manipulators:** Experiment 1 uses three manipulators in the testbed (both PUMAs and the Robotic Transporter), whereas experiment 2 uses only a single PUMA robot.

- **Sensor data for manipulation:** Force/torque sensors, joint encoders, and possibly the vision system are used for feedback control in the first experiment. While the second experiment makes use of the joint encoders and the vision system, only the former is used to maintain the manipulator's position. In this case, the vision system is used to acquire data for a separate calibration process.

- **Supervisory control:** Either of the experiments could be supervised by an automated process, human interaction, and/or pre-planned instructions that are simply read from a file. The supervision instructions could range from setpoints for the joint controllers to high-level commands for setting the operational mode of the system.

- **User proficiency:** For the first experiment, the user of the MCS requires a system that is highly configurable in terms of control laws, communication rates, and processor allocation—the system cannot be constraining in these ways or any other that would hinder the experimental verification of the theory. Contrasting this requirement of unlimited flexibility with that needed for the second experiment, the usefulness of a system that can provide basic functionality without assuming extensive knowledge of control, kinematics, etc., by the user, is also evident.

- **Data logging:** Data logging, i.e., the amount, quality, and speed of storing experimental data differs between experiments. In our examples, the second experiment requires no data logging, or, minimally, the actual setpoints achieved by the manipulator. The first experiment, on the other hand, may require storage of many state variables (both, those maintained by the system and those maintained by the user), raw sensor data, etc.

- **Experiment run-time:** Camera calibration can be a long and tedious process, requiring the MCS to operate for hours. While it is hopeful that the first example experiment could last indefinitely, the realistic running time is on the order of minutes.

- **Inter-processor relationships:** Consider the first experiment, whereby task decomposition involves servoing the arms independently, performing grasps of the object, and then the dual arm cooperation. The MCS relies on other processors for this task decomposition. The MCS may be subordinate to the task decomposition process, or it may execute as a stand-alone entity. If vision is added to the experiment for position verification, the MCS would probably serve as the master of the vision processor. Contrastingly, in the second experiment, the MCS serves in a slave role to the vision processor, and depending on how setpoints are generated, may execute the experiment without relying on the computational resources of other processors.

### 1.6.2 Functional Requirements

Considering the issues listed above, the purposes of the testbed, and the variety of manipulators, the design and implementation of the MCS is a comprehensive task. Prior to writing the first line of code, a MCS Design Team was formed. After a review of the literature on current robotic systems, consultation with other organizations conducting robotic experimentation, and conversations with potential users of the CIRSSE Testbed, detailed functional requirements for the system were generated. These requirements reflected a good understanding of advantages and disadvantages of current systems, the experimental requirements of researchers, and speculation on future equipment in the testbed.

One of the foremost intentions of the design was to have a system that would not have to be scrapped because of changes in processors, additional manipulators, or unforeseen experiments. It is, however, impossible to design a system that is entirely useful for today and also for the unpredictable future. Therefore, the approach taken was to split the design between hardware dependent and hardware independent modules. This, in itself, is standard software engineering. However, using this discrimination, generic services, rather than a complete solution to all the manipulation problems, were designed. Examples of services include inter-process communication and high-speed task synchronization.

The longevity of the services will be far greater than the complete solution. Complete solutions become useless with the addition of equipment or changes of processors. Service design does not have this drawback—regardless of changes, services such as communication, synchronization, etc., will be needed. Additionally, limiting the core of the MCS to services increases the hierarchical nature and modularity of the design. This core functionality is called the CIRSSE Testbed Operating System (CTOS), and its implementation is complete. The following list describes the services included in CTOS:

- Bootstrap mechanism to load and initialize software modules on the various processors.

- Task identification mechanism to uniquely reference all software modules on the various processors.

- Message passing mechanism to provide communication between software modules.
- Synchronization mechanism to provide high frequency, low latency execution of software modules.
- Inter-processor blocking mechanism to provide efficient scheduling of software modules.

The MCS, however, consists of much more than CTOS—other components include joint controllers, trajectory generators, user interface, and system state management. Since space limitations prevent a discussion of these components, the interested reader is referred to the MCS/CTOS system documentation, [57], [58], [59], [60], [61], [62], [63], [64], available from CIRSSE. The remaining discussion provides more details about the MCS core services, i.e., CTOS. The utility of these services is conveyed via an overview of their purposes, functionality, and flexibility.

**Bootstrap**

Experiments will be coded by user programs that may be distributed throughout the CIRSSE Testbed processors. Specifically, even simple experiments for measuring the quality of multi-joint PID algorithms may require more computational power than provided by a single CPU in the Motion Control Cage. More elaborate experiments may use complex task planning, trajectory generation, error detection and collision avoidance algorithms, thereby requiring computational resources on several of the Motion Control Cage CPUs and/or the SUN Workstations. A mechanism whereby the user specifies which software modules will be loaded on the various processors is necessary. Additionally, a method of allocating data storage in shared memory and making these allocations known to all the processors, is needed. This mechanism is also advantageous in that it provides a central place, for modifying the way in which the software is loaded, disjoint from the software to implement the algorithms necessary for the task. Thus, software modules are easily redistributed to balance CPU loads, and modifications to the testbed that result in improved CPU performance or an increased number of processors are easily accomplished.

The bootstrap mechanism does more than load software on the various processors. System and user software modules typically require initialization. This is particularly true given the distributed and user-configurable environment of the MCS. To meet this requirement, the bootstrap software broadcasts three phases of initialization messages (message passing is discussed below). Each phase of initialization guarantees that all software modules have completed the previous phase. This allows the software modules to conduct initialization in isolation, followed by inter-module initialization, followed by a synchronization message.

### Task Identification

System and user software modules can be distributed by the user via the bootstrap mechanism. Message passing, discussed below, is a means for software modules to communicate with one another. To facilitate this communication, the tasks associated with the software modules have to be uniquely identified.

CTOS provides a task identification service that assigns to each task a number specifying the computer system, the CPU board, and of course, the task. The assignments are made by the bootstrap when the tasks register themselves by a unique symbolic name. A task can find another's identification by referring to the task's symbolic name. Thus, the task identification service resolves at run-time the distribution of software modules. The identification numbers of tasks are used by the message passing service to resolve routing issues.

### Message Passing

Object-oriented programming is a state-of-the-art software development paradigm. This paradigm includes many features that make software development, integration, and enhancement/replacement more reliable. To make an analogy, object-oriented software modules are compared to hardware chips—not only are they modular, but they maintain their own state variables, and provide only a limited, yet well-defined, interface for communication of data and/or commands. It is only these commands that can cause state changes. The object-oriented paradigm also includes object classes, inheritance, and message passing. A disadvantage of the object-oriented paradigm is its tendency for slow execution speed. Available object-oriented programming languages include C++, Objective C, and SmallTalk.

As mentioned above, message passing is part of the object-oriented paradigm. The inefficiency of using an object-oriented language for the full implementation of a real-time system can be prohibitive. However, an interesting alternative is to use an object-oriented approach for the higher levels of the system (typically less time critical), and retain useful aspects of the paradigm at the lower, more time critical levels. This approach was taken in the development of the MCS/CTOS. Specifically, message passing, as an isolated tool, provides a flexible interface between distributed software modules and a compatibility to higher levels.

The message passing mechanism is implemented in CTOS as one of its fundamental services. The contents of a message are similar to those found in a network packet: the sender of the data, the receiver(s) of the data, the data itself, and various status flags (e.g., to indicate that a reply to the message is expected). Any software module can construct a message and send it to any other module. Routing messages to modules on the same CPU, on the same VME-Cage, or on another machine on the network, is handled by the message service manager, and is virtually transparent to the user.

Minimal effort is required by the user to write modules to send/receive messages. Specifically, an "event handler" is written in a standard form. Thus, the arrival of a message is considered as an event, and the message service manager calls the appropriate event handler. This handler can then choose to ignore a message, let the system respond with a default action, or process the message by calling user-supplied functions.

The user is free to declare and use new messages in addition to those provided by the MCS/CTOS. Thus, improvements to algorithms require only the programming of the new functionality (and possibly the addition of new messages)—the integrity of other software modules remains, and a consistent interface is maintained via message passing.

### Synchronization

Effective control of the testbed manipulators requires periodic sampling of sensors and commanding of actuators. For example, typical PID control loops for the PUMA joints have periods of a few milliseconds (i.e., 200–1000 Hz). These control loops contain several software modules, including, hardware interfaces, joint controllers, and trajectory generators. Some modules (e.g., hardware interfaces) are system modules, while others (e.g., joint controllers) are user-implemented or chosen by the user from a library of standard modules. Communication between these software modules needs to occur regularly, at high frequency, and with low latency.

Each CPU board in the Motion Control Cage contains a clock chip that can generate periodic interrupt calls to a user-specified software module at a user-specified frequency. This mechanism provides local synchronization of a software module on each CPU. The distributed nature of the MCS/CTOS requires synchronization of software modules running on (possibly) different CPUs. The difficulty of having all the CPU boards initialize their interrupts at precisely the same time, and the benefit of having a centralized source of interrupts for system start-up, shut-down, and emergency stops, make it desirable to have one source of interrupts that cause responses on all the CPU boards.

To provide a centralized source of interrupts, a synchronization service is included in the CTOS. One of the Motion Control Cage CPUs is designated as the synchronization master. Using its auxiliary clock chip, this CPU maintains the MCS clock and generates bus interrupts, which are received by the remaining CPUs. The synchronization service manager on each CPU then handles local synchronization issues. Software modules can, therefore, be synchronized simultaneously regardless of their distribution. The use of interrupts provides high frequency, low latency synchronization.

Synchronization of a software module can be made impossible by aperiodic hardware or software error conditions, communication delays, or algorithmic anomalies. This type of situation may indicate a non-recoverable failure in the synchronous software module or an aperiodic "hiccup" that can be handled gracefully. Upon detecting that a synchronous software module has not completed its task prior to the subsequent synchronization period, the service manager disables synchronization of that particular module and executes a user-supplied exception handler.

Using motion simulation software instead of moving the actual manipulators, is valuable for testing and improving the performance of controllers, trajectory generators, etc. However, often the simulation cannot perform its calculations in real-time. To accommodate this requirement, the synchronization service manager provides a mechanism to scale the MCS clock in a manner that is virtually transparent to the user.

Thus, this service performs global synchronization of software modules, and provides a paradigm that permits the overruns to be detected and handled.

### Inter-processor Blocking

Because VxWorks is effectively a single-CPU, single-user, multi-task operating system, task scheduling and memory management issues are effectively handled *on* each CPU, but there is no consideration of these issues *between* CPUs. One requirement for any software system is the efficient use of the available CPUs. High priority software modules should indicate to the operating system that other modules can run while they await data, (called blocking). Additionally, the operating system needs to permit execution of a software module to continue when the necessary data becomes available, (i.e., the operating system unblocks the module). VxWorks, like many operating systems, provides constructs called *semaphores* to manage the blocking and unblocking of software modules. However, blocking/unblocking is only considered locally on each CPU.

To maximize the CPU efficiency in the distributed environment of the MCS, an "inter-processor semaphore" is needed. Semaphores are useful for more than blocking/unblocking, and implementation of the full capabilities of semaphores between processors would be difficult. Fortunately, analysis of the requirements indicates that the functionality of an unblocking mechanism that is available between processors was sufficient. In other words, a software module running one CPU can cause the operating system on another CPU to unblock a specified software module (which is, of course, running on that CPU). This is done with low latency, so as to minimize the time between the decision to unblock and actually performing the unblock.

To provide the required functionality, an inter-processor blocking service was implemented in CTOS. A software module that registers itself with this service can block (awaiting some particular data) and subsequently be unblocked when this data becomes available, regardless of which CPU is running the unblocking software module. If software modules are distributed in such a way that they run on the same CPU, the service manager automatically uses the local semaphore capabilities of the operating system. Thus, this service gives the user a virtually transparent mechanism to block processes (and thereby maximize CPU efficiency) regardless of the distribution of the software modules.

### 1.6.3 Manpower Requirements

Development of the MCS is an extensive and on-going task. The following provides an estimate of the human resource requirements for the work completed to date. The work associated with supervising the design of the custom-built Robotic Transporter and outfitting the testbed is not considered in the following. It is assumed that a particular robotic testbed environment is in place, and that the reader is interested in designing and implementing a software control system similar to the MCS.

Preliminary design of the MCS resulted in an architecture for core services (i.e., CTOS). This architecture was designed and implemented by the MCS Design Team, which consisted of four people, each spending between one quarter to three quarters of their time on this project. By deciding on an architecture that permitted a set of core services to be designed and implemented first, the development process fit well with the academic environment. It was useful to have an overall system architecture in mind (subject to modifications and improvements), the architecture of the core services thoroughly developed, and a small design and implementation team for these services. The complete development of CTOS, (including a review of current systems, and consultation with potential users of the testbed), was accomplished in 30 weeks.

The completion of CTOS provided a bare-bones system that could not effectively control the robotic manipulators. The completion of CTOS coincided with the middle of the summer, thus making many more people available for design and implementation. The stability of the core services provided a foundation from which the larger group, with varying levels of programming proficiency, could build. In approximately 8 weeks time, control of all 18-DOFs in the testbed have been demonstrated. Additionally, demonstrations of gravity compensation, simple master/slave (i.e., position following), and self-motion control have been created, and libraries of controllers, kinematic routines, etc., are undergoing development.

## 1.7 High-Level Control of the CIRSSE Testbed

The role of the *Coordination Level* in the Theory of Intelligent Machines is to interpret and manage the plans from the *Organization Level*, add real-time details as appropriate, and communicate instructions to the *Execution Level* in order to coordinate operations. In addition, the Coordination Level monitors machine performance, providing feedback to refine short-term decision making and to assist replanning when the nominal plan is not successful, [21].

### 1.7.1 Coordination Level of Intelligent Machines

The Coordination Level is organized into a tree structure, as shown in Figure 1.2, with a *Dispatcher* at the root and multiple *Coordinators* at the leaves, [19]. All communication with the Organization Level is conducted through the Dispatcher, while the Coordinators interface with their corresponding Execution Level functions. Currently, the CIRSSE Testbed supports four Coordinators that manage manipulator motion, grippers, vision services, and path planning.

Viewed from the bottom up, the Dispatcher is the first component of the Intelligent Machine that deals with the machine as a whole. Thus, the Dispatcher can coordinate actions across execution-level functions, e.g., move a manipulator to a position determined by visual sensing. As its name suggests, the Dispatcher's primary function is to receive commands from the Organization Level and to dispatch these commands to the appropriate Coordinator for implementation. Organization-level commands often must be decomposed by the Dispatcher into a sequence of substeps and real-time values must be supplied for the abstract values used by the planner. This description is a simplification, however, as the Dispatcher is called upon to perform a number of other functions, such as supporting system-wide communications, scheduling resources, detecting and correcting intermediate-level errors, and performing some run-time planning.

A Coordinator can be considered as an expert in applying the deterministic functions of a narrow domain of the Execution Level, [65]. For instance, the Motion Coordinator may have several strategies for moving the tool frame of a manipulator into a requested position and orientation (e.g., employing redundant degrees-of-freedom), and can choose the strategy with the highest probability of success based on current constraints of the environment (obstacles) and time goals. While a nominal approach may be recommended by the organization-level planner, it is the Coordinator's responsibility to reliably accomplish the requested task in real-time. Errors are handled first by the Coordinator, and are passed up to the Dispatcher only when a local strategy is not adequate to resolve the condition. In some instances the Dispatcher must defer error resolution to the Organization Level, whereby operations may be replanned. Feedback is also provided through the Coordinators in order to update knowledge bases and improve the decision making of the Intelligent Machine.

### 1.7.2 Petri Net Implementation

Recent research at CIRSSE has employed a linguistic decision schemata approach to model the Coordination Level, [66]. In particular, Petri Net Transducers have been proposed, as a means to translate an organization-level command language into a sequence of coordination-level subtasks, [67]. Petri nets (PNs) are tools for modeling the dynamic behavior of discrete event systems that promise to be a useful method for both decomposing organization-level commands and for real-time control of the Intelligent Machine. Hence, current development work is aimed at implementing PN software that is integrated with the CIRSSE Testbed, and at designing the Dispatcher and Coordinator Petri Net Transducers.

Ordinary PNs are directed graphs with two types of nodes called *places* and *transitions*, which are connected by *arcs*, [68]. A PN graph uses circles for representing places and bars for transitions. A non-negative number is associated with each place to indicate the number of *tokens* it contains. Tokens reside in places, travel along arcs, and their flow through the net is regulated by transitions. A transition is *enabled* when each of its input places contains at least one token. An enabled transition can *fire*, whereby a token is removed from each of its input places, and a token is added to each of its output places. Between consuming input tokens and generating output tokens, the fired transition may call a function to perform additional processing, [69].

*Petri Net Transducers* (PNTs) are PN realizations of linguistic decision schemata that translate an input language to an output language. They are an extension of ordinary PNs in which transitions optionally have an additional enabling condition. This optional condition is read from an *input tape* composed of elements of the input language. When a PNT is used to model the Dispatcher or a Coordinator, the places represent the possible states, the tokens represent the validity of these states, and the transitions represent events that can cause state changes. The Dispatcher is a reconfigurable PNT defined by the Organization Level, while the Coordinators have a fixed structure.

A coordination-level PNT, i.e., the Dispatcher or a Coordinator, may be composed of many subnets that execute concurrently and are distributed among several of the SUN workstations and VME-Cages of the CIRSSE Testbed. This heterogeneous, distributed organization of an individual PNT is made possible by using the CTOS to communicate between the subnets. Because each subnet is a CTOS task, it executes independently and can be located on any computer reachable via the testbed's local network. The operations of subnets are coordinated by the message passing service described above.

In the example shown in Figure 1.3, subnet PNET-A is connected to subnet PNET-B. This is done by designating place `p1` of PNET-B as a "remote" output place of PNET-A's transition `t1`, and place `p2` of PNET-A as a "remote" output place of PNET-B's transition `t2`. When transition `t1` fires, a message is sent

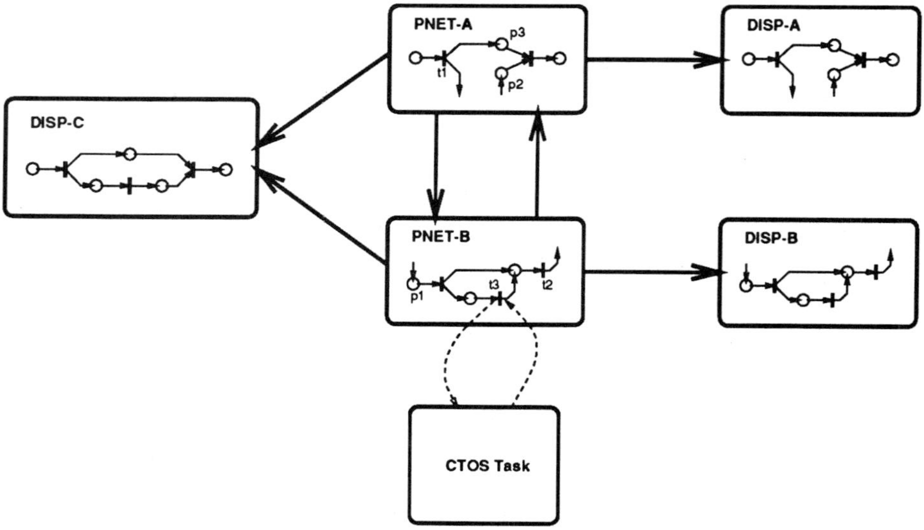


Figure 1.3: Distributed Petri Net

from task PNET-A to task PNET-B indicating that a token has been added to place `p1`; tokens are also written to any local output places (`p3` in this case). Messages may also be used to invoke an action when a transition fires. For instance, the figure shows PNET-B sending a message to a CTOS task when transition `t3` fires. The destination CTOS task may reside on any computer. With this architecture, a coordination-level PNT, e.g., the Motion Coordinator, may be implemented on a SUN Workstation, yet easily communicate with an execution-level function running on a VME-Cage, such as the Motion Control System.

In addition to allowing the composition of a PNT as an assembly of subnets, the CTOS architecture supports flexible displays of PNT execution. Figure 1.3 shows an example with two subnets attached to three displays. Two of the displays, DISP-A and DISP-B, have been configured to show their corresponding subnets. The remaining display, namely DISP-C, shows a subset of places and transitions from two different subnets. This capability of many-to-many connections between subnets and displays enables the construction of overview displays that summarize the operation of a complex PNT spanning many subnets. Operation of the Dispatcher and Coordinators can be monitored by means of these overview displays.

Another flexibility inherent to this design results from separating the subnet execution and displays into separate tasks. The display tasks must run on the workstations to access graphics services. The subnets, however, may execute on either the workstations or VME-Cages. Thus, a Coordinator PNT may be designed with a portion running on the same VME-Cage as the execution-level function with which it interfaces.

Underlying the decomposition and sequencing of commands performed by the Dispatcher and Coordinators is the flow of data required to implement these commands. To support this flow of data the Coordination Level employs a *Data Object Manager* task to maintain globally available data. The Data Object Manager's role is to receive data produced by coordination-level and execution-level functions, and to later retrieve this data on demand.

### 1.7.3 Coordination Level Research

Implementation of the Coordination Level is not complete and research continues in a number of areas. Research topics include the following: 1) integration with run-time organization-level functions to replan operations; 2) incorporation of a world model and a learning algorithm to dynamically improve decision making at the Coordination Level; 3) further development of Coordinator error recovery processes to enhance reliability and robustness; 4) adaptation of PN analytical techniques and PNT theory to guide synthesis of coordination nets from subnets; and 5) development of measures for complexity and analysis of the Coordination Level based on these measures.

## 1.8 Summary

Automation of in-orbit activities is beset with many challenging engineering problems and issues. While problems such as planning, control, and machine intelligence, present themselves in all complex robotic systems, they are complicated by the space environment and the types of tasks to be performed. Other issues, such as zero-gravity and adverse lighting conditions, are unique to space. Automated space-based assembly of structures provides an ideal experiment for researching these engineering problems.

Several facilities exist for robotic experimentation. The Center for Intelligent Robotic Systems for Space Exploration (CIRSSE) Testbed houses a unique set of hardware for researching space-based robotic engineering problems. The testbed includes two independent robotic manipulators mounted on a robotic transporter, totaling to 18-DOFs, and custom-designed sensing grippers. Other data are provided by force/torque sensors and the vision system, which includes static stereo cameras, arm-mounted stereo cameras, and a laser scanner. Several processing systems, ranging from single-board computers to workstations to real-time multi-processor VME-cages, are utilized to implement an experiment.

The need for a flexible control system for the testbed's manipulators has led to the development of the Motion Control System (MCS). This software system provides a vehicle to use or experiment with the components of robotic controllers. At the heart of the MCS is a set of core services, i.e., the CIRSSE Testbed Operating System (CTOS). These services—bootstrap, message passing, synchronization, and inter-processor blocking—provide mechanisms by which the user can readily take advantage of the multi-processor environment of the testbed.

A hierarchical approach has been proposed for the division of tasks within a complex robotic system. High-level control of the CIRSSE Testbed is being implemented by the intermediate level, i.e., the Coordination Level, of this hierarchy. Extensions to Petri net models are designed to provide control of lower functions and give the user a mechanism to observe the evolution of tasks.

Solutions to the space-based robotic engineering problems will draw from the efforts of several investigators and the integration of knowledge from several fields. The following chapters address some of the engineering problems identified in this chapter: truss assembly, planning, vision, and control.

## Acknowledgments

Funding for the development and operation of the CIRSSE Testbed has been provided by NASA under Grant NAGW-1333. The research conducted in the CIRSSE Testbed is the result of many investigators. Wherever possible, references were cited to acknowledge these efforts.

In addition to the research, much effort has been expended in accomplishing a service objective: bringing the CIRSSE Testbed on-line and making it a useful research tool. The on-going success of this monumental task is the result of the contributions of many people: the MCS Design Team—Keith Fieldhouse, Don Lefebvre, Steve Murphy, and Jim Watson—Linden Carmichael, Jay Cosentino, Kevin Holt, Amy Lin, Russ Noseworthy, Lance Page, Joe Peck, Art Ryan, Sanjeev Seereeram, Deepak Sood, Tim Southward, Dave Swift, Jodi Tsai, Ken Walter, John Weaver, and Lee Wilfinger.

## 1.9 References

[1] L. S. Homem de Mello and A. C. Sanderson, "AND/OR graph representation of assembly plans," *IEEE Transactions on Robotics and Automation*, pp. 1888–1899, April 1990.

[2] L. S. Homem de Mello and A. C. Sanderson, "A correct and complete algorithm for the generation of mechanical assembly sequences," CIRSSE Report 33, Rensselaer Polytechnic Institute, Troy, NY, June 1989.

[3] B. Armstrong, O. Khatib, and J. Burdick, "The explicit dynamic model and inertial parameters of the PUMA 560 arm," in *Proceedings of the*

*1986 IEEE International Conference on Robotics and Automation*, (San Francisco, California), pp. 510–518, April 1986.

[4] R. A. Howard, *Dynamic Probabilistic Systems, Vol. 1, (Markov Models)*. Wiley, 1971.

[5] L. Kleinrock, *Queuing Systems, Volume 1: Theory*. Wiley, 1975.

[6] R. Suri, "Perturbation analysis: The state of the art and research issues explained via the G/G/1 queue," *Proceedings of the IEEE*, vol. 77, pp. 114–137, January 1989.

[7] M. K. Molloy, "Performance analysis using stochastic Petri nets," *IEEE Transactions on Computers*, vol. C-31, pp. 913–917, September 1982.

[8] P. Ramadge and W. M. Wonham, "The control of discrete event systems," *Proceedings of the IEEE*, vol. 77, pp. 81–98, January 1989.

[9] C. A. Petri, *Kommunikation mit Automaten*. PhD thesis, University of Bonn, Bonn, West Germany, 1962.

[10] C. V. Ramamoorthy and G. S. Ho, "Performance evaluation of asynchronous concurrent systems using Petri nets," *IEEE Transactions on Software Engineering*, vol. SE-6, pp. 440–449, September 1980.

[11] M. A. Marsan, G. Conte, and G. Balbo, "A class of generalized Petri nets for the performance evaluation of multiprocessor systems," *ACM Transactions on Computer Systems*, vol. 2, pp. 93–122, May 1984.

[12] R. Y. Al-Jaar and A. A. Desrochers, "Performance evaluation of automated manufacturing systems using generalized stochastic Petri nets," *IEEE Transactions on Robotics and Automation*, vol. 6, pp. 621–639, December 1990.

[13] J. F. Watson, III and A. A. Desrochers, "Applying generalized stochastic Petri nets to manufacturing systems containing non-exponential transition functions," *IEEE Transactions on Systems, Man, and Cybernetics*, vol. 21, July/August 1991.

[14] J. B. Dugan, A. Bobbio, G. Ciardo, and K. Trivedi, "The design of a unified package for the solution of stochastic Petri net models," in *Proceedings of the IEEE International Workshop on Timed Petri Nets*, (Torino, Italy), pp. 6–13, July 1985.

[15] G. Chiola, "A graphical Petri net tool for performance analysis," in *Proceedings of the 3rd International Workshop on Modeling Techniques and Performance Evaluation*, (Paris), AFCET, March 1987.

[16] J. Robinson and A. A. Desrochers, "Performance analysis of a robotic testbed control architecture," in *Proceedings of the 1990 IEEE International Conference on Robotics and Automation*, (Cincinnati, Ohio), pp. 1782–1787, May 1990.

[17] G. N. Saridis, "Intelligent robotic control," *IEEE Transactions on Automatic Control*, vol. AC-28, pp. 547–556, May 1983.

[18] G. N. Saridis, "An analytic formulation of knowledge-based systems for intelligent machines," CIRSSE Report 3, Rensselaer Polytechnic Institute, Troy, NY, October 1988.

[19] G. N. Saridis and K. P. Valavanis, "Analytical design of intelligent machines," *IFAC Journal Automatica*, vol. 24, pp. 123–133, February 1988.

[20] G. N. Saridis, "On the revised theory of intelligent machines," CIRSSE Report 58, Rensselaer Polytechnic Institute, Troy, NY, June 1990.

[21] G. N. Saridis, "Architectures for intelligent machines," CIRSSE Report 96, Rensselaer Polytechnic Institute, Troy, NY, July 1991.

[22] G. N. Saridis and M. C. Moed, "Analytic formulation of intelligent machines as neural nets," CIRSSE Report 1, Rensselaer Polytechnic Institute, Troy, NY, October 1988.

[23] M. C. Moed, "The organizer: Planning tasks with an emergent connectionist/symbolic system," CIRSSE Report 42, Rensselaer Polytechnic Institute, Troy, NY, September 1989.

[24] F. Wang and G. N. Saridis, "Coordination model for intelligent machines," CIRSSE Report 15, Rensselaer Polytechnic Institute, Troy, NY, January 1989.

[25] F. Wang, K. J. Kyriakopoulos, A. Tsolkas, and G. N. Saridis, "A Petri net coordination model of intelligent mobile robots," CIRSSE Report 50, Rensselaer Polytechnic Institute, Troy, NY, January 1990.

[26] R. K. Mathur, "A hierarchical planner for space truss assembly," CIRSSE Report 57, Rensselaer Polytechnic Institute, Troy, NY, May 1990.

[27] A. C. Sanderson, H. Zhang, and L. S. Homem de Mello, "Assembly sequence planning," CIRSSE Report 36, Rensselaer Polytechnic Institute, Troy, NY, August 1989.

[28] F. J. Schima, III, "Two arm robot path planning in a static environment using polytopes and string stretching," CIRSSE Report 77, Rensselaer Polytechnic Institute, Troy, NY, December 1990.

[29] R. Munger, "Path planning for assembly of strut-based structures," CIRSSE Report 91, Rensselaer Polytechnic Institute, Troy, NY, May 1991.

[30] K. J. Kyriakopoulos and G. N. Saridis, "Collision avoidance of mobile robots in non-stationary environments," CIRSSE Report 63, Rensselaer Polytechnic Institute, Troy, NY, September 1990.

[31] K. J. Kyriakopoulos and G. N. Saridis, "An optimal control strategy for collision avoidance of mobile robots in non-stationary environments," CIRSSE Report 83, Rensselaer Polytechnic Institute, Troy, NY, February 1991.

[32] K. J. Kyriakopoulos, "A supervisory control strategy for navigation of mobile robots in dynamic environments," CIRSSE Report 86, Rensselaer Polytechnic Institute, Troy, NY, March 1991.

[33] R. B. Kelley and S. Bonner, "A representation scheme for rapid 3-D collision detection," CIRSSE Report 9, Rensselaer Polytechnic Institute, Troy, NY, August 1988.

[34] J. Tornero, "Spherical-Object representation and fast distance computation for robotic applications," CIRSSE Report 64, Rensselaer Polytechnic Institute, Troy, NY, September 1990.

[35] V. J. Ree, Jr. and L. K. Lauderbaugh, "Issues in the safety of complex systems," CIRSSE Report 10, Rensselaer Polytechnic Institute, Troy, NY, November 1988.

[36] T. D. Montgomery, "A matrix safety frame approach to robot safety for space applications," CIRSSE Report 14, Rensselaer Polytechnic Institute, Troy, NY, December 1988.

[37] D. J. Minnick, H. Kaufman, and G. W. Neat, "Expert hierarchical adaptive controller for robotic systems," CIRSSE Report 27, Rensselaer Polytechnic Institute, Troy, NY, July 1989.

[38] R. Steinvorth, H. Kaufman, and G. W. Neat, "Model reference adaptive control of flexible robots in the presence of sudden load changes," CIRSSE Report 68, Rensselaer Polytechnic Institute, Troy, NY, October 1990.

[39] R. Steinvorth, "Model reference adaptive control of robots," CIRSSE Report 87, Rensselaer Polytechnic Institute, Troy, NY, March 1991.

[40] S. H. Murphy and J. T. Wen, "Stability analysis of position and force control for robot arms," CIRSSE Report 53, Rensselaer Polytechnic Institute, Troy, NY, May 1990.

[41] J. T. Wen and K. Kreutz-Delgado, "Motion and force control of multiple robotic manipulators," CIRSSE Report 88, Rensselaer Polytechnic Institute, Troy, NY, March 1991.

[42] J. T. Wen and S. H. Murphy, "Simulation and analysis of flexibly jointed manipulators," CIRSSE Report 56, Rensselaer Polytechnic Institute, Troy, NY, May 1990.

[43] F. Wang and J. T. Wen, "Nonlinear dynamical model and control for a flexible beam," CIRSSE Report 75, Rensselaer Polytechnic Institute, Troy, NY, November 1990.

[44] L. Lanari and J. T. Wen, "A family of asymptotically stable control laws for flexible robots based on a passivity approach," CIRSSE Report 85, Rensselaer Polytechnic Institute, Troy, NY, February 1991.

[45] S. H. Murphy, J. T. Wen, and G. N. Saridis, "Simulation of cooperating robot manipulators on a mobile platform," in *Proceedings of the 1990 IEEE International Conference on Robotics and Automation*, (Cincinnati, Ohio), pp. 1190–1195, May 1990.

[46] S. H. Murphy, J. T. Wen, and G. N. Saridis, "Analysis of cooperative robotic manipulators on a mobile platform," in *Symposium on Advances in Intelligent Systems*, vol. 1387, pp. 14–25, SPIE Proceedings, November 1990.

[47] J. E. McInroy and G. N. Saridis, "Reliability analysis in intelligent machines," CIRSSE Report 39, Rensselaer Polytechnic Institute, Troy, NY, August 1989.

[48] D. Sood, M. C. Repko, and R. B. Kelley, "Using multiple sensors for printed circuit board insertion," CIRSSE Report 17, Rensselaer Polytechnic Institute, Troy, NY, January 1989.

[49] M. C. Zhou and F. DiCesare, "Adaptive design of Petri net controllers for error recovery in automated menufacturing systems," *IEEE Transactions on Systems, Man, and Cybernetics*, vol. 19, pp. 963–973, September/October 1989.

[50] K. E. Nicewarner and R. B. Kelley, "Vision-guided gripping of a cylinder," CIRSSE Report 93, Rensselaer Polytechnic Institute, Troy, NY, forthcoming.

[51] J. L. Cosentino, "Development of a control system for a pair of robotic platforms," CIRSSE Report 62, Rensselaer Polytechnic Institute, Troy, NY, August 1990.

[52] A. Lin, "Calibration of Aronson 6-DOF robotic platform," Master's thesis, Rensselaer Polytechnic Institute, Troy, New York, August 1991.

[53] R. B. Kelley, J. W. Tsai, J. D. Bethel, and J. Peiffer, "Gripper for truss structure assembly," CIRSSE Report 70, Rensselaer Polytechnic Institute, Troy, NY, October 1990.

[54] J. W. Tsai, "M68HC11 gripper controller software," CIRSSE Report 90, Rensselaer Polytechnic Institute, Troy, NY, May 1991.

[55] J. D. Bethel, "M68HC11 gripper controller electronics," CIRSSE Report 92, Rensselaer Polytechnic Institute, Troy, NY, forthcoming.

[56] J. R. Noseworthy and L. A. Gerhardt, "Three dimensional vision requirements and applications in a space environment," in *Symposium on Advances in Intelligent Systems, SPIE Proceedings*, vol. 1387, pp. 26–37, SPIE Proceedings, November 1990.

[57] K. R. Fieldhouse, K. Holt, D. R. Lefebvre, S. H. Murphy, D. Swift, and J. F. Watson, III, "Lecture materials for the MCS/CTOS introductory course," CIRSSE Report 97, Rensselaer Polytechnic Institute, Troy, NY, August 1991.

[58] J. F. Watson, III, "Using the CIRSSE Testbed synchronous service," CIRSSE Technical Memorandum 4, Rensselaer Polytechnic Institute, Troy, NY, May 1991.

[59] D. R. Lefebvre, "CTOS tasks and bootstrap process," CIRSSE Technical Memorandum 5, Rensselaer Polytechnic Institute, Troy, NY, forthcoming.

[60] D. R. Lefebvre, "User guide to CTOS message passing," CIRSSE Technical Memorandum 6, Rensselaer Polytechnic Institute, Troy, NY, forthcoming.

[61] K. Holt, "MCS controllers," CIRSSE Technical Memorandum 8, Rensselaer Polytechnic Institute, Troy, NY, forthcoming.

[62] D. Swift and K. Holt, "MCS channel drivers," CIRSSE Technical Memorandum 9, Rensselaer Polytechnic Institute, Troy, NY, forthcoming.

[63] K. R. Fieldhouse, "CTOS inter-processor blocks," CIRSSE Technical Memorandum 10, Rensselaer Polytechnic Institute, Troy, NY, forthcoming.

[64] K. R. Fieldhouse, "The MCS state manager," CIRSSE Technical Memorandum 11, Rensselaer Polytechnic Institute, Troy, NY, forthcoming.

[65] F. Wang and G. N. Saridis, "A coordination theory for intelligent machines," *IFAC Journal Automatica*, vol. 26, pp. 833–844, September 1990.

[66] G. N. Saridis and J. H. Graham, "Linguistic decision schemata for intelligent robots," *IFAC Journal Automatica*, vol. 20, pp. 121–126, January 1984.

[67] F. Wang and G. N. Saridis, "A formal model for coordination of intelligent machines using Petri nets," in *Proceedings of the 3rd International Intelligent Control Symposium*, (Arlington, VA), 1988.

[68] J. L. Peterson, *Petri Net Theory and the Modeling of Systems.* Prentice-Hall, 1981.

[69] J. E. Peck, "Coordination level Petri net software," CIRSSE Technical Memorandum 7, Rensselaer Polytechnic Institute, Troy, NY, forthcoming.

# Chapter 2

# An Automated Assembly System for Large Space Structures

by: Ralph Will
Marvin Rhodes
William R. Doggett
Catherine L. Herstrom
Carolyn Grantham
Cheryl L. Allen
P. Daniel Sydow
Eric G. Cooper
NASA Langley Research Center
Hampton, VA
Cuang C. Quach
Marion A. Wise
Lockheed Engineering and Sciences Company
Langley Program Office
Hampton, VA

This chapter presents a research program that is being conducted at the NASA Langley Research Center to evaluate methods for automated telerobotic assembly of large space truss structures. The research program, facility, and hardware are described, and a summary of the results obtained to date is presented. The research is intended to identify the specific problems and considerations that must be dealt with when the total assembly process is addressed in detail. The assembly operation is structured to focus on a generic truss structure rather than a particular mission or design. However, the assembly issues being investigated are typical of many automated in-space processes. The program was initiated to study automated truss assembly operations and no other objectives were included in this initial investigation. The structure, although not space qualified, was designed to be representative of that required for future antennas and observatories.

Implementation of this research effort has required the integration of a strong interdisciplinary team composed of specialists in mechanical and structural design concepts, automation and robotics, controls, software development, configuration management, and electronic design and development. Many support personnel have contributed to the success of this research effort by their dedication and hard work. Having a well defined focus has enhanced the capability of the team to work together and meet the goals and milestones in a timely manner.

## 2.1 Introduction

A number of proposed space missions both to and from planet Earth require large truss structures to provide a stiff and stable platform for experimental measurements, observation antennas/telescopes and habitats/shelters. An example of a planned mission to planet Earth is shown in Figure 2.1 and an example of a mission from planet Earth is shown in Figure 2.2. Figure 2.1 is a large antenna, and Figure 2.2 is an aerobrake concept for a Mars mission vehicle. The truss structures for these, as well as other future missions may involve the assembly of thousands of members. Recent studies, [1], [2], have elevated the concern regarding the use of astronauts to perform in-space construction operations. Therefore, alternative assembly techniques to those that traditionally rely on astronaut Extra Vehicular Activity (EVA) must be explored. The reference studies have also recommended that astronaut Internal Vehicular Activity (IVA) time be minimized, which may limit the use of astronauts as teleoperators for assembly operations that may involve many hours of repetitive operations. One attractive alternative is to design an automated telerobotic system that can utilize either an astronaut or earth-based operator as an executive monitor who is called upon to intervene only when the automated system encounters a problem and requires assistance. This mode of operation, known as *supervised autonomy*, holds the most promise for the accomplishment of large or complex assembly and construction tasks with the limited crew resources available on orbit. Supervised autonomy has the additional advantage that it can be monitored from the ground if they involve no time-critical control functions.

To date, very little work has been directed toward interdisciplinary research to develop automated robotic assembly methods for large erectable truss structures. The current program was initiated within the past several years at the NASA Langley Research Center by merging basic technology in robotics and truss structure design and assembly. The program focuses on the actual au-

Figure 2.1: Space-Based Antenna

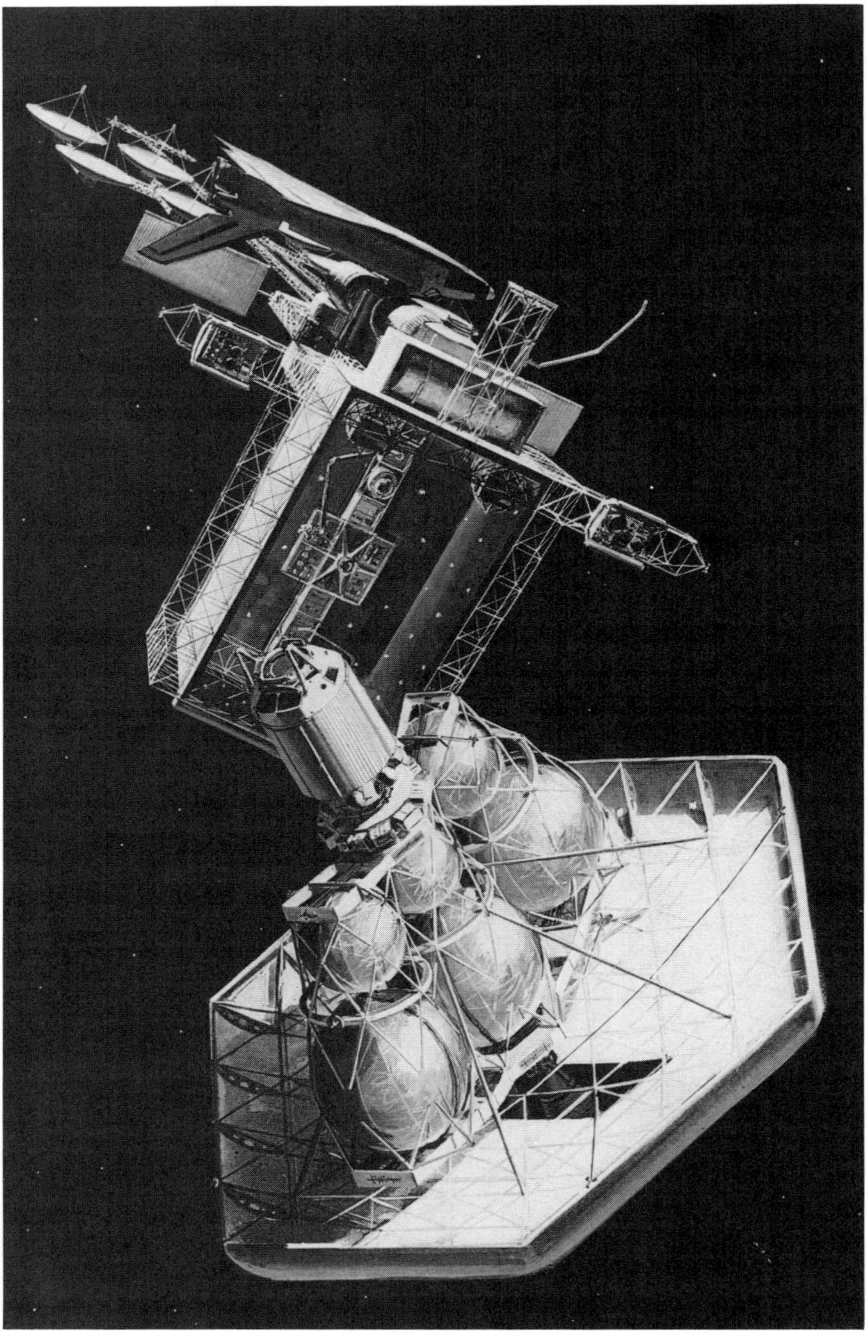

Figure 2.2: Truss-Supported Aerobrake Configuration

tomated assembly of a generic structural unit that serves as the basic element for very large structural systems. The specific objectives of the program are to determine what types of joining and end-effector mechanisms are suitable for telerobotic operation, to develop a software architecture capable of reliably performing a complex assembly task that incorporates realistic system errors, and to provide an operator interface compatible with the volume of internal information necessary for successful operation of the automated system. This effort provides practical experience in the assembly of truss hardware designed for automated telerobotic operations. An additional objective is to collect sufficient information on actual assembly operations to develop time lines characterizing automated telerobotic truss assembly and construction. It also has the potential to provide an operational testbed for many aspects of more advanced assembly techniques such as automated path and sequence planners, and machine vision guidance operations. The hardware testbed can also be used to measure the effectiveness of the operator interface and display system in communicating with operators who have only modest levels of training.

This chapter describes the system design considerations, facility and system hardware, and the software and operator interface functions involved in implementing the hardware testbed. The test facility is operational and the system executive program and error recovery software is performing all assembly operations in a fully automated mode. Several end-to-end assembly and disassembly sequences for the complete 102 member truss structure have been performed and information on timing, error frequency, error causes, and recovery techniques has been collected. The sequence for the assembly tests conducted to date will be discussed along with the research directions toward a more robust system.

## 2.2 Facility and Hardware Description

At the initiation of this program several ground rules were established to guide the development of the component design. These ground rules were: 1) existing "off-the-shelf" components would be used directly, or modified for use, where possible to hasten implementation; 2) design and fabrication would incorporate inexpensive materials and simple systems so that they could be modified easily as experience and future research requirements dictate; 3) passive guidance features would be incorporated wherever possible to aid alignment for minimizing position errors; 4) all assembly operations would be reversible so that truss repair and error recovery procedures could also be automated and; 5) system hardware and software design would be based on total system automation with operator intervention rather than enhancing a teleoperated system with automated functions. The current automated assembly facility is shown schematically in Figure 2.3 and a photograph of the actual hardware system is shown in Figure 2.4. Important aspects of the various components are discussed below.

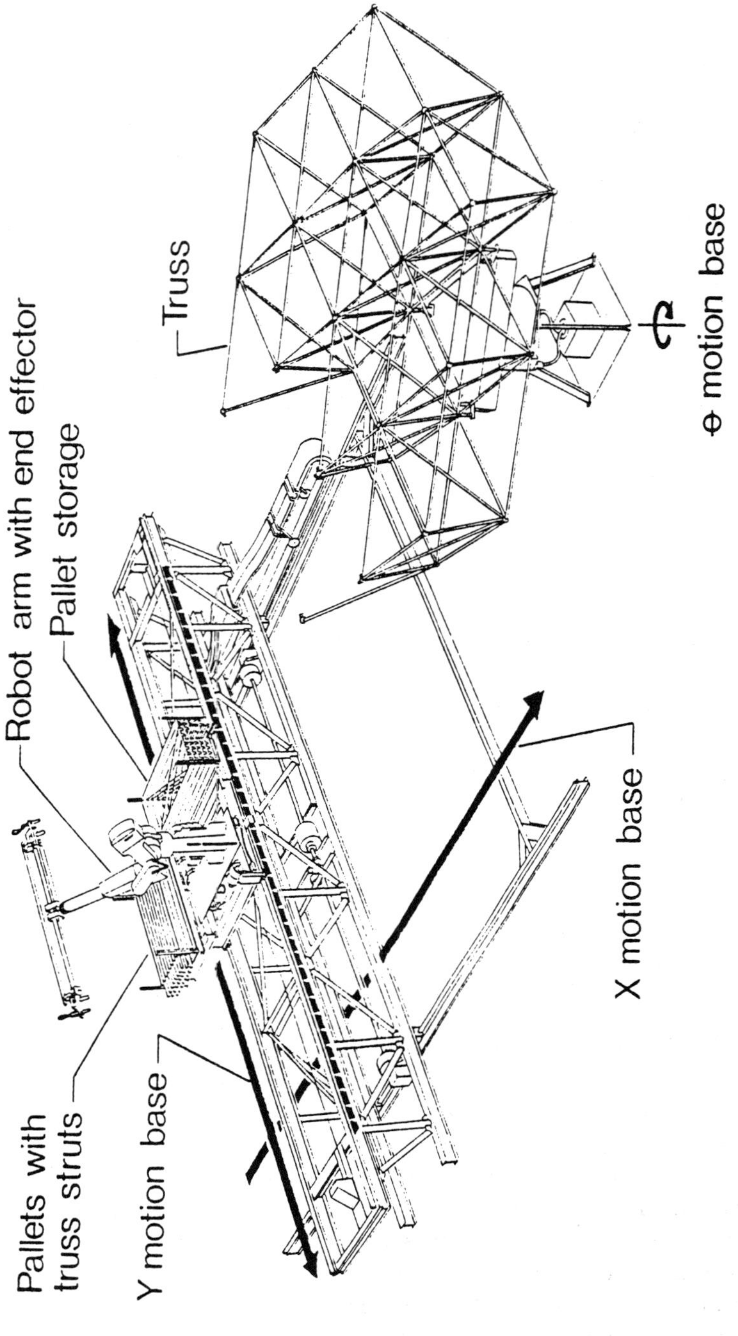


Figure 2.3: Schematic of Automated Structures Assembly Laboratory

Figure 2.4: Photograph of Automated Structures Assembly Laboratory

A detailed description of the performance characteristics is presented in [3].

### 2.2.1 Robot and Motion Bases

The robot arm shown in Figures 2.3 and 2.4 is a six-degree-of-freedom, electrically driven, industrial robot selected for its reach envelope 1.52 m (5 feet), payload capacity 89N (20 pounds), positioning repeatability ±0.010 cm (0.004 inches), and reliability. No modifications have been made to the robot over those supplied as options by the manufacturer. The robot is mounted on an x-y Cartesian translational motion base that has approximately 6.1 m (20 feet) of travel in both x and y directions and has a measured positioning accuracy of ±0.005 cm (0.002 inches) using linear encoders. The truss structure is assembled on a rotating motion base with ±3 revolutions from the nominal reference location and a positioning accuracy of ±0.25 mm (0.01 inches) at a radial distance of 6.1 m (20 feet). The motion bases are designed to minimize deflections that induce position errors due to the increasing mass of the truss during assembly and repositioning of the robot. Motion base drive motors on all three axes are controlled by an 80286 microprocessor-based personal computer. The robot computer is based on an M-68000 microprocessor and all robot motions are programmed in a modified Basic programming language supplied by the manufacturer.

### 2.2.2 Truss and Joining Mechanism

The truss structure selected for this study is the regular tetrahedral configuration shown by the model in Figure 2.5. This configuration was chosen for several reasons: 1) it is representative of supporting trusses required for large antennas, reflectors, and aerobrakes; 2) it is composed of regular hexagonal rings as shown in the figure, therefore, the developed assembly procedure can be applied to construct large multi-ring units; and 3) the truss has some special geometric characteristics that provide desirable access paths for the installation of struts. The truss hardware used in this investigation has members that are nominally 2 meters (6.6 feet) long and 2.64 cm (1.04 inches) in diameter. The two ring test model has a diameter of 8 meters (26.25 feet) and the complete structure has 102 struts and 31 nodes.

The special geometric characteristics noted previously can be seen in Figure 2.6. The truss geometry includes a natural set of orthogonal axes and there are sets of planes normal to this axis system. Every member in the truss is oriented to lie within one of the sets of planes and the members in a given plane form square geometric patterns (Figure 2.6). The squares provide the maximum area for collision-free manipulation and access for installation of strut members. The squares and their planar orientation are hereinafter referred to as the member's insertion plane. Assembly of the truss involves maneuvering the strut along a predefined path into the insertion plane, aligning it parallel

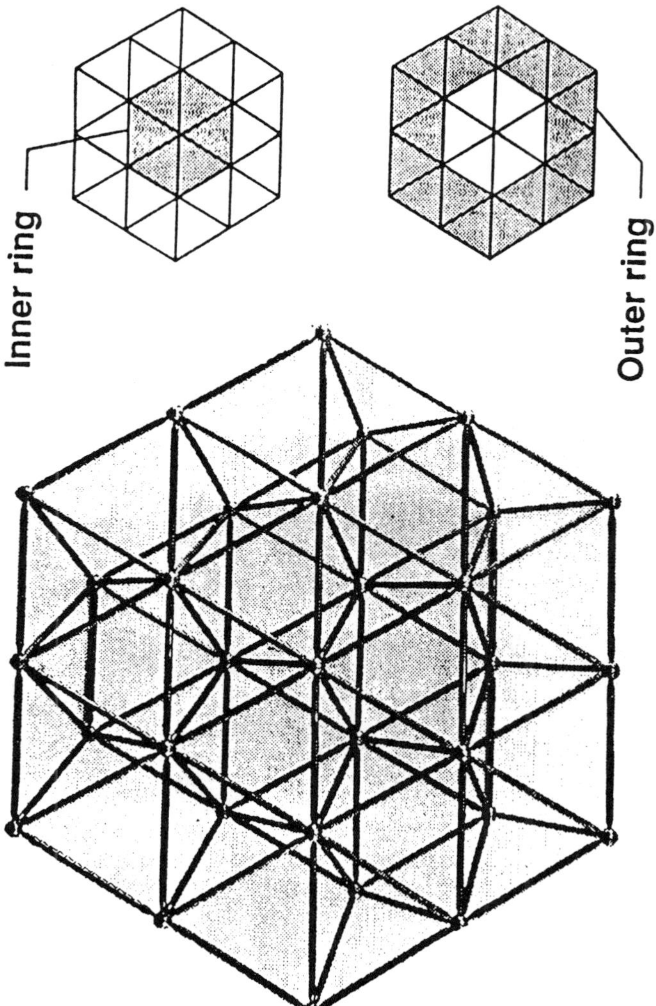


Figure 2.5: Top View of Two-Ring Tetrahedral Truss Structure

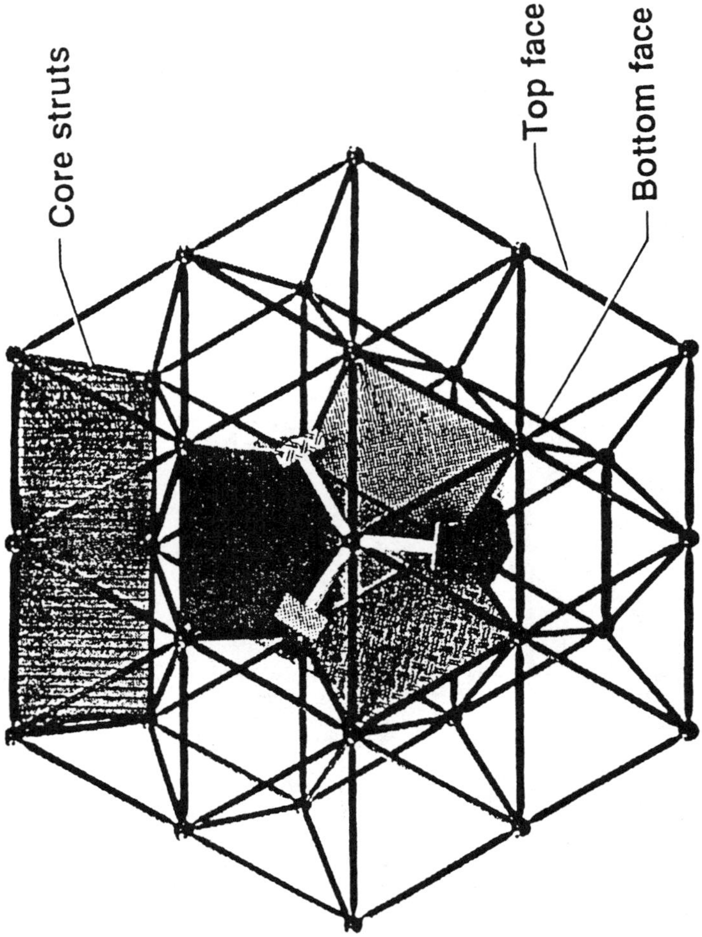


Figure 2.6: Tetrahedral Truss Structure Showing Strut Insertion Planes

to the installed orientation approximately 10 cm (4 inches) in front of the installed position, and then moving it directly to the installation position. The assembly process begins by connecting members to three pre-mounted nodes on the rotating motion base to build the first hexagonal ring. Additional detail on assembly operations will be presented in a subsequent section.

The joint connectors for space-truss structures have a number of requirements that must be met for the truss to be assembled and have a predictable structural performance. The connectors must have a predictable linear load-deflection response for reverse tension-compression load cycles and have no "free play". The joints must permit the strut members to be installed by side entry between two nodes a fixed distance apart, because the truss is structurally redundant and it is desirable to minimize disassembly to replace or repair a damaged member. Structural redundancy also requires the length of the member to be accurately set, and ramped entry guides on the joint are desirable to aid the insertion operation. Additional requirements are defined in [3].

The truss joints and nodes developed for this test assembly are shown in Figure 2.7. Each node must be capable of connecting a total of nine struts that are aligned with the node center. Six of the connecting struts lie in the plane of the face and the remaining three are core members that connect the top and bottom faces of the truss. The joint is composed of two parts: a connector section that is bonded to the graphite epoxy tube to form a strut, and the receptacle section that is mechanically attached to the node. The connector section is inserted into the node-mounted receptacle and a locking nut is turned to draw the connector plunger toward the strut, securing the strut to the receptacle. The locking mechanism applies approximately 890 N (200 lbf) preload to the joint connection to eliminate "free-play". The joints are fabricated from aluminum and have a mass of 134g (0.296 lbm). Additional details concerning the mechanism and its operation can be found in [4].

Following fabrication of the truss struts and node components, they were assembled and measured to evaluate the accuracy of the total truss structure. The truss was manually assembled for this series of tests with the design torque closure limits applied to the joints. Three sets of photogrammetry measurements were made on a set of high contrast retro-reflective targets centrally positioned on each of the 19 upper surface nodes. The test measurements indicated that the rms deviation between the 19 upper surface nodes and the best fit plane was about 0.015 cm (0.006 inches) and the rms positioning repeatability for two assembly tests was about 0.005 cm (0.002 inches). Additional information regarding these tests can be found in [4].

### 2.2.3 End-Effector

At the start of the program a number of truss joints were evaluated for possible implementation and/or adaptation to the automation process. The current

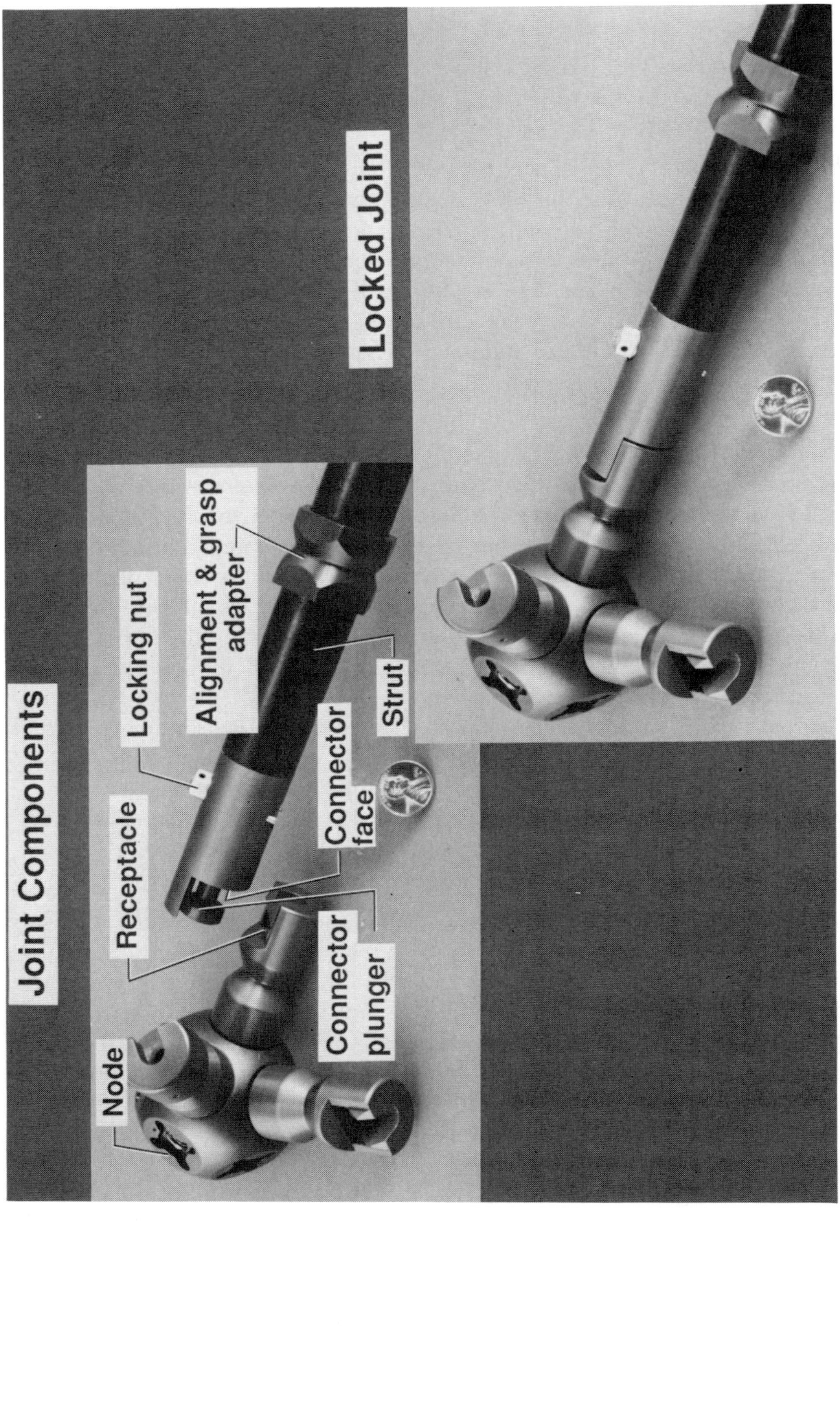


Figure 2.7: Strut/Node Joint Connector Hardware

joint was selected because, in addition to meeting the structural requirements, it could be easily adapted to operate with a simple end-effector. The development of the joint and end-effector as a total design aided in the development and success of the project. The end-effector is a specialized tool that performs all operations required for strut installation or removal including the locking of the joint connector. The end-effector's operational functions are shown schematically in Figure 2.8 and in a photograph of the hardware mechanisms shown in Figure 2.9. The strut is grasped by a set of strut holders which close around a hexagonal alignment and grasp adapter that is bonded to the strut tube (Figure 2.9). The hexagonal shape aids in circumferentially aligning the strut with the end-effector and is a configuration that can be easily grasped by the end-effector jaws. The center of the hexagonal adapter has a machined vee-groove that mates with a protrusion in the end-effector jaws to passively position the strut axially in the end-effector. These passive alignment features are representative of the many such details that are incorporated in all of the components where alignment is critical for connecting or mating.

To insert a strut in the truss, the end-effector is moved to a predetermined ("taught") position in the insertion plane. At this location the fingers are closed on the receptacle. The fingers are configured so that during closure they are passively aligned with the receptacle if the receptacle is at any location within a cylindrical envelope 5 cm (2 inches) in diameter by 1.5 cm (0.6 inches) long. Both ends of the strut are pushed forward by platforms that insert the connector into the joint receptacle. The strut is held in place while small gear-head motors lock the joints. The strut holders are then unlatched, the platforms retracted, and the receptacle fingers opened to complete the installation. The end-effector is designed to permit operation either with a node preattached to either end of the strut or to insert a strut into two nodes already in the structure.

All end-effector mechanisms and actuators are equipped with simple sensors such as microswitches or proximity probes so that a computer program can monitor the operation and notify the operator if a problem occurs. Small video cameras (Figure 2.9) are mounted on each end of the end-effector to permit operator monitoring of end-effector mechanisms and to verify the sensor function. The total mass of the end-effector including a strut is about 6 kg (13.5 lbm). A commercial force/torque load cell is mounted between the end-effector and the robot wrist to provide compliant move capability during both strut pick-up and installation operations. This aspect of the assembly operation will be discussed in a subsequent section.

### 2.2.4 Strut Storage

The requirements defined for storage of the struts that are in preparation for truss assembly included the following: 1) all struts be placed at a convenient location within the reach envelope of the robot without interfering with other assembly operations; 2) the precise location and orientation of each strut be

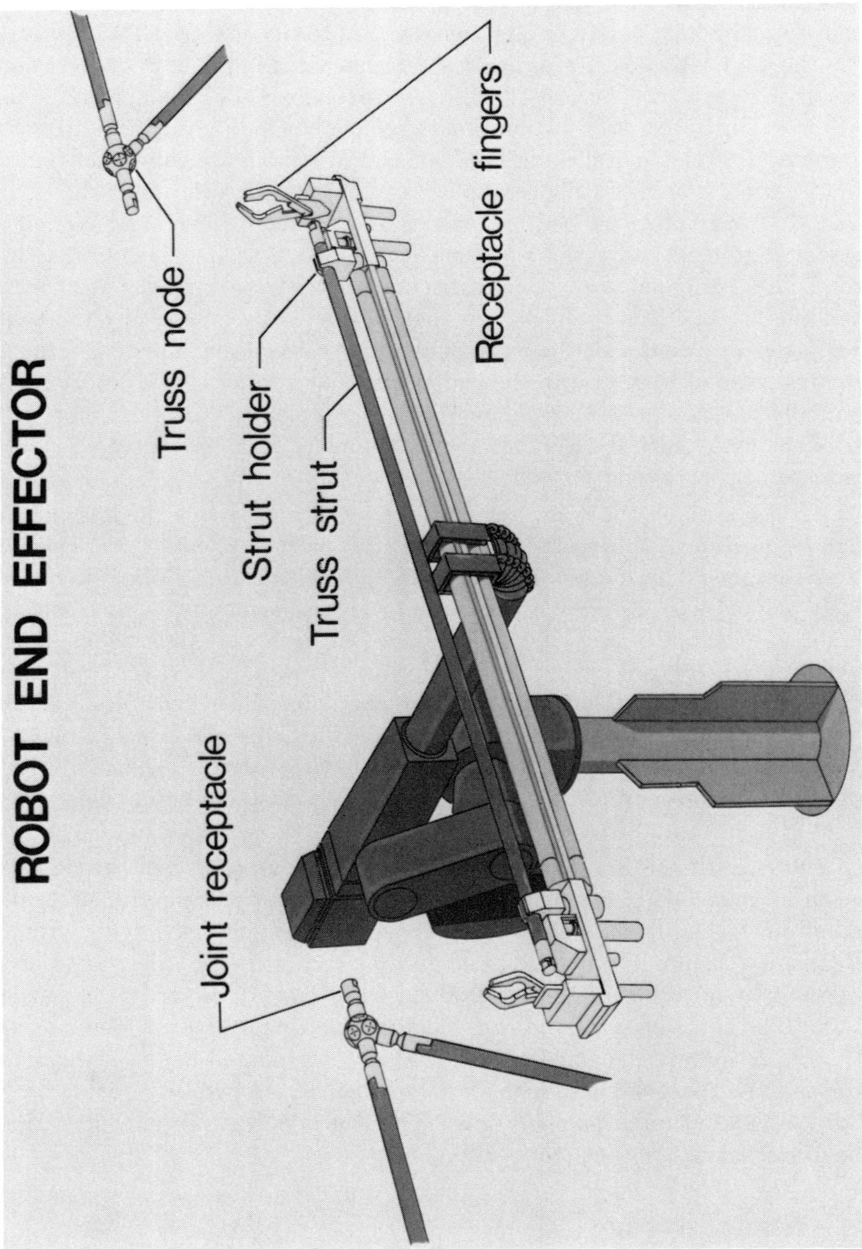


Figure 2.8: Schematic of End-Effector and Strut Insertion Concept

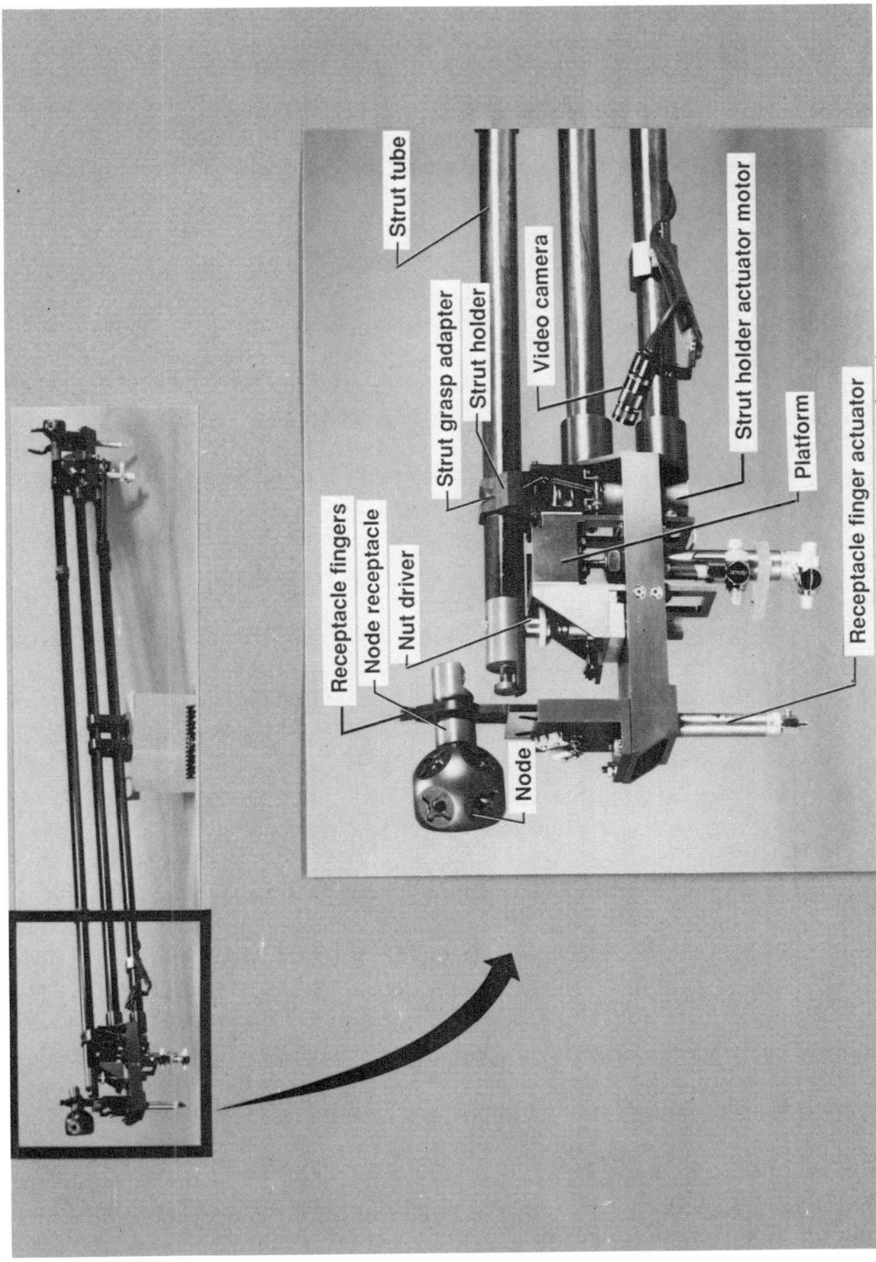


Figure 2.9: End-Effector and Actuator Mechanisms

fixed; 3) the struts be packaged in the smallest volume possible and the individual members still be readily accessible; 4) each member be restrained but capable of being removed or installed in the storage container by a small latching force. Several concepts were initially considered including a device that would present each strut to the end-effector at a common location. Implementing this concept appeared difficult and the packing efficiency appeared to be poor. An alternative concept, having an additional advantage of being a passive storage system, was developed. It consists of a series of trays which are stacked in a rack behind the robot arm (Figure 2.3). The trays, shown in Figure 2.10, are aluminum frames that hold thirteen struts each, between vertical positioning pins. The positioning pins have spring loaded pin plungers on both sides to hold the struts in place. A small force is required to extract each strut from its storage slot. Alignment and grasp adapters like those used to pick up the strut with the end-effector are used to interface with the positioning pin. The flat sides on these hexagonal adapters are trapped between the tray above and the base of the holding tray to prevent the struts from rotating. The nodes are preattached to selected struts in accordance with the assembly sequence. The protrusion of the receptacles from the preattached nodes imposes a sizable storage space requirement and a special stacking arrangement had to be developed and coordinated with the assembly scenario. The stacking arrangement is shown in the sketch of Figure 2.11. The arrangement has four pre-noded struts in tray 1 at the top of the stack. The pre-noded struts are located in the first and last tray slots with two other noded struts equally spaced in-between, separated by three unnoded struts. The second tray also has four pre-noded struts at the same slot locations, however, these nodes are all placed in the opposite end of the tray to those of tray 1. Tray 3 has three pre-noded struts with the nodes staggered to fit in slots between those of tray 1. Tray 4 is similar to tray 3 except that the nodes are on the same end of the tray as those of tray 2. Tray 5 is identical to tray 1 and the four tray pattern is repeated. Therefore, all even numbered trays have struts with alternating nodes of four and three on the same tray end, and likewise with the odd numbered trays. Due to the efficient packaging of struts in the trays, the receptacle fingers on the end-effector must be closed to pick up an unnoded strut so as not to interfere with the receptacle on adjacent pre-noded struts. The complete 102 member truss is packaged in 9 trays with several positions in selected trays left empty to accommodate the assembly sequence. Each tray has cylindrical handles on each end with hexagonal positioning and alignment adaptors identical to those on the struts. The trays are picked up by the end-effector as they are emptied, and transferred to a storage canister located to the left side of the robot arm (see Figures 3 and 4).

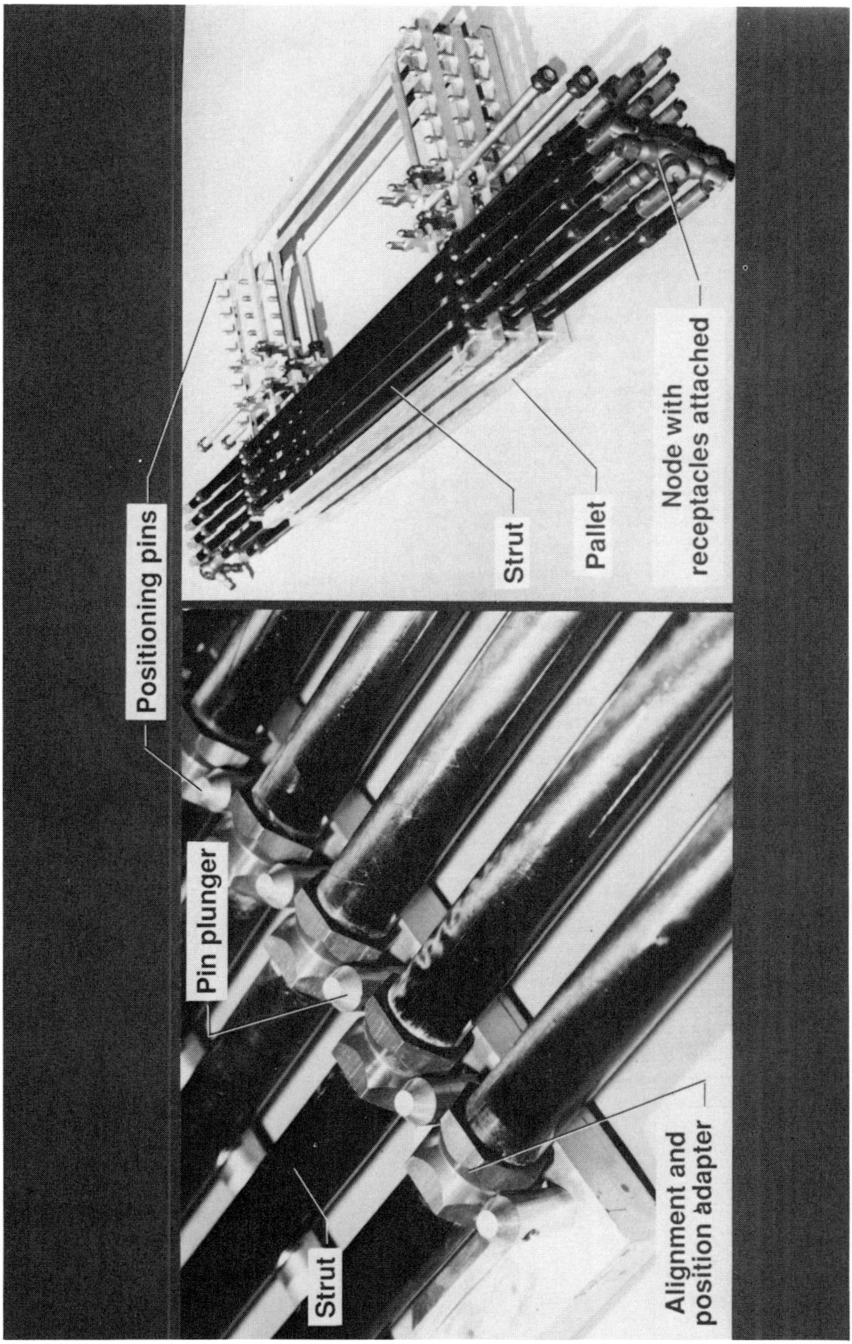


**Figure 2.10: Strut Storage Tray Pallet Details**

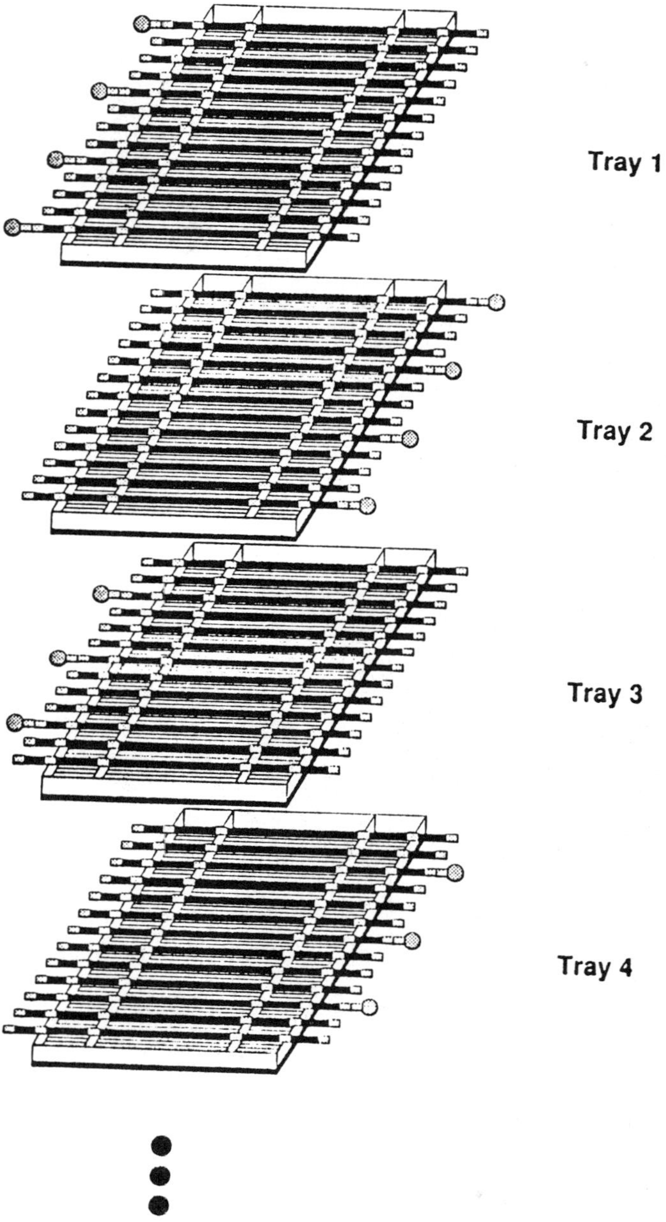


Figure 2.11: Strut Storage Tray Stacking Arrangement

### 2.2.5 System Control and Communication

The entire system is managed by several digital computers serially connected through RS-232 communication lines as shown schematically in Figure 2.12. Overall system executive and operator interface functions are performed on a micro-VAX workstation and implemented in FORTRAN. Also located on the VAX are the robot arm path control logic and the assembly system error recovery algorithms. The robot carriages are controlled by an indexer board associated with an 80286 personal computer (PC). Commands to this board are generated by a driver program on the PC written in BASIC. The robot arm motions are controlled by a modified BASIC programming language developed by the robot manufacturer. The robot program is executed on the robot's M-68000 processor in response to commands from the executive program. The robot processor also includes local data which define pretaught paths from the canister to the installation position of each strut. This minimizes the amount of information that must be passed between the processors during execution.

### 2.2.6 Operator Control

The operator's workstation is shown in Figure 2.13. The operator has four basic sources of information available to monitor system operation. These are: 1) the computer CRT menu display; 2) a video system display; 3) the CRT display from the robot computer and; 4) a panoramic view through the control room windows. All operator command inputs are entered via a standard terminal keyboard. The operator's role is to initiate and monitor the automated assembly process using a series of hierarchical menus that are displayed dynamically. The automated system is directed by a command file to install struts in a predetermined order. The operator monitors all operations via the displayed menus that highlight the current system command status. If an error occurs, a brief summary of the error condition and a menu of operator interventions for recovery from the error are displayed. The operator uses the video displays to verify the nature and determine the severity of the problem. All error recovery interventions consist of keyboard commands and menu selections. The assembly will not proceed until the error is corrected. If the error cannot be resolved, the system reverses the sequence and rolls back to the state where the last command file directive was issued. Additional discussion of the operator's display and procedures is provided in the section on Software Structure.

The video system consists of four fixed position cameras: one located on each end of the end-effector to provide a close-up view of the end-effector mechanisms; one located on the Y carriage behind the manipulator arm to provide an over-the-shoulder view of the strut installation and a direct overhead view of canister operations; and one camera fixed to a wall approximately 6.1 m (20 feet) high and 40 feet from the rotational motion base to provide surveillance of assembly operations from the right rear quarter of the robot. Both the

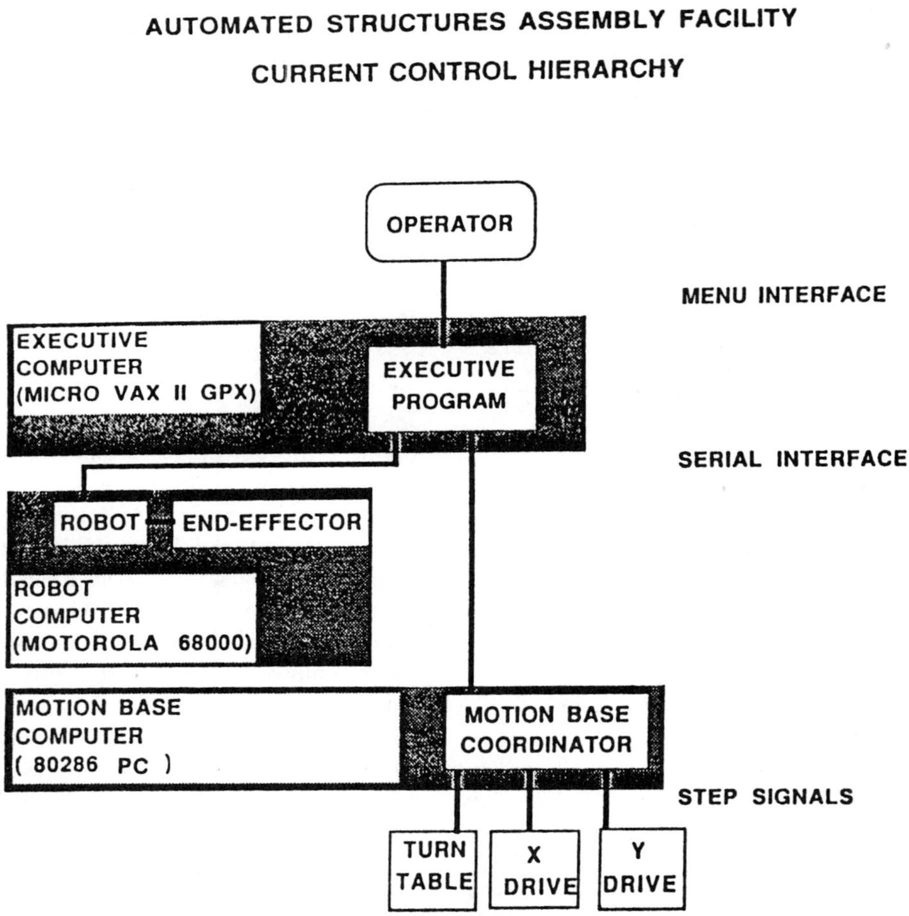


Figure 2.12: Schematic of Facility Computer Control System

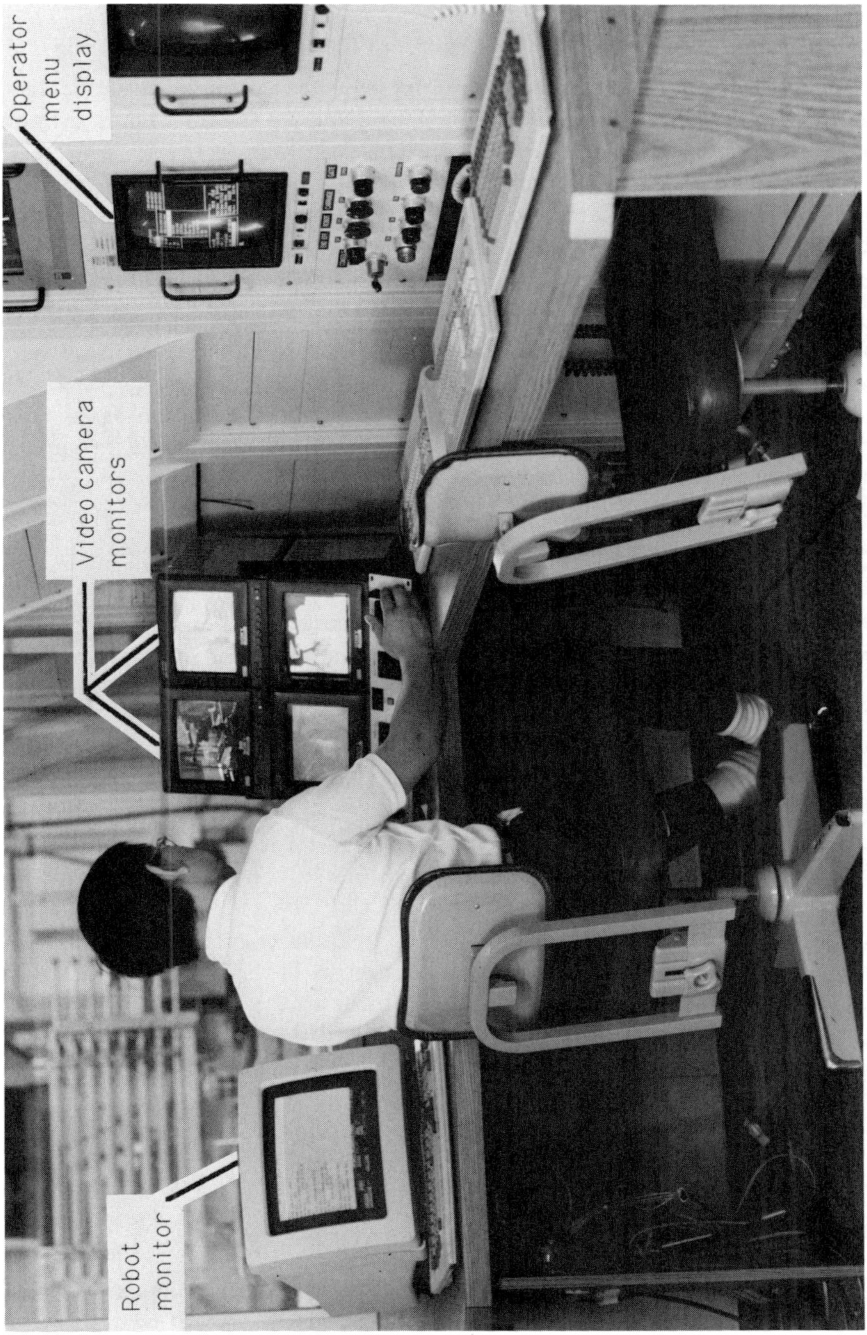


**Figure 2.13: Operator's Workstation**

over-the-shoulder and wall mounted cameras are equipped with pan/tilt/zoom capability that is controlled by joysticks from the operator's workstation.

The CRT display for the robot computer provides the operator with a visual cue of the robot commands and states. This display provides backup information to the menus associated with the executive program to aid in assessing error sources and recovery operations that involve robot motion. Also, the forces and moments measured using the force/torque load cell are displayed on the robot CRT display.

The panoramic view that the operator has through the control room window does not provide substantial information to correct detailed errors, because the control room is about 35 feet from the turntable, which is too far to observe detailed mechanical operations. The view is conveniently located to permit the detection of potential collisions between the truss and the motion bases. Tests during which blinds were drawn over these windows resulted in no adverse effects on operator performance or comfort.

## 2.3 Assembly Operations

Each strut in the truss assembly has a specific name based on an identification convention developed for this investigation. When a strut is selected for installation, the system database is checked to establish the status of the requested strut and determine if it is being installed in an acceptable order. A number of factors may affect the order. For example, connecting struts required to support the member must have been previously installed and a collision-free path must be available. If all factors are satisfied, the installation sequence is initiated. The first operation is to acquire the strut from the canister tray and hold it in the end-effector with the arm in a rest position just above the canister. The motion bases are positioned so that the arm can move the strut along a predefined collision-free path to the installation point in the insertion plane. The end-effector inserts the strut connectors in the joint receptacles and locks the joint connector. The arm then returns to the rest position over the canister. The following subsections give a detailed account of each of these operations and outline some of the checks and procedures that are followed if an error is detected. The initial assembly sequence developed for this investigation, including the rules that were followed during the development, is listed in Appendices A and B.

### 2.3.1 Strut Identification Convention

Since the tetrahedral truss has a regular geometric pattern composed of basic sub-elements, it was desirable to have a naming convention that would be common for those members with similar orientations and positions. A convention that was suitable for any "n" ring configuration was also desirable. A sketch illustrating the adopted convention is shown in Figure 2.14. The figure shows a top platform view of a large planar truss. The struts in the top face are represented by lines of medium width, the struts in the bottom face are represented by lines of narrow width, and the core struts that connect the top and bottom faces are represented by the dashed lines. The wide solid lines outline the $n$ hexagonal rings that are used in this chapter to describe the truss. Each ring is composed of $6(n\text{-}1)+3$ cell units and the members within each cell are defined by their end positions that fall on the even numbered locations of a conventional clock face, noted on the figure as the operator's convention. Any strut in the truss can, therefore, be specified by its ring number, cell number and clock position. An example of the notation used for a member of ring 3 and cell 1 whose ends are at clock positions 10 and 2 is R3 C1/10_2. Using this convention, all top and bottom face struts have a unique designation. However, since the core struts lie on the boundary of the individual cells, they can be identified by two cell designations.

The naming convention defined above permits an operator to locate any member in a large multi-member truss with respect to an origin located at a specific reference node. The convention is based on a fixed orientation of the operator and the truss. The reference node for the automated assembly experiment is the pivot node on the axis of the turntable. Since end positions of the struts change with the angular orientation of the turntable, the designation of a member from the vantage point of the robot may be different than the designation from the vantage point of the operator. From the vantage point of the robot, the truss contains 12 unique positions if the turntable is positioned in 120 degree increments.

### 2.3.2 Strut Pick-up From the Canister

Acquiring a strut from the canister (and strut return to the canister during disassembly) is accomplished using information from a system database described in section 2.4. This information includes the strut's assigned tray number and the slot number within that tray. The database also contains information regarding preattached nodes if the strut is a core strut. The end-effector is moved from the rest position to a location directly above the tray slot. A logic algorithm is executed to close any receptacle fingers that may hit receptacles on adjacent struts with preattached nodes.

There are two techniques that may be used to define the location of points for a robot arm. One technique is to store an absolute location of tool tip coordinates, and the robot controller always uses the same kinematic solution for that point, thus achieving an accurate repeatability. The second technique is to specify a point in terms of an offset (a point that is relative to an absolute point). Although this method may not be as precise as specifying an absolute location, it has generally provided adequate positioning.

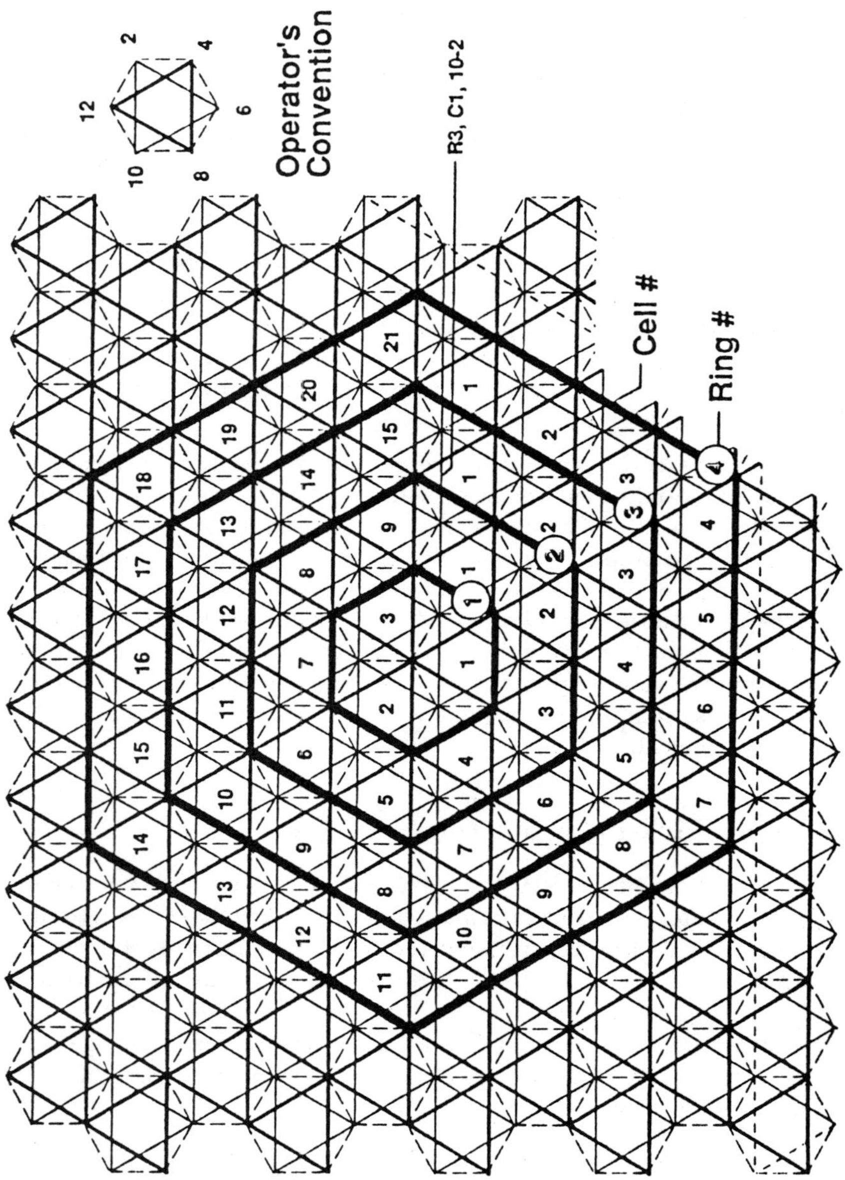


Figure 2.14: Schematic Illustrating Strut Naming Convention

To move to the tray slot, the location of the tray and the location of the slot are computed as an offset from a taught central point in the vicinity of the tray system. This approach was used because the computation storage required to keep each individual strut location would have been extensive and the time and effort to teach each of the locations would have been prohibitive. The slots are numbered from 1 to 13, beginning with the slot closest to the robot, and the trays are numbered from 1 to 9 beginning with the tray at the top of the stack. The central (reference) position is located in the center (slot 7) of the bottom tray (tray 9).

For strut acquisition, the end-effector is lowered to the level of the tray containing the desired strut and the end-effector platform is extended. At this position, the strut grippers on the extended platform lightly contact the grasp and alignment adapter on the strut. The arm is then moved incrementally under the control of a force-feedback loop until either a sensor on the gripper jaws indicates closure or a vertical load of 89N (20 lbs.) is applied by the end-effector to the strut. During this operation, all loads and moments in the plane of the tray, are maintained at levels below 3.6N (0.8 lb.) and .56N-m (5 in-lbs.), respectively. When either of the above two conditions are met, the gripper jaws are commanded to be latched and the platform is retracted, extracting the strut from the tray.

Since the total truss assembly was a major goal of the program, the assembly scenario had to include retrieval of each strut from the canister and transfer of the empty trays. The struts were placed in the trays as noted previously to facilitate efficient packaging. The strut removal sequence could not be sequential across a tray. Also, the packaging efficiency and interference between the end-effector and the preattached nodes required that the noded struts be removed prior to an adjacent unnoded strut in the same tray. The empty trays are moved from the supply canister to the storage canister by using the end-effector to grasp adapters on the tray handles in the same manner that the struts are grasped. Although most trays were substantially filled, several strut positions in selected trays were left vacant to satisfy all assembly requirements and eliminate the need to transfer trays with struts in place.

### 2.3.3 Motion Base Moves

The carriage and turntable positions (X , Y , $\Theta$) necessary for the installation of a particular strut are predetermined and stored in a data array in the executive program. The relative location of each motion base position with respect to an axis reference point is given in the scenario included in the appendix. The positions were established empirically to permit installation of each strut within the reach envelope of the arm and not be hampered by arm singularities. When a strut is called for installation, the strut name is parsed to determine the motion base location associated with that strut. The motion base repositioning commands are transmitted, one at a time, to the PC-based program with the command order determined by a collision avoidance routine. The PC program verifies each new position and the system status database is updated at the completion of each move. All motion base moves are performed with the robot arm at the rest position and the end-effector located above the strut canister. This position was selected to minimize the distance that the arm protrudes toward the structure.

The carriage collision avoidance algorithm prevents the robot arm and the motion bases from colliding with any part of the structure currently on the turntable, and at the same time seeks to minimize the moves, and thus the time, required by the motion bases. Two points on the carriage are considered to be the potential collision points: the robot elbow, and the handles of the empty trays in the storage canister located on the left side of the robot. Collisions can occur during carriage Y-axis moves and turntable rotations if the carriage and robot arm are positioned too close to the turntable. Because of truss geometry, only the core struts, or those that connect the top and bottom faces of the truss structure, are potential obstructions. A database that includes these potential obstructions is included in the VAX executive program to facilitate these checks.

The collision avoidance algorithm computes the radial distance from the center of the turntable to the point on the strut where a collision can occur, (e.g., a point at the height of the robot's elbow in its rest position) and the angle between this point's radius and the turntable zero reference. If this radius of any installed strut is greater than the radius of either of the robot or tray handle collision points in both their initial and final positions, then that strut is considered as a candidate for collision. The Y-positions of both the candidate obstruction and the carriage collision points are then compared to determine if interference actually occurs. For motion base moves involving both Y-carriage moves and turntable rotations, the situation is more complex. Here the sequence of the moves must be considered and the possibility that a collision may be avoided by making the moves in a particular order must be checked. If a collision is determined to be unavoidable, then a minimum distance to move the X-carriage away from the turntable to avoid collision is calculated and that avoidance move is commanded first.

### 2.3.4 Robot Paths and Capture Sequence

With the motion bases properly positioned, the robot arm then begins to traverse a predetermined path to transfer the strut from the canister rest position to the insertion plane where it is to be installed. Each path consists of a series of points that are executed in a sequential order. Most of these points are absolute points and the arm is moved from point-to-point in a sequential order without pausing between moves. This technique is used because the robot controller defines the actual robot path as straight line segments from point-to-point and the system has no method for specifying a continuous path. Many of the paths contain common segments because the actual location of the arm is not significant. The primary consideration is that the member does not collide with previously installed members. Near the end of the path, the speed of the arm is reduced at a location called the "approach point". At this point the strut is aligned in the insertion plane 10.16 cm (4 inches) directly in front of its location in the truss. The location of the approach point is critical to final positioning, as is the point where the strut is inserted, (insertion point). Therefore, the approach point is defined relative to the insertion point to minimize the number of taught points.

There are three different truss/strut conditions that must be accommodated when members are inserted. Therefore, the final portion of the path from the approach point to the insertion point must be tailored to the particular condition. The first condition is a direct assembly that entails placing a strut between two fixed nodes already in the structure. For this case, the end-effector moves directly to the insertion point and the receptacles on both nodes are grasped for installation. The second condition involves the installation of a core strut (one that fits between the top and bottom planes of the structure) with a preattached node at one end. Here, the end-effector also moves directly from the approach point to the insertion point, however, only the receptacle fingers (and locking mechanism) at one end of the end-effector operate. After installation the strut is left cantilevered from the truss. To minimize the effect of gravity induced strut deformations, only the core struts were installed in a cantilever condition with a node preattached. The third condition involves the installation of a member that connects the free end of a cantilevered member with the remainder of the structure. For this condition the end-effector must move from the approach point to the deflected position of the cantilevered core strut, grasp that receptacle, and then move to the insertion point before grasping the other receptacle. The intermediate (cantilever) receptacle capture points for this condition were defined relative to the insertion points and the repeatability of the deflected struts have made this is a successful procedure. Obviously, for a zero gravity space condition this would not be required. For this, as well as several other conditions influenced by gravity, the procedures developed for this demonstration test may be more challenging than would be required for space operation.

The accuracy and repeatability of the motion bases and the commercial robot are individually adequate to perform the structural assembly demonstration. However, the sum of the positioning errors and the relatively high stiffness of the truss, can generate significant loads when the receptacle fingers of the end-effector are attached to the joint receptacles for strut installation. To minimize the effect of these loads and assist with error free operation of the end-effector mechanisms, a wrist mounted force/torque load cell was installed between the end-effector and the robot. All loads are measured with respect to a reference frame located midway between the centers of the receptacle fingers.

The force/torque cell is nulled at the approach point for each strut. After the end-effector fingers close on the receptacle, the loads are checked to determine if they are above predetermined threshold limits of 3.6N for loads and 0.56 N-m for moments, (typically, the loads do exceed these limits). The arm is then repositioned using controlled incremental moves. The passive guidance features designed into the truss and the end-effector, guide the repositioning to the exact insertion point. The balancing algorithm consists of calculating the displacement necessary to eliminate the largest force or moment using empirically derived stiffness gains. Also, empirically derived limits are set so that a calculated move that substantially exceeds the limit, and may potentially damage the mechanisms, cannot be executed. After each move the loads are evaluated and the process repeated until the loads are below the threshold limits. The number of cycles generally required to reduce the forces/torques below the threshold values is between 10 and 30 and the final arm displacements are typically in the range of thousandths of a centimeter. Following the arm repositioning, the strut is moved forward by the end-effector platforms inserting the joint connector into the receptacle.

The attachment of the receptacles must be coordinated with the assembly sequence. A core strut may be installed in the insertion plane in either of two directions. To maintain uniformity, the receptacles on all nodes were identically aligned and were rotationally symmetric with respect to the axis of the node. When located in the insertion plane during the approach to the node receptacles, clearance between the end-effector and the adjacent struts is approximately 1.9cm (0.75 inches).

Approximately 19 taught paths were required to install all members instead of the 12 cell positions discussed previously. The 7 additional positions were the result of different robot base locations necessary to avoid collision situations and the orientation of the nodes in the trays (stacking requirements for the struts in the trays and the limitation of attaching nodes to only core struts).

### 2.3.5 End-Effector Installation Operations

The platform is extended to insert the strut into the receptacle and the locking nut is turned by a small gear head motor, locking the strut into place. The strut latches are released, the platform retracted, and the receptacle fingers are opened to release the structure and complete the installation process. Sensors mounted on the end-effector are used to monitor the success or failure of each operation and the sequence does not proceed until the previous operation is successfully executed.

After the strut is installed the end-effector is moved back to the approach point. For those struts installed to fixed nodes, the move back to the approach is direct. For core struts installed in the cantilever condition, the move back to the approach point requires that the end-effector be rolled slightly to compensate for the gravity induced deflection, otherwise, the joint receptacle will frequently drag against the fingers and substantially deflect a strut before being released. From the approach point the robot moves the end-effector back along the same path to the canister rest position.

### 2.3.6 Operator Pause and Reverse

The need for a pause and reverse capability was identified early in the hardware checkout phase of the assembly tests. The ability to temporarily stop the assembly process to consider the safety or success of the ongoing operation, reduced operator apprehension and established confidence in the system. The capability to pause and reverse is required for system checkout, but is also used frequently during assembly, particularly where the operation involves close clearances or a video camera must be adjusted for better coverage. The increased control accorded to the operator significantly reduces the stress associated with the operator's monitoring task. Pause represents a nonintrusive level of intervention which, if combined with a graphics predictor simulation and display, would provide a powerful interface tool for the transfer of internal knowledge to the operator. The "hold and evaluate" feature is a key element in a supervised autonomy environment.

### 2.3.7 Error Recovery

Error conditions detected by sensors are reported to the executive program and then to the operator for selection of error recovery procedures. These may involve further end-effector action, such as simply cycling the actuator, or movement of the robot arm to reposition the end-effector, thus, minimizing interference and permitting the component to function properly. The robot arm motions are either in the form of manually commanded adjustments in arm position, as determined from the video displays, or by the force/torque balancing algorithm. For the manually commanded position adjustments the operator is prompted to enter the displacement and rotation that the end-effector is to be moved in the end-effector coordinate reference frame. This requires the operator to estimate these distances from the visual information provided by the video cameras. Inexperienced operators may require supplemental support to aid in this task, such as models of the truss hardware or computer generated graphics displays.

All error recovery actions are commanded by the operator as selections from menus that are specific to each error. The only exception to the operator intervention is a force/torque balance that is automatically invoked by receptacle finger closure errors. This was automated because these errors are easy to correct and it is not necessary to involve the operator. The error recovery menu also includes two selections that do not involve corrective actions. One is an ability to ignore the indicated error ("go on anyway") when the operator can determine that the indicated malfunction is due to a sensor failure or is deemed not to be serious enough to warrant action. In this case, the operator assumes responsibility and the assembly sequence proceeds as if the action was successful. Another operator option is to "give up" trying to correct the error. In this case, the system is rolled back to the beginning of the command it was currently executing and stopped. This is done so that the operator is not required to remember any operational sequences.

### 2.3.8 Tray Transfer Operations

The transfer of empty strut trays to and from the storage canister is intended to be a simple operation involving no logic other than the pause and reverse capability. The robot arm always traverses a fixed path from one canister to the other. Whether the operation involves getting or storing a tray determines the order in which the canisters are accessed. The tray pick-up with the end-effector is identical to acquiring a strut from a tray, and in fact, utilizes the same end-effector routines. The tray release function, however, differs from the strut replacement in the tray. Here, when the end-effector reaches the release point, the strut latches are simply opened and the arm is commanded to move down in small increments until a downward force of 156N (35 lbf.) is being applied to the tray. This sets it properly in the spring-loaded pins that hold it in place.

The height of the tray acquisition and release points within the canisters is a function of the tray number and is calculated in increments relative to the taught point representing the bottom tray position in the respective canister. This positioning scheme is identical to that used for the struts and described in a previous section. The trays traverse up and down within the canister against nylon slides on the corner posts. The initial tray transfer operations were conducted with the trays fully supported by the strut latches on the end-effector. The tray sides would occasionally rub against the nylon slides, which, imposed a large reactive moment on the end-effector mechanism because of the width of the trays. To aid in counteracting this moment, a fork-shaped bracket was designed and installed on each tray which fits on either side of the end-effector tubular structure. The bracket also aids in aligning the trays during the transfer and stacking operation.

### 2.3.9 Truss Disassembly

To remove a damaged strut from the truss or to follow an alternative sequence due to some failure, the system must be capable of disassembling the truss. Therefore, each component of the system was designed to be capable of removing the struts from the truss structure and storing them in the canister trays. The empty trays are also retrieved from the storage canister when required and placed in the working canister to be filled. To disassemble the truss the struts are removed in precisely the opposite order from the assembly sequence. It should be noted, however, that the sequence of component operations required to remove struts is not a simple inverse of the original install sequence. For example, the cantilever conditions of a strut are not the same when it is installed as when it is removed resulting in a significant change in the operational sequence of the end-effector. Therefore, separate logic routines are invoked for the disassembly operations. Force/torque positioning is used to guide the end-effector for the removal from the structure and insertion of the strut into the tray slot. Operator monitoring, pause and reverse operations, and error recovery are identical to the installation operations.

## 2.4 Software Structure

### 2.4.1 Executive Program

The executive software program was designed in a top-down manner and is shown in Figure 2.15. The program structure is hierarchical to support the "supervised autonomy" mode of assembly operations. The program was developed around the operational scenarios and the requirements of the hardware actuators and sensors. Projected scenarios were reviewed and terminology to clearly differentiate the various components was developed and modified several times. However, once they were established they were strictly adhered to during code development. This resulted in a modular program that reflects the total system's architectural and functional capabilities. The software structure (Figure 2.15) is divided into five levels: planning, truss element, device, component, and verification. Commands may be entered at any level so that the system will operate in either a completely automated mode, where high level commands are automatically decomposed into lower level commands, or in a manual mode where any level may be entered directly. Commands can only call routines which are lower in the hierarchical structure and in the same branch. The highest or planning level is designed to be coupled with an automated assembly sequence planner which is not currently implemented. The automated planner output is represented by a file of "FETCH and CONNECT" text commands. These are processed by the second level (truss element) that decomposes them into individual device commands in the following order: position the motion bases, position the robot arm, and operate the end-effector. The truss element level also offers the operator the ability to install or remove individual struts at any time.

Device level commands for the carriages, robot arm, or the end-effector address the subsystems individually and make it possible for the operator to direct each subsystem manually. The device level commands are also decomposed into sequences at the component level, which actually perform the specified operation. Although commands at this level are device specific, their successful ex-

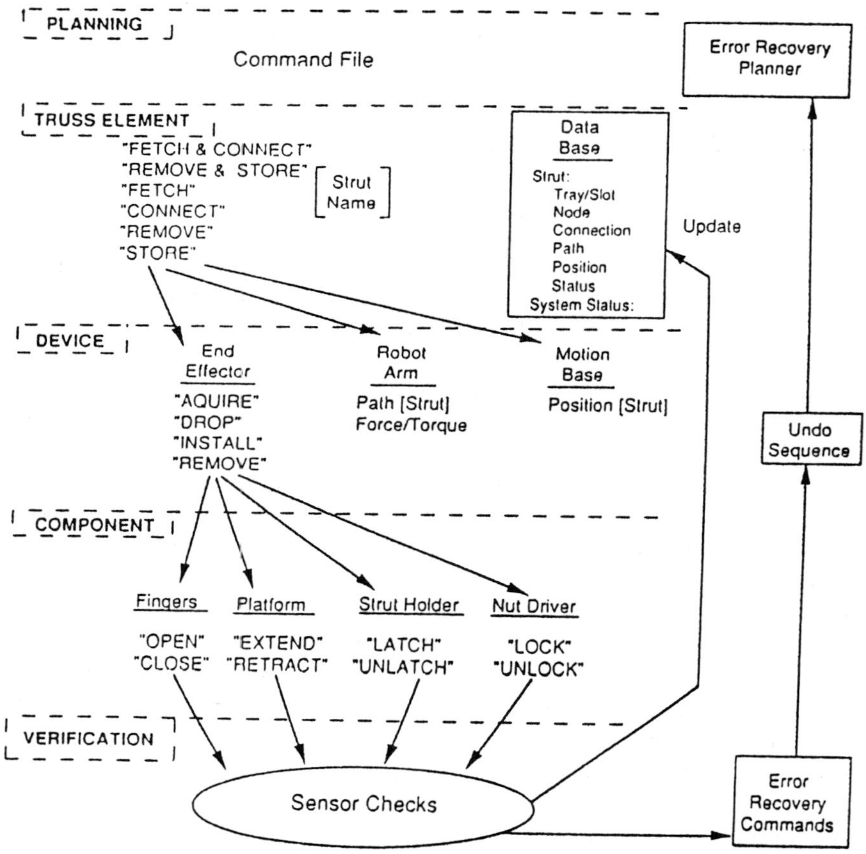


Figure 2.15: Automated Assembly System Software Hierarchy

ecution may require interaction with a second device whose execution involves another processor. For example, arm repositioning may be required during the "INSTALL" sequence to reduce loads on the end-effector mechanisms.

The command decomposition sequence at any level is not fixed, but varies with the type of strut being installed and any special conditions determined from the database information for that strut. Database checks are performed to ensure that the specified strut's status is consistent with the operation to be performed, i.e., a strut to be removed must be currently installed in the structure. The database is kept current by status updates as each low level command is completed and verified by sensor checks. Unsuccessful resolution of a problem results in initiation of an inverse sequence to roll the assembly process back to a state where replanning may be performed. The failure information is passed back up through the software hierarchy, displayed on the operator's console, and the assembly process is paused.

The operator's menu display, shown in Figure 2.16, exhibits the same hierarchy as the executive software. The commands currently being executed, including the decomposition of high level commands, are highlighted on each menu. This allows the operator to follow the action and observe the trace to the current status. Special windows are also displayed to identify the strut being manipulated and the status of the device-level components. Dialog windows provide a running description of the success or failure of the component currently operating and prompts the operator when a menu selection can be made. An error menu is displayed when a problem is encountered and a dialog message gives additional information on the error condition. Operational experience has demonstrated that the hierarchical command/display structure is very successful in keeping the operator abreast of the assembly process and in dealing with the problems encountered. This is a result of designing the hierarchical command/display structure around the operational requirements of the hardware.

The system software was developed around the hardware requirements and the defined operational scenarios using standard software engineering practices. No consideration was given to structuring the software around the NASA Standard Reference Model (NASREM) architecture described in [5]. However, a subsequent comparison of the resulting program structure with the NASREM architecture has proved interesting. The NASREM architecture is depicted in Figure 2.17 and the automated assembly executive software structure in Figure 2.18. As can be seen, the automated assembly hierarchical structure corresponds closely with the four lowest levels of NASREM. For example, the NASREM primitive level corresponds with the automated assembly device level that includes the robot arm, the end-effector, and the motion base. The NASREM element move level corresponds to strut operations (fetch, connect, remove) in the automated assembly hierarchy. Figure 2.18 includes only those functions at each level that were needed in the automated assembly application. Other

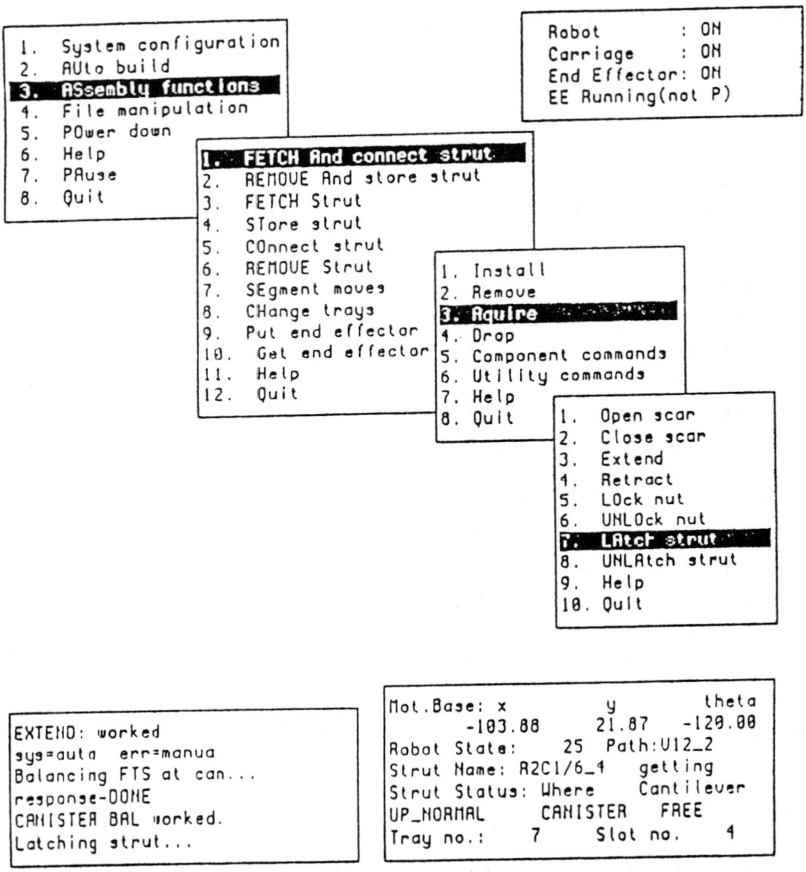


Figure 2.16: Operator's Menu Display

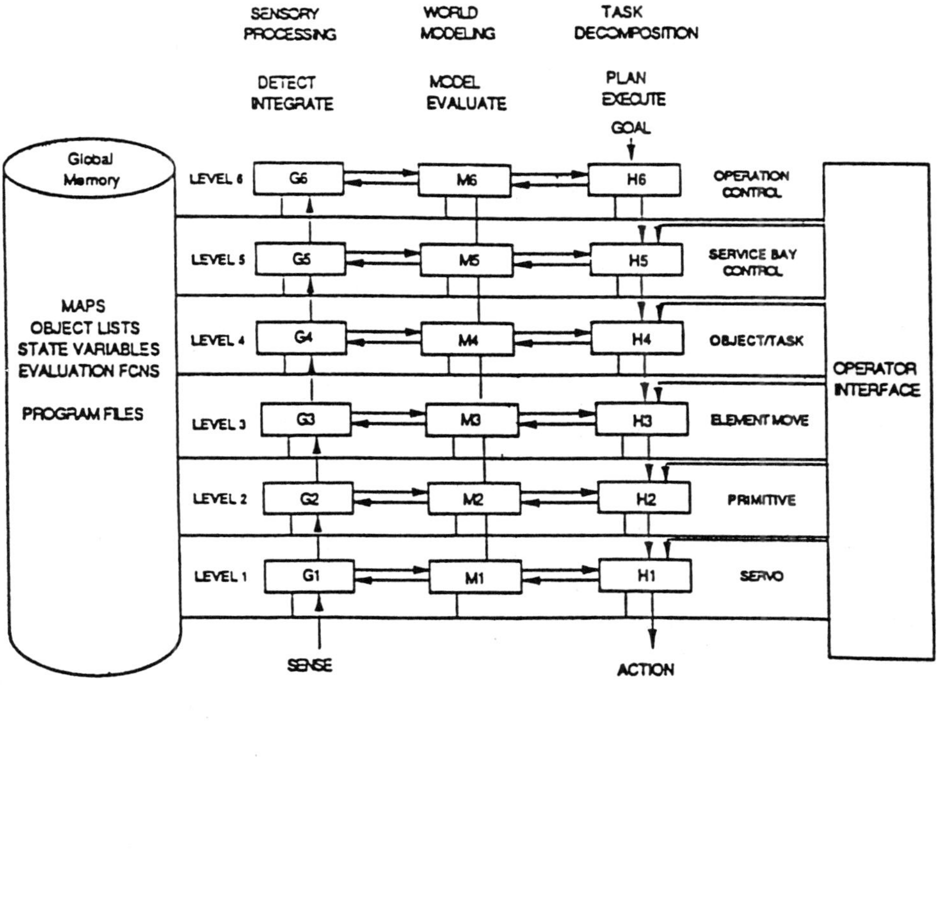


Figure 2.17: NASREM Standard Reference Model Architecture

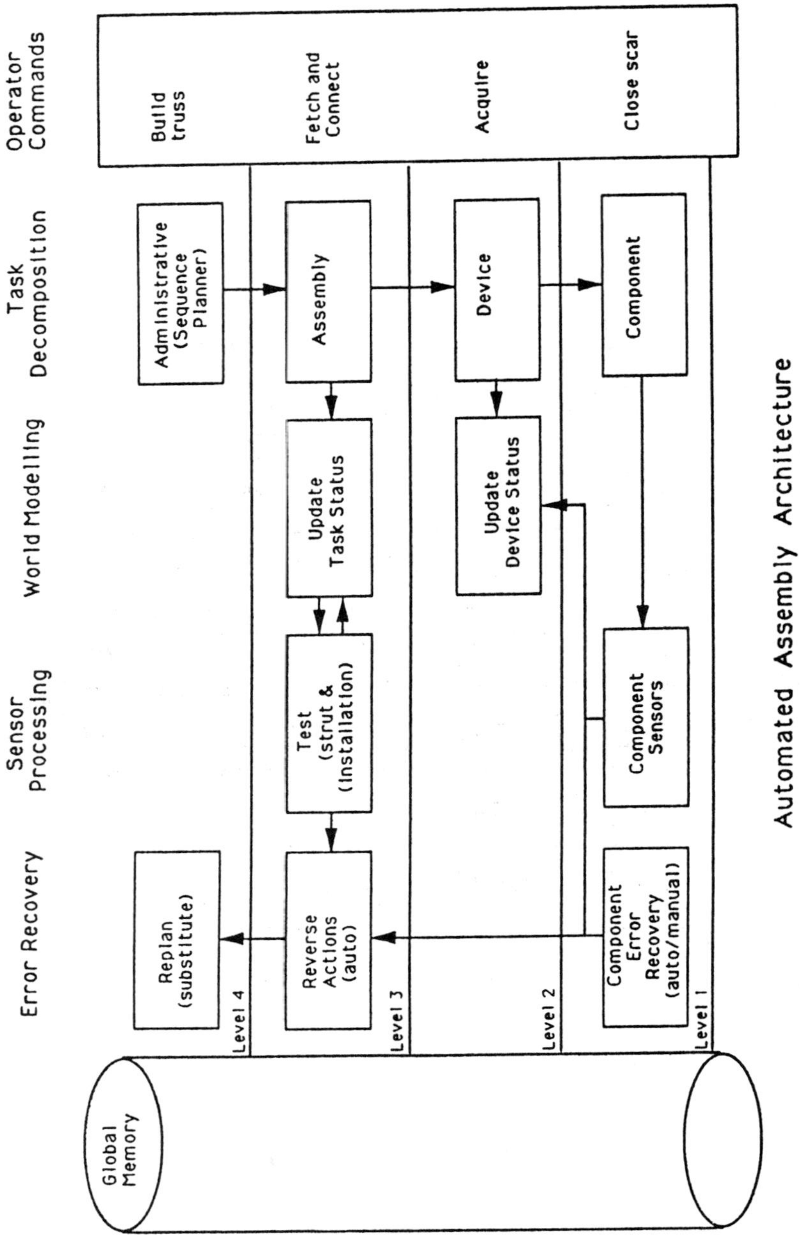


Figure 2.18: Automated Assembly Program Architecture

activities, such as typical operator commands at each level and error recovery, are included for completeness. Operator command capability is provided at all levels, however in practice, it is seldom used below the element move level (fetch and connect), except for system checkout.

All hardware actions and sensor processing occurs at the component (NASREM servo) level. Also, all error conditions are resolved by either operator intervention or automated actions at the component level. Unresolved errors are passed back up the hierarchy initiating an automatic reversal of the tasks performed at each level. For the assembly task, alternative actions are available only at the planning level, which take the form of substituting struts for failed members. This requires replanning the assembly sequence. Aside from the component level, the only other testing is performed at the element move level. These tests involve physically exercising and testing the locking mechanism immediately after a strut is removed from the canister. Another test has been performed immediately after locking a strut into the structure by attempting to retract the strut platform before unlatching to verify the integrity of the joint lock. A failure of either of these tests, results in the selection of a substitute strut and a replanning of the assembly sequence. The world model information base (global memory) is updated at only two levels: the device level and the element move level. At the device level, the end-effector status model is updated at the completion of each component action. At the element move level, the truss structure model and the storage canister status is updated with the installation or removal of each strut.

The NASREM architecture provides good conceptual agreement with the automated assembly application, although not all activities have an entry at every level. The hierarchical model does provide a particularly concise display for operator visualization. The hierarchical structure is capable of supporting several assembly operations by providing a standard interface between the levels. For example, standardization of device-level primitives for several end-effector configurations used for different assembly operations is showing promise by using the same executive to direct different tasks. Device-level commands will be carried out on a microprocessor mounted to the end-effector. A microprocessor-based device at the component level has the added advantage of freeing the operator from the details of the end-effector mechanisms. Analysis of the automated assembly software has demonstrated that the results are very similar to the NASREM architecture.

The executive software program was initially implemented in FORTRAN to verify the operation and structure of the specified command hierarchy. The FORTRAN prototype was used to check out and refine the assembly system operation, particularly the robot state and path sequence logic, and the error recovery procedures. When the assembly system logic and procedures became stable, work began toward developing an operating command system based on an expert system building tool. The results thus far, indicate that considerable savings in both development time and code size are possible. The expert system provides an inference engine for rule evaluation and a framework for the information bases. The FORTRAN program will still be used to provide the menu displays and I/O interfaces to the operator, the robot processor, the carriage control software, and the end-effector program. The information base resides in a shared permanent storage on the VAX. Information included in this permanent storage is restricted to system status that must be retained from one assembly test to another, along with information that is shared between the FORTRAN code and the expert system.

### 2.4.2 Robot Program

The program to control the robot motion is written in an expanded version of BASIC provided by the robot manufacturer and is executed on the robot's M68000 processor. The software program structure is simple and consists of a short execution loop that calls a series of subroutines that contain the instructions to perform specific commands. Command functions to setup, initialize, configure, and move the arm are examples of some specific commands. Utility subroutines were developed to perform special tasks such as repositioning the arm to reduce forces/torques indicated by the load cell or operator directed position adjustments.

The geometric data for the 19 possible robot paths and the tray storage path segments are stored in a point definition library or PDL file on the robot processor. The data are stored according to the path names specified by the command string, so that execution of the utility routines concatenates the path name onto the robot's "MOVE" commands and the desired path is traversed. When the arm has completed execution of the specified command, a "DONE" is returned to the executive program. If a robot motion error occurs during the routine, an "ERROR" message that includes the robot system state is returned. Interruption of the robot servo power is used to pause the system during arm motions. The power shutdown is detected by the robot processor and a "PAUSED" condition is returned to the executive program. A separate monitor function in the executive program precludes sending of any robot command while the servo power is shut-off. The robot program software has been deliberately configured to represent a low-level-of-knowledge system operation. Its primary function is a repository of the points that constitute the geometric positions for the 19 robot paths and the tray path storage segments. This is consistent with the hierarchical structure of the automated system.

### 2.4.3 System Data Structures

The software program's knowledge of system conditions resides in a shared database that contains two basic types of information: 1) the current status of all elements of the assembly system and structure, and 2) predetermined conditions/positions that are used to direct and control the robot and motion bases. The current status information is maintained continuously to represent the physical state of the system at any point in time and thus, ensure continuity of system operations. This is initialized externally and is updated automatically during test runs. The predetermined position information for the robot and motion bases are points that are associated with the installation of individual struts. The predetermined position information also describes the collision-free "taught" paths by which the arm moves between the canister and the various installation positions in the truss. Table 2.1 illustrates the data section which is broken down by the following elements: motion base position, strut type, robot status, tray status, current truss status, and end-effector status.

The MOTION_BASE_POSITION element stores the x, y and $\theta$ values (X_-Car, Y_Car, Turntable) for the pre-determined motion base locations that establish the positioning relationship between the robot base and the truss. Due to the reach capacity of the robot, many of the struts are installed with the robot base situated at the same motion base position.

The STRUT_TYPE element contains all the data necessary to describe in detail the installation and storage conditions for each of the 102 strut members. The struts are identified and accessed by an alphanumeric designation (Name). The current location of the strut (Where) is accessed by the system before any strut operation can be initiated. The system must know if the strut is currently in its tray, installed in the structure, or held by the end-effector. When a strut is selected for installation, the system refers to a list of struts (Connect_To) which defines those struts that must be installed in the truss prior to installation of the selected strut. This is a safety feature that ensures the required initial conditions for installation of the selected member are satisfied. The secondary reference to the location status (Where) of each strut on this list certifies that all conditions are met. The installation position (Loc_In_Cell) identifies which of the 19 predetermined robot paths is to be followed in installing or removing a strut. If a node is preattached to a strut the record (Node_End) indicates which nut driver on the end-effector must not be operated while installing that strut. If the end-effector must capture another node, (Cap_End) specifies which end. The condition of the installed strut (Cantilever) is used to establish predefined modifications for the robot path that must occur during the capture sequence. Due to tray packing requirements, a preattached node may not be located on the correct end associated with a direct path entry. The record (Flip) identifies this condition and initiates a robot command to rotate the strut 180 degrees at an intermediate point in the installation path. To remove or insert a strut in the tray requires that the assigned tray and slot position (Tray, Slot), be supplied. As noted previously, each strut is installed along 1 of 19 predefined paths. Each state in the path is defined by a robot position record (X, Y, Z, Roll, Pitch, Yaw) which are the "taught" points that are used to define the path. The collision avoidance algorithm requires that the end position of the core struts be defined

| Record | Variable | Description |
|---|---|---|
| MOTION_BASE_POSITION | | Motion base configuration |
| | X_Car | Carriage x position |
| | Y_Car | Carriage y position |
| | Turntable | Theta angle in degrees |
| STRUT_TYPE | | Strut description |
| | Name | Strut name |
| | Where | Current strut location |
| | Connect_To | Struts needed for support |
| | Loc_In_Cell | Installation position in truss |
| | Node_End | End with node |
| | Cap_End | End to capture: |
| | Cantilever | End conditions of strut |
| | Flip | Flip indication |
| | Tray | Tray number containing the strut |
| | Slot | Position in tray |
| | State_Pos | Position for each robot state |
| | X, Y, Z, Roll, Pitch, Yaw | |
| | X_End, Y_End | Core strut end points for collision avoidance |
| ROBOT_STATUS | | Manipulator arm current mode |
| | State | Robot path location |
| | Cond_State | Sub state point in a path |
| | Strut_Now | Name of strut in arm, if any |
| | Strut_Getting_Now | Name of strut in process of retrieving |
| | Strut_Just_Had | Name of last strut installed |
| TRAY_STATUS | | Current tray locations and mode |
| | Tray_State | Path location |
| | Tray_Mode | Current tray operation |
| | Current_Tray | Tray on top |
| | Working_Ap | Robot position for working approach point |
| | X, Y, Z, Roll, Pitch, Yaw | |
| | Storage_Ap | Robot position for storage approach point |
| | X, Y, Z, Roll, Pitch, Yaw | |
| TRAY_HANDLE_LOCATIONS | | |
| | Storage_Loc | Robot position in storage canister |
| | X, Y, Z, Roll, Pitch, Yaw | |
| | Working_Loc | Robot position in working canister |
| | X, Y, Z, Roll, Pitch, Yaw | |
| CURRENT_STRUT | | Strut currently in arm |
| | Left_Seat | Nut driver alignment status |
| | Right_Seat | |
| | Left_Nut | Nut status |
| | Right_Nut | |
| CURRENT_MOTION_BASE_POSITION | | Motion base configuration |
| | X_Car | Carriage x position |
| | Y_Car | Carriage y position |
| | Turntable | Lazy susan theta angle in degrees |
| END_EFFECTOR | | End effector status |
| | Left_Receptacle_Finger | Left receptacle fingers status |
| | Right_Receptacle_Finger | Right receptacle fingers status |
| | Platform | Status of platform |
| | Latch | Strut holders status |
| | Storage_Pos | Robot position for storing |
| | X, Y, Z, Roll, Pitch, Yaw | |

Table 2.1: Automated Assembly Data Structure

for computation of potential collision conditions. The record (X_End, Y_End) is used to supply this information with the strut element.

The ROBOT_STATUS element contains the path and current positioning point on the path (State, Sub_State) for all strut paths and the tray transfer path, as well as the current strut/tray (Strut_Now, Getting_Now). This element also defines the name of the last strut/tray installed/removed by the robot (Strut_Just_Had).

The TRAY_STATUS element maintains all information pertaining to the strut storage trays. The robot path identifier is (Tray_State), and (Tray_Mode) denotes if the objective of the move is to store or retrieve a tray. The number of the tray that struts are being removed/stored in is noted as (Current_Tray).

The CURRENT_STRUT element contains information pertinent to the status of the end-effector for the strut that is currently in the end-effector. The status variables indicate whether the nut driver sockets are seated to lock/unlock the joint connector (Left_Seat, Right_Seat) and the current status (locked or unlocked) of the joint (Left_Nut, Right_Nut).

The END_EFFECTOR element maintains the current status of the various components on the end-effector. The records for the position of the receptacle fingers at each end of the end-effector are (Left_Scar, Right_Scar) and indicate whether they are open or closed. The position of the strut insertion platform (Platform) and the condition of the strut grippers (Latch) are also maintained.

Data examination is available to the operator through the main menu. This provides a direct method for accessing the status of any component and determining if initialization conditions are correct. If the operator determines that a value is incorrect or desires to manually override the indicated status he can do so, but only through password entry into the database. This protects the system from haphazard modifications by an inexperienced operator, while permitting flexibility in control of variables for system set-up and debug. If a user modified variable affects other data items, the operator is responsible for making these changes also. For example, if the location of a strut is changed from the tray to the end-effector, the operator must change the status of the end-effector strut latches from open to latched.

### 2.4.4 End-Effector Software

The end-effector software has evolved on an empirical basis. The initial task of generating actuator commands and monitoring sensor output was increased by the need to provide effective error recovery. This was developed as the end-effector mechanism error modes were identified through actual assembly tests rather than preliminary bench testing and the solutions were developed through experience. The need for a pause and reverse capability for the operator and the ability to roll back following an unresolved error complicated the end-effector sequencing algorithm and its software.

As noted earlier, the end-effector software evolved into a modular, hierarchical structure resembling the lower levels of the NASREM model. The software program to implement the end-effector command hierarchy was initially located in two places. The lower level end-effector component commands such as "LATCH" , "LOCK" , and "CLOSE FINGER" and the actual interface to the actuator and sensor hardware were implemented on the robot processor in a dialect of BASIC. This was done because the robot processor was well equipped to communicate with external equipment and sensors. This provided an excellent testbed environment for the development of the suite of end-effector sensors, operational sequences, and error recovery procedures. The higher level end-effector component commands such as "INSTALL" , "REMOVE", "ACQUIRE", and "DROP" were programmed in FORTRAN on the VAX executive processor. These commands were decomposed into the lower level sequences, which were then passed to the robot processor for execution. Error recovery routines and menus were programmed on the VAX in FORTRAN.

## 2.5 Tests

Although the system was designed in a top-down manner, implementation has moved from the bottom-up by subsystem development, testing, and integration. This has resulted in complete end-to-end assembly (and disassembly) of the 102 member truss structure in a supervised autonomy mode. The assembly tests conducted to date have followed a manually developed assembly sequence based on taught paths and strut locations. The objectives of the end-to-end tests are: 1) identify through experience the type of errors likely to be encountered during assembly of space-truss structures; 2) examine the reliability of the system to determine if the concepts implemented are suitable for space application; 3) evaluate the effectiveness of the operator in resolving problem situations; 4) quantify the time required for the various portions of each task and, 5) assess the operator time required to resolve the errors encountered. The testing has resulted in a database that can be used to estimate the effect that changes in individual operations will have on the entire assembly. The database also maintains a compilation of some of the errors that may be encountered along with a number of potential solutions.

Each test is directed by a command file that contains a predetermined strut sequence for either assembly or disassembly. The system executes the stream of commands as they are read from the command file. If an error occurs the system will execute preprogrammed commands to recover. If the error recovery commands fail to correct the error, then control is turned over to the operator and an error recovery menu is displayed. The operator then verifies the problem using the video system and selects the appropriate recovery action. When the error is resolved the system will resume executing commands from the command file at the step following the command where the error occurred.

When resolution of the problem from the console is unsuccessful the operator's last resort is to enter into the assembly area with the system disabled and manually intervene to correct the problem.

The operator uses a personal computer spreadsheet program to record elapsed times for all assembly operations, the type of errors, and the recovery technique employed. The operator's objective is to resolve the problems from the console using the menu display and video information. The spreadsheet data are analyzed to identify systematic problems in the component hardware and difficult operator interactions. This is used to refine the mechanisms and operational procedures to minimize problems that require operator attention. A complete end-to-end assembly test is not conducted in one single continuous time period since it requires 12 to 15 hours. Instead, it proceeds in four to six hour blocks so that operator fatigue is not a factor.

An assembly test is initiated with three nodes mounted to the turntable motion base and all of the struts in the trays of the supply canister. The robot is commanded to a predefined calibration position by the operator and placed into a vendor supplied self-calibration routine. The reference position of all motion bases is checked to verify their calibration. After all system checks have been made and visually verified by the operator the assembly sequence is started. The timing begins with the robot arm located at a rest position near the top of the supply canister. There are seven timing segments that are performed during the assembly of a strut. These segments are illustrated in the sketch of Figure 2.19. The first time-segment involves the movement of the arm from the rest position to the strut acquire position, immediately above the strut. The second segment is the time required for the end-effector to remove the strut from the tray and involves several force/torque directed moves and many of the end-effector component operations and sensor checks. The third segment is the movement of the arm back to the rest position. The motion bases are then moved (fourth time segment) to the desired location as directed by the collision avoidance algorithm. The arm is then moved along the preplanned path (fifth segment) to the approach point which is 10.2 cm (4 inches) immediately in front of the installed location in the truss. The sixth segment is the actual install operation and involves the 10.2 cm of arm movement to the install location, force/torque repositioning, end-effector operations including locking of the joint, followed by arm motion back to the approach point. The last timing segment involves the return of the arm along the preplanned path to the rest position just above the supply canister. Several of these segments, e.g., arm motion to the install point and arm motion back to the canister rest position might appear to be redundant, however, there are differences in these individual segments that can affect the total time. If an error occurs during any segment that requires operator intervention, the time required for the operator to correct the problem, as well as the source of the problem, is recorded.

A total of four complete assembly sequences and four complete disassembly

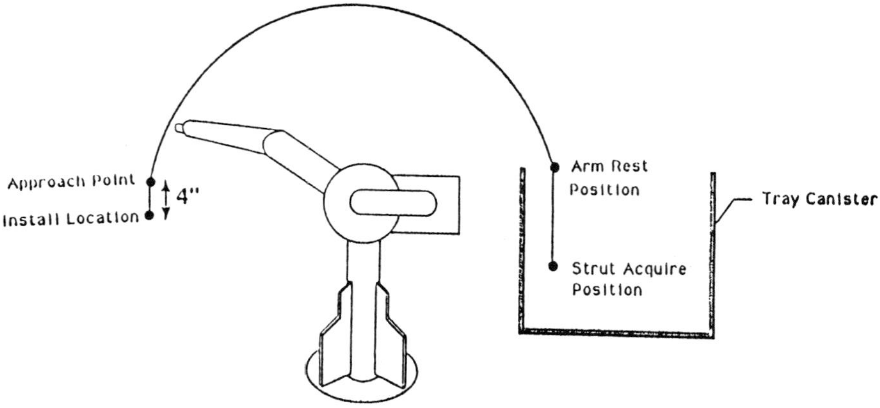


Figure 2.19: Robot Arm Path Segments

sequences have been conducted, although, only the results from two of each have been used for data analysis. The first two tests were introductory assemblies and a substantial number of changes were implemented in both the software and the hardware as a result of the initial test findings. Also the timing sequences were substantially modified following the introductory tests. The quantitative results presented for system timing, identification and evaluation of errors are intended to characterize the current performance of the automated assembly system.

## 2.6 Current Test Observations and Results

### 2.6.1 Preliminary Assembly Results

A summary of the results from the two assembly and disassembly tests are presented in Figure 2.20. The results are in the form of pie charts and the sectors of the pie are the timing segments identified previously. These results are the total times for all of the 102 members and therefore, they can be used to determine representative averages for the total process as opposed to values for a particular strut. The average time for installation is slightly over 9 minutes per strut and the removal time is approximately 8.5 minutes per strut. This difference is primarily due to operations associated with removing and replacing the struts in the storage tray. The time to acquire the strut in the assembly

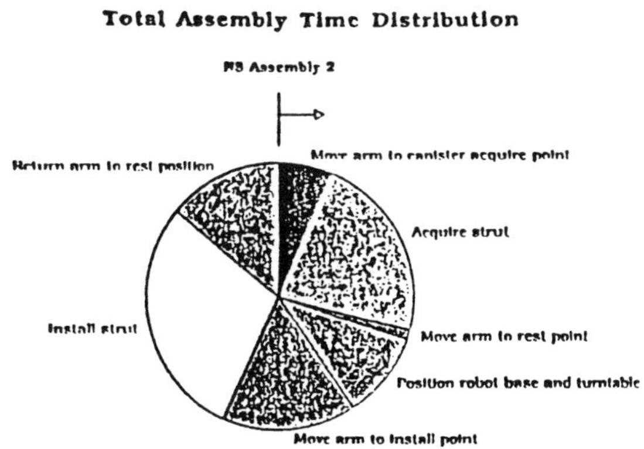


| | Min. | Percent |
|---|---|---|
| Move Arm to Canister | 61.42 | 6.55 |
| Acquire Strut | 209.47 | 22.34 |
| Move Arm to Rest Position | 11.50 | 1.23 |
| Position Motion Base | 97.55 | 10.40 |
| Arm Approach | 150.95 | 16.10 |
| Install Strut | 275.78 | 29.41 |
| Arm Retreat | 131.13 | 13.98 |
| Total: | 937.80 | |

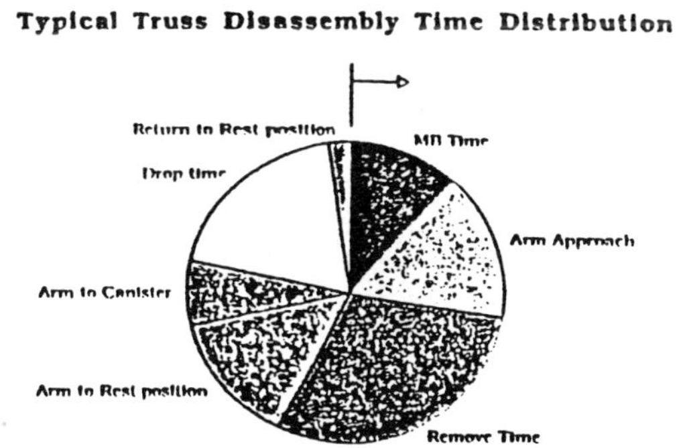


| | Minute | Percent |
|---|---|---|
| MB Time | 107.60 | 11.83 |
| Arm Approach | 142.22 | 15.64 |
| Remove Time | 272.31 | 29.95 |
| Arm to Rest Position | 126.83 | 13.95 |
| Arm to Canister | 64.26 | 7.07 |
| Drop Time | 173.80 | 19.11 |
| Return to Rest Position | 22.28 | 2.45 |
| | 909.29 | 100.00 |

Figure 2.20: Average Time Breakdown for Complete Truss Assembly and Disassembly

sequence is longer because this procedure involves a force/torque balancing to move the arm into position on top of the strut closing the strut latches. This is not required for disassembly. Of particular interest is the relatively large amount of time necessary for end-effector operations during "INSTALL", "REMOVE", "ACQUIRE", and "DROP". These segments are relatively long because they typically involve many actuation commands and sensor checks. A number of the actuation commands are relatively slow, such as the joint locking and latch closing commands. The actuators and sensors also involve a significant amount of communication between computers that typically require about 1 second each for the simple serial communication link currently being used. The time shown on Figure 2.20 does not include the time required to transfer the strut trays from canister to canister. The tray transfer time required about 4.75 minutes per tray or a total of about 38 minutes per test for both assembly and disassembly.

A simple time-motion analysis of the assembly was conducted for which every operational command was identified and an estimated time to complete each command was assigned. This analysis indicated that the assembly time could be significantly improved and that it could be expected to be performed within a range of 3-5 minutes per strut as opposed to the current time of over 9 minutes per strut. For the time-motion analysis all actuation commands were considered to be driven by electrical motors that required 3-5 seconds to complete as opposed to the current pneumatic actuators that take a much shorter time. The time required for arm motion, which is associated with the robot speed, was not reduced because the operator must have adequate time to react in order to prevent collisions. The assessment of the current system operator is that monitoring the arm while operating at speeds higher than those currently being used would be stressful and tiring. The major improvement in the assembly time occurred in end-effector operations associated with acquiring the strut from the tray and installing it in the truss. A microprocessor and associated software to command actuators and poll sensors has been developed and has been implemented in a second generation end-effector. This new capability is discussed in [6]. Retrofitting the current end-effector to be controlled by this microprocessor could easily reduce the time for the strut installation by a factor of two, and it has the potential for a reduction of up to five. In developing the time-motion estimate it was also assumed that a distributed control system with parallel execution of some operations could be implemented. The main advantage is the elimination of the time required for motion base moves (nearly 1 minute), since they could be accomplished while the robot was moving to the tray and acquiring the strut in the end-effector.

This preliminary timing information may be representative of what is possible with an automated assembly system and it may appear relatively slow in comparison to astronaut assembly times of about 30 seconds per strut reported in [7]. However, increasing the speed of assembly operations is not the major objective of this research program. Of much greater concern is the overall reliability of the system in completing the assembly task and the ability of the system monitor to successfully correct unplanned errors. Since no EVA is planned as part of these assembly operations and the system monitor could supervise operations from a terrestrial-based control room, all that is required is that the system proceed at a rate adequate to accomplish the desired task in a reasonable (as yet undefined) time.

The number of errors that occurred during the two complete assembly and disassembly tests ranged from a high of 74 for one of the assembly tests to a low of 28 for a disassembly test. For this preliminary test set, more errors occurred during the two assembly tests than during the two disassembly tests. In all tests, over 90% of the errors that occurred were the result of displacements that either prohibited the end-effector from properly functioning or caused difficulty in removing/inserting a strut in the tray. All positioning problems were corrected by the operator using either the automated force/torque balance routine or the adjust capability to slightly reposition the arm. The adjust moves are typically in the range of 5.1 mm (0.2 inches). The average time required for the operator to confirm the problem, make an independent assessment, and complete the correction was about 2 minutes per error. This is indicative of the effective error recovery procedures that have been developed. Although the timing profiles discussed previously do not include this error time, it does not contribute significantly to the total assembly time. The remaining errors were the result of a variety of mostly minor problems, several of which were so inconsequential that the operator was able to determine that it was safe to proceed without taking corrective action. Only about 1% of the struts required the operator to manually intervene and most of these were associated with errors in the length of the truss members that caused difficulty with member insertion. The truss is a redundant structure and although the length of each member is nominally the same, there are small differences in the member lengths due to fabrication errors. The joints were designed to compensate for some error in length and to have passive guidance features that would assist with misalignments. These guidance features, however, did not have adequate ramping and modifications to the basic design have been initiated.

The quantitative results for both assembly times and the errors encountered are presented more for interest and completeness than as a prediction of the achievements capable with a significantly more sophisticated space-based system. The system is in an evolutionary development process, therefore, these results must be considered as a "snapshot" in time during the preliminary phase and not the best that can be achieved with an automated system. As the evolution proceeds, the assimilation of advanced techniques (discussed in a subsequent section) such as machine vision are likely to significantly affect both the installation time and the number of errors that will be encountered.

### 2.6.2 Test Observation Results

Normally in a discipline oriented research program, a systematic approach is taken to implement improvements and assess the resulting changes in performance against some established reference or baseline. That approach is unsuitable for this study for several reasons. First, this program is structured around the development of a generic technology with application to a number of proposed missions, as opposed to being developed for a particular project for which a set of guidelines and requirements have been established. Second, identifying specific requirements, communicating them effectively, and assessing the impact of a proposed approach on the requirements is more difficult when the research crosses discipline lines. Third, there is no database or reference to judge progress or improvements against. Since there is no established methodology, the research is very dynamic with small perturbations occurring constantly. For example, in the process of evaluating an approach which is either unsuccessful or marginally successful, an approach that may be very successful will become evident. Therefore, the most important and potentially useful results at this stage of system development are qualitative results and observations. In addition to the specific items noted below, a major realization from this test experience is that most of the problems encountered and resolved would not have been evident using only graphic simulations or bench tests of separate components. The following is a compendium of the observations and results obtained to date:

(1) Integrating robotics into the initial design has been a key element in the success of this program, as opposed to retrofitting existing hardware developed for astronauts, or other assembly methods to automated telerobotic assembly operations.

(2) The development of test hardware that is designed around the use of simple components that are commercially available has aided in the ability to initiate tests quickly and to overcome problems through hardware modifications as well as software adaptation.

(3) The concept of having the end-effector insert the strut into a captured node has proven to be operationally successful. Using the robot arm to push the strut into a receptacle, which was considered early in the development, simply would not have worked. Any drag from friction that occurs during insertion, or misalignments, will move the receptacle and, thus, compound the problem. Also, for many truss structures that require assembly, the direction of insertion cannot be aligned to lie along the axis of a supporting member that reacts to the insertion load.

(4) Reliable strut installation was achieved only through compliant moves provided by a wrist-mounted force/torque feedback system. Misalignments, (although they may be small) that occur through normal positional

errors, cause high loads to be induced in the end-effector. A stiff robot arm attached through the end-effector to a stiff truss makes some form of compliance necessary to relieve the loading on the end-effector operating components. During force /torque repositioning, the final positioning of the arm has been observed to involve motions as small as 0.002 cm.

(5) Passive guidance features designed into the end-effector and joining mechanisms have significantly aided capture of the joint receptacle by the end-effector fingers and installation operations. Modifications to the passive alignment features of the joint receptacle, as indicated earlier, are required to overcome installation difficulties.

(6) Positive actuation of all capture and release functions is mandatory. Spring actuated release mechanisms were initially incorporated into the design of the end-effector and they proved to be unreliable due to loads caused by misalignment. Also solenoid-actuated latching devices were not effective. A robust positive force capability is needed for all power driven actuation devices.

(7) Full instrumentation of the end-effector is essential, particularly for automated operations. Checking each step, using simple instrumentation sensors, made end-effector operations very reliable. Approximately 33 tests are performed by the end-effector in removing each strut from the canister and installing it into the structure.

(8) Video camera coverage is necessary for observing end-effector component operations and verifying the end-effector instrumentation.

(9) Surveillance and over-the-shoulder video camera coverage is essential for operator viewing of system operations. However, the complex paths, the lack of depth perception, and the number of camera positions required to ensure collision-free paths, appear to make teleoperation for any task other than error recovery difficult. Teleoperation capabilities for complex assembly paths and tasks, similar to those examined in this investigation, need to be evaluated.

(10) The capture and installation of gravity-deflected cantilevered struts presented fewer problems than originally anticipated, due to the repeatability of the strut's deflected position.

(11) The assembly process must be under computer control. It is difficult for even highly experienced operators to remember system status and operational sequences. All manual commands and operator intervention should be verified by knowledge-based tools to ascertain their advisability and safety before execution.

(12) Complete database information is required on the status of the system to make decisions and resolve conflicts. The database information must be checked and updated by system sensors at program initialization and kept current by sensory input at the lowest level possible. The database update must be done automatically rather than being left up to the operator. All levels of the command hierarchy check the database prior to issuing each command to avoid unnecessary or conflicting commands.

(13) The software was designed primarily from the point of view of the operator and his role in controlling the system and handling errors. The resulting modular hierarchical format has remained virtually unchanged during the evolution of the system and it has proven to be very successful.

(14) Errors detected at the component level must be resolved before the assembly sequence can continue. The error condition is passed up through the command hierarchy and, in some cases, can be resolved by alternate actions at a higher level.

(15) The dynamic hierarchical menu displays provide a concise summary of what the automated system is doing and why, and are essential to the operator's understanding of the system's current status and its progress in the overall assembly sequence.

(16) A manual "adjust" capability to command small changes in the end-effector position is a desirable and useful feature. The adjust range should be limited by the software to avoid collision in the event of operator entry error. A teleoperator mode may be useful and is being considered for implementation and evaluation purposes.

(17) For the operator's convenience, every operation must be interruptible, or able to be paused at any point, to give the operator the capability to survey conditions and decide if the process is proceeding in an acceptable manner. From the paused condition, every operation must be reversible to permit the operator to roll back the operation to a known or successful condition.

(18) Clear, concise naming conventions for the truss members, the assembly system devices, and the component commands is vital to the operator's ability to monitor and intervene in the automated operation in an expedient manner. The name designations that finally evolved are simple but the process was surprisingly difficult and many of the names were modified several times before agreeing upon a suitable convention.

(19) Bench testing of strut joining is not a substitute for "in situ" operations. The problems that surfaced during actual assembly were more complex than those that are likely to have been encountered in controlled situation bench testing. Also, tests with real hardware are time consuming

and expensive, however, it permits many of the real problems to surface. Graphic simulation, while necessary and effective, will not substitute for test hardware experience.

(20) In the event of an error it is very desirable to be able to automatically roll back a composite command to the initiation point prior to handing control back to the operator. Otherwise, the operator is required to recall the sequence of primitive commands at the point where intervention is necessary and he must be prepared for any error situation.

(21) Receptacle presence sensors were mounted in the receptacle fingers to detect the joint receptacle. This relieved the operator from having to monitor the capture of cantilevered struts and eliminated many of the initial error conditions.

(22) Experience has led to the incorporation of several automated error recovery features. One of these is a force/torque balance for receptacle finger closure errors so that the operator is not bothered by easily corrected problems.

(23) The force/torque balance algorithm has been limited to 30 iterations to safeguard against limit cycling.

(24) The receptacle fingers are the most vulnerable end-effector component and several sets have been distorted due to robot position errors and loss of robot and/or motion base calibration. Since errors are likely to occur, the most vulnerable end-effector component must be capable of taking a substantial portion (or potentially all) of the full robot force capability without failure. An alternative would be to use the failure value of the most vulnerable component as a robot control limit.

Subsystem testing was helpful in the early development phase, however, it became apparent that tests of this type would not identify many of the problems encountered in the full assembly tests. Most of the problems encountered in the full assembly tests were problems that could not have been anticipated. For example, bench testing of strut installation with the end-effector would not simulate all conditions encountered in the actual in situ installation of a strut in the truss. It quickly became apparent that few struts could be installed, even with the passive guidance that had been designed into the hardware, without the use of force/torque feedback positioning. Most end-effector operations were initially performed manually by initiating commands at the component level. This demonstrated the need for additional sensor feedback and helped to develop the scenario for the correction of errors and the need for automated roll back to the condition that existed prior to the error. As the higher levels of automation were integrated, the value of the dynamic hierarchical menu display was recognized along with the need for a “Pause” and “Reverse” capability. A significant portion of the system capability, especially with regards to software requirements, was empirically developed from the problems that were encountered, most of which could not have been anticipated without the hardware experience.

The merging of basic technology in robotics, structures, and mechanical design into an interdisciplinary effort that has used a generic building block as a research focus has been successful in advancing the technology in a relatively short time. Addressing the total integrated task, as opposed to component testing, has forced all aspects of the task to be evaluated. No problems have been encountered which would indicate that automated assembly is not a viable option for in-space construction. All of the fundamental processes have been addressed and the system is fully functional in a supervised, automated mode. This hardware testing indicates that the basic methodology developed in the areas of mechanisms design, assembly techniques, software structure, and operator interface will apply directly to a space-based automated assembly system.

## 2.7 Future Research Opportunities

The research conducted to date has provided insight into additional studies that are needed to advance the current state of technology. Most of the topics listed in the subsequent sections have been initiated and the research is in differing stages of development. The intent is to identify these activities and thereby demonstrate the breadth of the research activity.

### 2.7.1 Vision System

The practice of operating a structural assembly system that relies totally on the taught points for strut capture and installation is neither flexible or robust enough for space operations. Therefore, work on a sensor guidance system has been initiated to develop and evaluate the capability to locate and guide the robot to a passive target mounted on the joint receptacle. Another benefit of a sensor guided system is its ability to aid in the implementation of the second generation end-effector that is capable of installing struts one end at a time.

Several alternatives for locating the joint receptacle and guiding the arm have been considered and a variety of sensor options are available: infrared, laser-guided, and optical. Two of these options, infrared sensors and optical machine vision, have been evaluated. The use of infrared sensors was ruled out because initial tests indicated the need for a large servo range to accurately track distant targets and the implementation would have required the use of multiple sensors for the total guidance operation. Therefore, attention has recently focused on the potential use of a vision system. Incorporation of a vision system into the assembly operation was delayed until sufficient progress had been made in the vision development and the assembly scenario requirements.

The telerobotic delivery of a strut from the canister to installation in the structure is projected to be made up of three phases: 1) an automated path planner that uses geometric information on the structure, robot arm, and carriage positions to guide the arm to a position where; 2) a camera mounted on the end-effector can locate targets on the node to which the strut is to be attached and guide the arm to a position where the end-effector fingers can grasp the joint receptacle. At this location; 3) force/torque feedback can be used to precisely align the end-effector for strut insertion. Although a path planner for the automated assembly system has not been evaluated, it is estimated that such a tool should be able to guide the arm to a point within approximately 0.46 m (18 inches) of the strut's installation point and be located within about ±5.1 cm (2 inches) of the desired location in each of the three translational tool tip coordinates. This point is denoted as the Vision Approach Point. At this point the vision system would discriminate and locate the target, then align and guide the robot to a point near the joint receptacle. The targets must be specifically designed, accurately fabricated, and positioned on each node receptacle to provide the information necessary to guide the robot end-effector to the proper alignment position with the node receptacle at about 13 cm from the node for strut insertion/removal.

A system has been designed and preliminary target acquisition tests have been conducted in the assembly test facility. Implementing the vision system introduced a number of hardware issues. One major concern is the limited space available for the location of the camera on the end-effector and the target on the node receptacle. Also, the camera had to be aligned with the target and be nonobtrusive so as not to affect the assembly scenario or end-effector component operations. Therefore, the video images for these tests were produced by miniature Charge Coupled Device (CCD) video cameras similar to those already mounted on the end-effector. Since these miniature video cameras do not have the capability for remote focusing and iris setting, the range of operation and lighting conditions must be bounded.

The primary advantages of the vision system are the capability to acquire the desired target with little or no servoing, to provide a real-time view of the image scene to the system operator, and to process the images at a rapid rate. The current image processing system consists of a commercially available system configured with processor boards to perform the tasks of image capture, storing, thresholding, finding centroids of regions (blobs), and displaying graphics. It is hosted by the executive computer. Controlling software has been developed to process images through the various processor boards and to provide for operator interaction.

Figure 2.21 shows the receptacle target that measures approximately 1.3 cm x 2.5 cm. The light colored dots on the flat black background are made of a

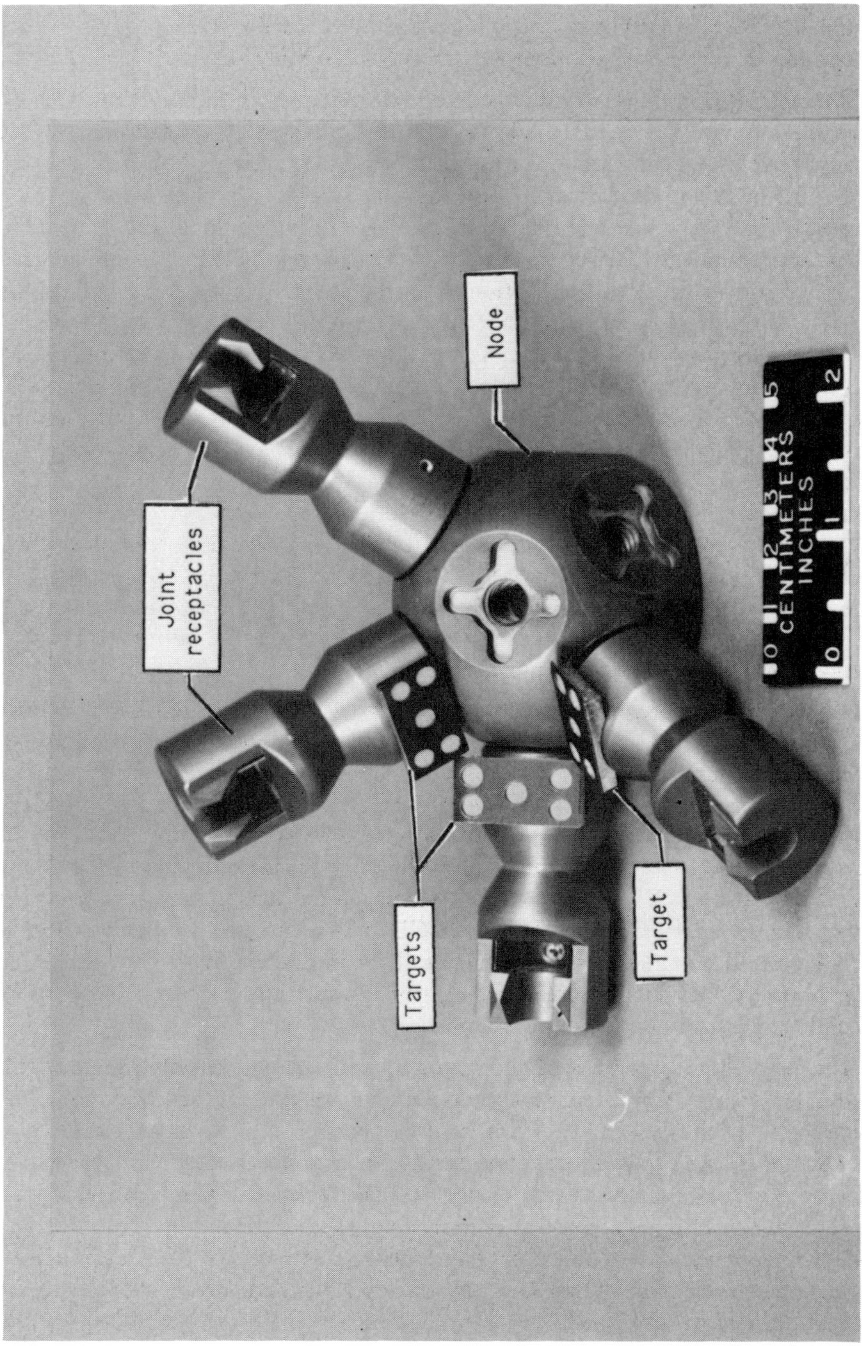


**Figure 2.21: Targets for Machine Vision System**

retro-reflective material and are arranged in a distinctive pattern to facilitate target recognition by an image processing algorithm. The current target design satisfies a variety of requirements: it is small so as not to interfere with the fingers of the end-effector; simple and accurate; capable of being mass-produced; unique enough to be distinguished in a varying and highly-cluttered background; and provides range information in addition to planar positioning, so that the arm can be guided to the target in three-dimensional space.

To simplify fabrication the target is constructed in layers. The bottom layer is a commercially available retro-reflective tape that has a high reflectivity for light rays that are incident parallel to the optical axis of the video camera. The top layer is a thin sheet of flat black anodized metal to reduce its reflectivity, thereby providing a high contrast between the target dots (blobs) and the surrounding background. This black surface acts as a mask for the bottom layer and reduces the possibility of dots blending with each other or the background in the processed image. The domino configuration of five dots is simple enough to be accurately mass-produced and its geometry makes it relatively easy to discriminate from most background clutter. Only four dots are needed to determine the target's position and orientation; using five dots introduces some redundancy and allows for the possibility that one dot may be obscured. The target is illuminated by a small light source mounted adjacent to the video camera lens as shown in Figure 2.22.

Conditions in the laboratory have not been altered to simplify the image recognition process since the goal of the vision system is to enhance the robustness of the overall assembly system. Since the desired outcome of the image processing is to extract the correct target from its background, a combination of processing tasks that take advantage of the target dot configuration is employed. The first task is to acquire an image from the video camera and store it as a gray level digital image in a frame buffer. A technique that utilizes histogram information is used to determine the appropriate gray level threshold, between zero (black) and 255 (white). This reduces the amount of information in an image that must be processed, and tends to eliminate much of the background clutter. The next step is to identify regions of contiguous pixels, called blobs, that have nonzero gray level values. Checks are made to evaluate the circularity of the blobs. The centroid of each circular blob is defined in terms of pixel coordinates relative to the processing board. At the end of this screening process the individual centroids of each target dot are sent to a four-point quadrangle algorithm. This algorithm uses the centroid information of four of the five target dots located in the image plane to compute the three-dimensional location and orientation of the target relative to the end-effector. This information is passed from the vision control program to the robot arm command program that determines the corrections necessary to align the end-effector with the node receptacle. This procedure is performed iteratively until the end-effector is properly aligned with the node receptacle. If during the image processing

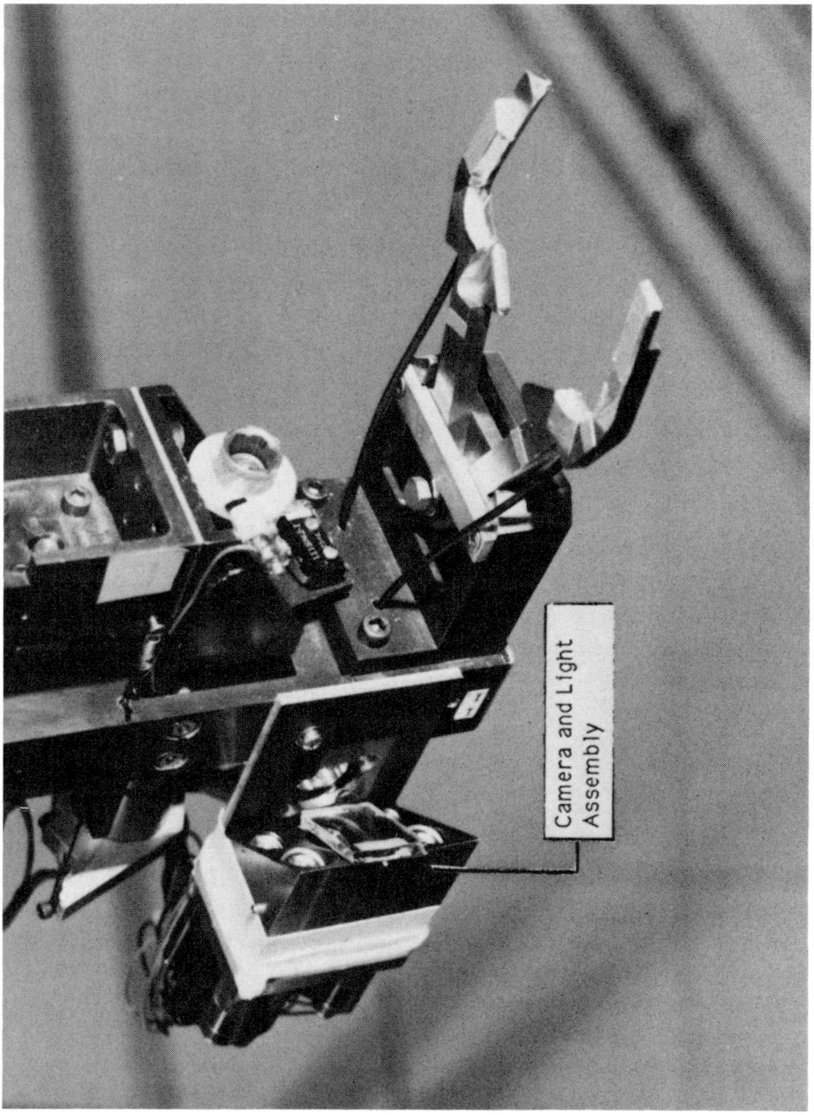


Figure 2.22: Machine Vision System Camera and Lighting

cycle the target dots evade detection, either by incorrect thresholding or inaccurate calculation of the centroids, tolerances placed on the various checks are automatically adjusted until either the target is correctly identified or the system operator chooses to intervene. If a situation occurs that prevents the vision system from identifying the target, the system operator may identify the target, thus, enabling the vision system to proceed. Additional development tests are anticipated to evaluate the vision systems performance under a broad range of environmental conditions. The interface between the vision control program and the robot command program is underway.

## 2.7.2 Installation of Panels

The assembly of a bare truss that has no instruments or surface, represents only a part of the task of assembling useful spacecraft. In general, the parts of such systems will be both diverse and complex. To accomplish a host of assembly operations there is a question whether this should be performed by developing a general purpose manipulator tool or using a special tool for each operational phase. To gain some experience in the application of interchangeable end-effector tools, development is underway to expand the assembly task to include the installation of reflector-type panels. An end-effector has been fabricated that is capable of acquiring a 2 meter wide hexagonal panel at three vertex points and attaching it to the nodes on the top surface of the truss. This operation will require a coordinated strut and panel assembly scenario, [8]. The procedure involves using the strut end-effector to assemble part of the truss, parking this end-effector and acquiring the panel end-effector to install several of the twelve panels. End-effector change operations occur several times during the integrated assembly and are accomplished using a quick-change device mounted between the robot wrist and the end-effectors.

The robot arm tasks for panel installation are relatively simple compared to strut installation. The path and procedures for the panel end-effector are identical for all panels, only the position of the motion bases are different for the twelve panels. Some of the strut installation procedures have aided developing panel installation procedures. The panel is removed from a canister that is adjacent to the X motion base and moved to a location that is just above the nodes on the top of the truss. Fingers on the end-effector grasp two of the truss joint receptacles to establish position and alignment . The panel is rotated around these receptacles and aligned with the third node. Actuators lower the panel onto the three support nodes simultaneously where it is latched in place. The latches are lock-bolt type units that release the panel from the end-effector at the same time it is attached to the truss. After the panel is attached to the truss, the end-effector is withdrawn from the truss to acquire the next panel. The panel end-effector will be instrumented with sensors to monitor each function and the installation will be controlled by a microprocessor similar to the one developed for the second generation strut end-effector (to be discussed later). Panel assembly tests are anticipated to begin soon.

### 2.7.3 Sequence Planning

The sequence for the automated assembly tests conducted to date was developed manually and required several modifications before the final sequence was established. It was a laborious task. For larger space-trusses that may involve thousands of members or for the development of repair scenarios, even trained human planners frequently fail to detect dead-end sequences until a lot of time and effort has been wasted. The development of a scenario for replacing a member in a damaged truss or replanning a sequence because of an unexpected error cannot be done in advance because it is virtually impossible to anticipate every failure. Speed and reliability will be very important in these replanning efforts. Therefore, a cooperative program with the Jet Propulsion Laboratory is underway to develop an assembly sequence planner.

The technique is based on representing the set of all assembly plans by either directed graphs or by AND/OR graphs. The nodes in the directed graphs correspond to states of the assembly process and the arcs in the AND/OR graphs correspond to assembly tasks associated with joining two components or subassemblies. Feasibility conditions and cost functions are the tools used to develop and gauge the quality of each solution. The algorithm for generating a sequence develops it in a backward manner. It begins with the state goal that has all of the components connected together and ends at the initial state with all of the components disconnected. Details of the operations and procedures for the approach to this assembly sequence planning can be found in [9], and in Chapter 4 of this book.

### 2.7.4 Path Planning

The use of predetermined, taught points and paths for delivering struts into the structure for installation is simply not robust enough for a viable flight system.

The complexity of the truss geometry, the limited reach of the robot arm, the total degrees-of-freedom available for positioning, and the length of the struts presents a challenge for the determination of collision-free paths. By the same token, the regularity of the truss geometry, and the rod shape profile of the struts, makes an automated path planning tool easier to implement. Therefore, an effort is anticipated to implement an on-line automated path planner that incorporates the database system developed for the assembly operations. Several approaches to the path planner may be explored. One technique is to define the geometry of the truss and use a brute force combination of the geometry and vector algebra. A second technique is to utilize a simple potential field model, several of which are discussed in [10], and in Chapter 4 of this volume. A path planner would be expected to deliver the strut along a collision-free path to a location where the machine vision system, discussed previously, could

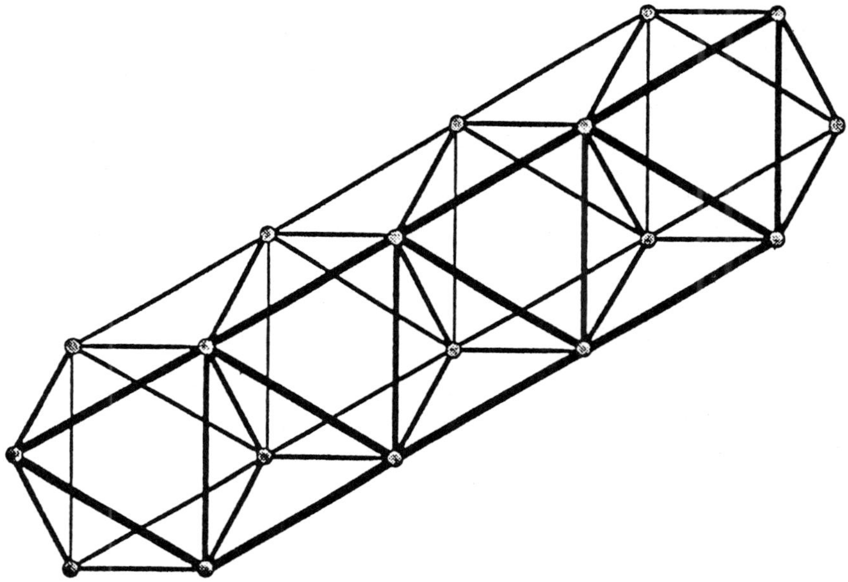

Figure 2.23: Tetrahedral Beam Configuration

assume control at the vision approach point. The path planning tool will be evaluated with the graphics simulation (to be discussed later) prior to tests with the assembly hardware.

### 2.7.5 Assembly of a Linear Truss Beam

One desirable objective is to evaluate the feasibility and difficulty of constructing other truss configurations using the same assembly system internal knowledge. This possibility has only been recently explored and the initial indications appear promising. The truss configuration under consideration is a linear tetrahedral beam as shown in Figure 2.23. The beam uses the same struts and nodes as the ring structure, but rather than building complete rings, struts are added to the ends of a beam formed on opposite sides of the turntable. Detailed analysis of this scenario reveals that the automated system already contains all the information required for the installation of these struts. All that is needed to produce a different truss configuration is an appropriate change in the assembly sequence that drives the automated system. The assembly sequence for the beam has been generated, but not yet tested.

The arrangement of noded and unnoded struts in the canister is fixed by geometry and node receptacle interference considerations. The development of a planning algorithm that uses a "node or no node" request during the assembly sequence to locate and select struts from the canister is relatively simple to implement. Current plans call for actually assembling a linear tetrahedral beam to verify that the system changes are indeed minimal and to develop and test a simple strut search and selection algorithm. As a further extension of this effort, a fixture is being developed for the turntable that will enable the beam to be constructed from one end and then pushed through a set of tracks so that each bay can be indexed forward after it is assembled. This fixture will permit the assembly of beams that have a range of member length from one to two meters.

### 2.7.6 Second Generation End-Effector

A second generation end-effector that is capable of acquiring a strut at one end only and installing it into the truss has been designed and fabricated. This end-effector will enhance the versatility of the system by providing the capability to assemble contoured trusses that have members of different lengths and to install system payload modules. Implementation of this end-effector will be challenging because the end-effector will be required to capture and connect struts that are cantilevered from the truss. Operations are not anticipated to begin until the path planning and vision systems have been fully developed and demonstrated.

The second generation end-effector hardware is very similar to one end of the current end-effector. Two strut grippers are located on the strut insertion platforms and a set of tube-gripping fingers has been added inboard of the strut grippers to aid in stabilizing the strut. Actuator and sensor functions are also similar to the current end-effector. Operational sequences will necessarily be more complex because every strut will have to be inserted at one end and then captured in a cantilevered condition and installed at the second end. This will also increase the installation and removal time. However, this new end-effector will incorporate an onboard dedicated microprocessor to command actuators and verify successful operation through the interrogation of sensors. The motivation for development of an onboard microprocessor is to reduce the number of signal lines from the end-effector to the robot controller and to incorporate a distributed computational capability that is capable of supporting concurrent operations. An 80535 microprocessor has been procured to control the end-effector and to communicate with the executive program upon request via RS232 serial lines. All end-effector software code and logic is implemented in "C". This includes both sequence control and sensor monitoring functions. While the microprocessor decision was driven primarily by communications considerations, an added benefit is the ease of accommodating several interchangeable end-effectors with the same executive program and command interface. This effectively makes the end-effector a stand-alone device, capable of performing standard high-level commands, monitoring its own sensors, and providing diagnostics and operator information for its error recovery. With this capability in place, an end-effector can perform standard functions such as

"INSTALL" and "REMOVE" autonomously, so the executive program will not have to handle the detail of how a function is performed, only whether it was successful. These commands may be general so they refer to whatever object the particular end-effector is dealing with and thereby the executive software can be reduced significantly. The speed of performing end-effector operations will also be increased because the processing of sensor checks will be based on interrupts. This will offset, to some extent, the increased operational time for installation and removal. The microprocessor-based end-effector control has the added potential of speeding up system operation by making motion base moves and end-effector operations concurrent. An example of this would be the capability to move the motion base carriages into position while the end-effector picks up a strut from the canister.

### 2.7.7 Graphics Simulation

The large amount of internal information involved in the automated assembly system, the need for clear, concise transfer of this information to the operator, and a desire to minimize operator training and experience requirements make 3-D computer-generated graphics displays extremely attractive. The ability to graphically simulate what the system is trying to accomplish, will allow the operator to accurately anticipate the operation about to occur, and permit a visualization of any intervention operation that may be contemplated. Also, the ability to depict mechanisms and assembly operations from any, (even a physically impossible) perspective could save in the assembly system's video coverage and lighting requirements. Another benefit of graphics simulation is the capacity to develop and check out software and system algorithms in a safe, non-hardware environment where all parameters are under simulation control. This would speed-up software development and thereby significantly reduce the cost. A final graphics simulation feature is the ability for rapid prototyping of system concepts and configurations.

A graphics simulation and display system has recently been procured and the development of a simulation is underway. One aspect that will be evaluated in initial simulation studies is the need for real-time simulation operations. It is not presently clear that the operator displays are required to operate at real-time rates or if sufficient information can be obtained, and a simulation can be effective that operates at other than a real-time rate.

### 2.7.8 Operator Training and Human Factors Evaluation

The control system was designed to perform the truss assembly with minimal operator intervention, however, there are many instances (particularly in dealing with system errors) when the computer software must present information to the operator and elicit a response. When the software was initially developed, some consideration was given to the selection of names for commands that would be identified with hardware function and error names that elicit the appropriate response. No formal evaluation of these terms was conducted, however, and personnel skilled in human factors response were not consulted. In light of the current state of the system and the opportunity for using it as a testbed for other studies, it became apparent that unskilled operator training and issues associated with human factors should be formally considered. Issues that may be evaluated are the type of information and the kind of visual aids that will be most helpful to the operator. One operator aid currently being developed is computer controlled video coverage (camera selection and pan/tilt/zoom) which will ease the task of monitoring operations.

To examine these issues, it is anticipated that personnel skilled in human factors studies will be consulted to evaluate the current system and recommend an assessment plan. It is anticipated that this will involve some testing and evaluation using unskilled test subjects as assembly system operators. These operators will be provided with a common level of training and instruction on the hardware and software menus. This may be in the form of a video training supported by hands-on-training at the controls with a skilled instructor. For the actual tests, the assembly tasks will be seeded with a planned sequence of errors so that all operators will be required to deal with the same problems under the same set of conditions to accurately assess their performance.

## 2.8 Conclusions

Techniques for construction and assembly in space other than those that traditionally rely on astronaut Extra-Vehicular Activity (EVA) must be explored. One alternative that is being examined at the NASA Langley Research Center is the use of an automated telerobotic system that uses an Earth-based, or a space-based, executive monitor to track the operation of robotic truss assembly and intervene only when the automated system encounters a problem and requires assistance. To examine the operations and obtain some practical experience a hardware testbed has been developed. The testbed was designed to incorporate off the shelf hardware and existing components to quickly obtain some practical operational experience. Several end-to-end assembly and disassembly tests have been conducted to date. The success of the research conducted in the testbed provides a very encouraging outlook for the realization of a space based operational system in an orbital, lunar, or planetary environment. One key item is the basic experimental approach which was developed around the detailed study of a focused problem. This proved to be essential in identifying the real issues and directing the effort toward well defined requirements rather than general concepts. Few of the significant problems and test observations for the automated assembly process were anticipated prior to this effort and many initial concerns never materialized. The viability and performance projections of automated system concepts have little relevance unless they are based on detailed consideration of the task or on practical experience.

One key element is that the program was developed around a cooperative, interdisciplinary approach to the designs of both the assembly system and the structure being assembled. All aspects of the automated operation were developed from a total system viewpoint. Experience has shown that the impact of automation on the design of vehicles and systems is just as important as the functional design of the system itself and close coordination between the required functional requirements and the automation aspects must be maintained. An important factor also, is the development of a system software and program structure based upon the needs and the role of the operator in error resolution. This led to effective error resolution by the operator and contributed to the success of the supervised autonomy mode.

The testing experience has highlighted the necessity for "in situ" development and demonstration in a complete and realistic testbed of all advanced automated system technology, as opposed to simple bench testing of individual components and end-effector mechanisms. The current Langley Research Center (LaRC) automated structures assembly facility provides both a baseline assembly system and modification potential to accommodate this need in many technology areas encompassing both automation and structural design.

It is recognized that much remains to be accomplished before such a system will be ready for flight test operations. For the current program, however, it was decided to delay development of advanced operations anticipated to be required for a flight system until a practical experience base could be developed that would provide a clear definition for advanced research tasks. The experience gained to date has now provided that direction. The reliance on the use of taught points is not adequate for space operation. Therefore, a preliminary machine vision system that is capable of identifying and locating a passive target to guide the robot to the strut installation position has been developed. The incorporation of other tasks in the assembly operation, such as, the installation of reflector panels has been initiated. No viable automated system for space operation could successfully operate without both a knowledge based sequence planner to deal with unanticipated failures and a path planner. The testbed experience has provided a better definition of the requirements for these efforts, an excellent facility in which to evaluate their operation, and a reference database against which they may be evaluated.

## 2.9 Appendix A

### Initial Conditions and Guidelines for Development of Truss Assembly Scenario

The following rules and guidelines have been formulated to assist in the development of the tetrahedral truss assembly sequences and scenarios. They address the considerations for structural stiffness and integrity during assembly, strut and node packing in the canister trays, and the robot and motion base constraints. The intent is to formalize these rules into a knowledge base for a goal-directed, automatic task sequence planner. The rules have been categorized into those applicable to the structural configuration, the tray packing conditions, and the robot and motion base constraints. By modularizing the knowledge base the assembly sequences may be easier to change.

**Initial Conditions**

1. The truss has 102 members and 31 nodes (19 nodes in the top plane and 12 nodes in the base plane).

2. The three base support nodes are set in place prior to initiation of assembly. These base nodes are fixed and are used as anchor points for stabilizing the truss system.

3. The ring assembly starts by building the center tetrahedron, then proceeds to the first ring. This is followed by assembling the lower part of the second ring and the unit is completed by constructing the upper part of the second ring. This process was adopted to aid in tracking the process and avoid leaving a member out of the truss as opposed to a requirement. However, it does aid in reducing the amount of structure that may be assembled asymmetrically, thus reducing the time/distance required to move the X carriage so it will clear the robot when repositioning the turntable.

**Structural Configuration Considerations**

4. A truss member can only be inserted in the plane of the square base of the pentahedron.

5. The entrance receptacles of the core struts are rotationally aligned.

6. A tetrahedral unit is completed before additional members are added to start a second tetrahedron supported by the first.

7. Two tetrahedrons may be partially developed and connected if they are supported by a third unit that has been previously completed, (i.e., a pentahedron may be completed before the two adjacent tetrahedrons are completed.)

8. Assembly is developed in rings and the rings are broken down into repeating sub-elements. Each sub-element is completed before beginning the next sub-element. After all sub-elements in a ring are assembled they are interconnected to complete the ring.

9. Only core struts are cantilevered to minimize derformations due to gravity and these cantilevered struts each have a preattached node. A cantilevered strut must be anchored to a completed tetrahedron or to one of the three fixed base nodes.

10. After installation of a cantilevered core strut, a face strut was always installed that would tie the cantilevered end to a fixed node. One cantilevered member cannot be cantilevered from another.

**Tray Packing Conditions**

11. Nodes are preattached to the members in the canister and are not free to be moved from one strut to another. In fact, the end-effector is incapable of removing a node and placing it on another strut. However, a pre-noded strut could be removed and replaced by simply unlocking the joints if it has at least one other core strut attached when the strut is removed.

12. Each canister tray holds up to 13 struts.

13. Trays must be removed from the supply rack and installed in the storage rack in the order (and orientation) that they are stacked since there is no provision for temporary storage. The members are required to be oriented in the proper direction, i.e., the node on the correct end, when they are inserted in the truss. This is handled in the assembly sequence as a "Flip".

14. For the current assembly scenario all members of a tray are removed before a tray is stored in the tray holder, however, not all positions in the trays are filled. Some spare struts could be permitted to remain in a tray while it is moved. They should be kept close to the center of the tray to minimize the inertial unbalance of the tray. The center storage position of the struts must be open to permit the end-effector to access the arms that are used to pick up the tray.

15. When the end-effector fetches struts from the tray it cannot fetch an unnoded strut immediately adjacent to a noded strut with the noded strut in the tray because of interference between the scar gripper and the adjacent node.

**Robot and Motion Base Constraints**

16. The robot reach and the axis constraints must not be violated.

17. The number of predefined robot paths and taught points should be minimized.

18. It is desirable to insert as many struts as possible at a given carriage and turntable location within the range of the robot arm. The robot base positions (X and Y carriage position) were chosen to maximize the number of struts inserted at a given location.

19. The turntable is capable of ±3 revolutions from its starting position. To avoid hitting the limit, the rotation of the turntable was monitored and it was reversed regularly to avoid even approaching the limits.

20. The X carriage is always aligned after rotation with the 10/2 or 8/4 position of the clock.

21. The Y carriage travel limit is ±116 inches from the turntable rotation axis.

## 2.10 APPENDIX B

**Assembly Sequences for a Tetrahedral Truss**

Table 2.2 lists the manually developed assembly sequence for the two-ring, 102-member tetrahedral truss structure. This table is used by the operator for reference during the assembly/disassembly operations and contains information on strut names, robot arm paths, strut cantilever conditions, motion base positions, and tray assignment positions for each strut in the sequence. The purpose of this appendix is to describe the strut data and define the host of assembly operations/conditions that must be accounted for.

Figure 2.24 shows a diagram of the tetrahedral truss with the cells labeled and the strut sequence numbers indicated. All information in this table refers to Figure 2.24 except for the motion base positions (X, Y, Θ), which refer to the facility layout diagram in Figure 2.25.

The truss is assembled in basic sections, the center pyramid followed by the first ring. The second ring is assembled as two units, the lower section first followed by the upper outer portion. The outer portion of the second ring has member positions similar to the first ring. The following items define the notation on the sequence chart:

1. Strut designation is referenced to clock position with "12" being directly in "front" of the robot and moving clockwise around a circle looking down on the assembly, as shown in Figure 2.14.

2. There are two viewing designations noted: the operator designation and the robot designation. The operator designation is referenced to a scheme to locate a member anywhere in the truss by the use of a ring, cell and position location that is unique for that strut. The operator names are referenced to the clock designation at the bottom of the figure. The robot designation is referenced to the path and position the arm must assume to successfully insert the member. The robot name is referenced to the turntable angle because this is the perspective from the position of the strut. The robot path is a function of the motion base location.

| SEQUENCE NUMBER | OPERATOR NAME | ROBOT NAME | X/Y/θ | TRAY/SLOT NODE/END |
|---|---|---|---|---|
| Center pyramid | | | | |
| 1 | R1C1/10_2 | 10_2/FIXED | -205.647/-17.50/0.0 | 1/3 |
| 2 | R1C1/12_2 | 12_2/FREE/UP/R | | 1/5/L |
| 3 | R1C3/6_10 | 10_2/FIXED | -205.647/-17.50/120.0 | 1/4 |
| 4 | R1C3/8_10 | 12_2/CAP1 | | 1/6 |
| 5 | R1C2/6_2 | 10_2/FIXED | -205.647/-17.50/-120.0 | 1/7 |
| 6 | R1C2/6_4 | 12_2/CLOSE | | 1/11 |
| Center pyramid completed. | | | | |
| 7 | R2C7/6_8 | F10_8/FREE/UP/F | -197.647/-17.50/-120.0 | 1/9/L |
| 8 | R1C2/12_4 | 12_8/CAP1 | | 1/8 |
| 9 | R2C7/6_4 | 12_2/FREE/UP/R | -205.647/-96.18/-120.0 | 1/13/L |
| 10 | R2C7/8_4 | 12_4/CAP2 | -197.647/-96.18/-120.0 | 1/10 |
| 11 | R1C3/12_8 | 12_4/CLOSE | -197.647/-17.50/120.0 | 1/12 |
| 12 | R1C3/6_4 | F10_8/FREE/UP/F | -197.647/-17.50/120.0 | 1/1/L |
| 13 | R1C3/8_4 | 12_8/CAP1 | | 1/2 |
| 14 | R2C1/10_8 | F12_2/FREE/UP/F | -205.647/-96.18/120.0 | 2/5/R |
| 15 | R2C1/12_8 | 12_4/CAP2 | -197.647/-96.18/120.0 | 2/4 |
| 16 | R1C1/12_4 | 12_4/CLOSE | -197.647/-17.50/0.0 | 2/6 |
| 17 | R1C1/10_8 | 10_8/FREE/UP/R | -197.647/-17.50/0.0 | 2/9/R |
| 18 | R1C1/12_8 | 12_8/CAP1 | | 2/7 |
| 19 | R2C4/12_2 | F12_2/FREE/UP/F | -205.647/-96.18/0.0 | 2/13/R |
| 20 | R2C4/12_4 | 12_4/CAP2 | -197.647/-96.18/0.0 | 2/8 |
| 21 | R1C2/8_4 | 12_4/CLOSE | -197.647/-17.50/-120.0 | 2/11 |
| 22 | R1C2/12_8 | 8_4/FIXED | -172.0745/-17.50/-120.0 | 2/12 |
| 23 | R1C1/8_4 | 8_4/FIXED | -172.0745/-17.50/0.0 | 2/10 |
| 24 | R1C3/12_4 | 8_4/FIXED | -172.0745/-17.50/120.0 | 2/3 |
| First ring completed. | | | | |
| 25 | R1C2/10_8 | B6_4/FREE/DN/F | -200.88/-38.0/-120.0 | 2/1/R |
| 26 | R1C2/6_10 | 6_2/CAP1 | -198.88/-38.0/-120.0 | 2/2 |
| 27 | R1C2/10_2 | 6_10/CLOSE | | 3/1 |
| 28 | R2C6/6_4 | 12_2/FIXED | -137.454/-56.87/-120.0 | 3/5 |
| 29 | R2C7/10_8 | D6_4/FREE/DN/R | -198.88/-116.74/-120.0 | 3/3/L |
| 30 | R2C7/6_10 | 6_2/CAP1 | | 3/2 |
| 31 | R2C6/6_2 | 10_2/CLOSE | -137.454/-56.87/-120.0 | 3/4 |
| 32 | R2C8/10_8 | B12_2/FREE/DN/F | -137.454/21.87/120.0 | 3/7/L |
| 33 | R2C7/6_2 | V6_2/CAP1 | -141.0/-19.50/-180.0 | 3/6 |
| 34 | R2C7/10_2 | V10_2/CLOSE | -133.0/-17.50/-180.0 | 3/8 |
| 35 | R1C3/12_2 | D6_4/FREE/DN/R | -198.88/-38.0/120.0 | 3/11/L |
| 36 | R1C3/10_2 | 6_2/CAP1 | | 3/9 |
| 37 | R1C3/6_2 | 6_10/CLOSE | | 3/10 |
| 38 | R2C9/10_8 | 12_2/FIXED | -137.454/-56.87/120.0 | 3/12 |
| 39 | R2C1/12_2 | B6_4/FREE/DN/F | -200.88/-116.74/120.0 | 4/3/R |
| 40 | R2C1/10_2 | 6_2/CAP1 | -198.88/-116.74/120.0 | 4/1 |
| 41 | R2C9/6_10 | 10_2/CLOSE | -137.454/-56.87/120.0 | 4/2 |
| 42 | R2C2/12_2 | D12_2/FREE/DN/R | -137.454/21.87/0.0 | 4/7/R |
| 43 | R2C1/6_10 | V6_2/CAP1 | -141.0/-19.50/60.0 | 4/4 |
| 44 | R2C1/6_2 | V10_2/CLOSE | -133.0/-17.50/60.0 | 4/5 |
| 45 | R1C1/6_4 | B6_4/FREE/DN/F | -200.88/-38.0/0.0 | 4/11/R |
| 46 | R1C1/6_2 | 6_2/CAP1 | -198.88/-38.0/0.0 | 4/6 |
| 47 | R1C1/6_10 | 6_10/CLOSE | | 4/8 |
| 48 | R2C3/12_2 | 12_2/FIXED | -137.454/-56.87/0.0 | 4/9 |
| 49 | R2C4/6_4 | D6_4/FREE/DN/R | -198.88/-116.74/0.0 | 5/5/L |
| 50 | R2C4/2_6 | 6_2/CAP1 | | 5/4 |
| 51 | R2C3/10_2 | 10_2/CLOSE | -137.454/-56.87/0.0 | 5/3 |
| 52 | R2C5/6_4 | B12_2/FREE/DN/F | -137.454/21.87/-120.0 | 5/9/L |
| 53 | R2C4/10_2 | V6_2/CAP1 | -141.0/-19.5/-60.0 | 5/8 |
| 54 | R2C4/10_6 | V10_2/CLOSE | -133/-17.5/-60.0 | 5/7 |
| 55 | R2C5/2_6 | 10_2/FIXED | -137.455/21.87/-120.0 | 5/6 |
| 56 | R2C8/10_6 | 10_2/FIXED | -137.455/21.87/120.0 | 5/10 |
| 57 | R2C2/10_2 | 10_2/FIXED | -137.455/21.87/0.0 | 5/11 |
| This completes the lower part of the second ring. | | | | |
| | CW 360 TO STRAIGHTEN WIRES | | | |
| 58 | R2C4/10_8 | V12_2/FREE/UP/R | -153.79/-53.34/-30.0 | 5/13/L |
| 59 | R2C4/12_8 | V12_4/CAP1 | | 5/12 |

Table 2.2: Assembly Sequence

| SEQUENCE NUMBER | OPERATOR NAME | ROBOT NAME | X/Y/Θ | TRAY/SLOT NODE/END |
|---|---|---|---|---|
| 60 | R2C5/8_6 | F10_8/FREE/UP/F | -129.455/100.68/-120.0 | 5/1/L |
| 61 | R3C7/12_4 | V8_4/CAP2 | -88.5/-55.555/-60 | 5/2 |
| 62 | R2C5/8_4 | 12_4/CLOSE | -129.455/21.87/-120.0 | 6/3 |
| 63 | R2C7/12_2 | F112_2/FREE/UP/F | -153.79/-53.34/-150.0 | 6/1/R |
| 64 | R2C7/12_4 | V12_4/CAP1 | | 6/2 |
| 65 | R2C8/12_10 | 10_8/FREE/UP | -129.455/100.68/120.0 | 6/5/R |
| 66 | R3C12/8_4 | V8_4/CAP2 | -88.5/-55.555/180.0 | 6/4 |
| 67 | R2C8/12_8 | 12_4/CLOSE | -129.455/21.87/120.0 | 6/6 |
| 68 | R2C7/12_8 | 8_4/FIXED | -172.0745/-96.18/-120.0 | 6/7 |
| 69 | R2C7/12_10 | V12_10/FIXED | -90/14.87/-180.0 | 6/11 |
| 70 | R2C6/12_2 | 10_8/FREE/UP/R | -129.455/-56.87/-120.0 | 6/9/R |
| 71 | R2C6/12_4 | 12_8/CAP1 | | 6/8 |
| 72 | R3C11/8_4 | V8_4/CLOSE | -88.5/23.185/-180.0 | 6/10 |
| 73 | R2C5/12_2 | 10_8/FREE/UP/R | -129.455/21.87/-120.0 | 6/13/R |
| 74 | R2C5/12_4 | 12_8/CAP1 | | 6/12 |
| 75 | R2C6/8_4 | 12_4/CLOSE | -129.455/-56.87/-120.0 | 7/1 |
| 76 | R2C6/12_8 | 8_4/FIXED | -103.8837/-56.87/-120.0 | 7/5 |
| 77 | R2C5/12_8 | 8_4/FIXED | -103.8837/+21.87/-120.0 | 7/9 |
| 78 | R2C1/6_4 | V12_2/FREE/UP/R | -153.79/-53.34/90.0 | 7/7/L |
| 79 | R2C1/8_4 | V12_4/CAP1 | | 7/8 |
| 80 | R2C2/4_2 | F10_8/FREE/UP/F | -129.455/100.68/0.0 | 7/11/L |
| 81 | R3C2/12_8 | V8_4/CAP2 | -88.5/-55.55/60.0 | 7/12 |
| 82 | R2C2/12_4 | 12_4/CLOSE | -129.455/21.87/0.0 | 7/13 |
| 83 | R2C4/8_4 | 8_4/FIXED | -172.0745/-96.18/0.0 | 7/10 |
| 84 | R2C4/8_6 | V12_10/FIXED | -90/14.87/-60.0 | 7/6 |
| 85 | R2C3/10_8 | F10_8/FREE/UP/F | -129.455/-56.87/0.0 | 7/3/L |
| 86 | R2C3/12_8 | 12_8/CAP1 | | 7/2 |
| 87 | R3C6/12_4 | V8_4/CLOSE | -88.5/23.185/-60.0 | 7/4 |
| 88 | R2C2/10_8 | F10_8/FREE/UP/R | -129.455/21.87/0.0 | 8/3/R |
| 89 | R2C2/12_8 | 12_8/CAP1 | | 8/2 |
| 90 | R2C3/12_4 | 12_4/CLOSE | -129.455/-56.87/0.0 | 8/1 |
| 91 | R2C3/8_4 | 8_4/FIXED | -103.8837/-56.87/0.0 | 8/4 |
| 92 | R2C2/8_4 | 8_4/FIXED | -103.8837/21.87/0.0 | 8/5 |
| 93 | R2C1/12_4 | 8_4/FIXED | -172.0745/-96.18/120.0 | 8/9 |
| 94 | R2C1/4_2 | V12_10/FIXED | -90/14.87/60.0 | 8/13 |
| 95 | R2C9/6_4 | 10_8/FREE/UP/R | -129.455/-56.87/120.0 | 8/11/R |
| 96 | R2C9/8_4 | 12_8/CAP1 | | 8/10 |
| 97 | R3C1/12_8 | V8_4/CLOSE | -88.5/23.185/60.0 | 8/12 |
| 98 | R2C8/6_4 | 10_8/FREE/UP/R | -129.455/21.87/120.0 | 8/7/R |
| 99 | R2C8/8_4 | 12_8/CAP1 | | 8/6 |
| 100 | R2C9/12_8 | 12_4/CLOSE | -129.455/-56.87/120.0 | 8/8 |
| 101 | R2C9/12_4 | 8_4/FIXED | -103.8837/-56.87/120.0 | 9/3 |
| 102 | R2C8/12_4 | 8_4/FIXED | -103.8837/21.87/120.0 | 9/7 |

This completes the assembly sequence.

Empty Slots: <u>tray/slots</u>

3 /13

4 /10,12,13

9 /1,2,4,5,6,8,9,10,11,12,13

Table 2.2 (continued): Assembly Sequence

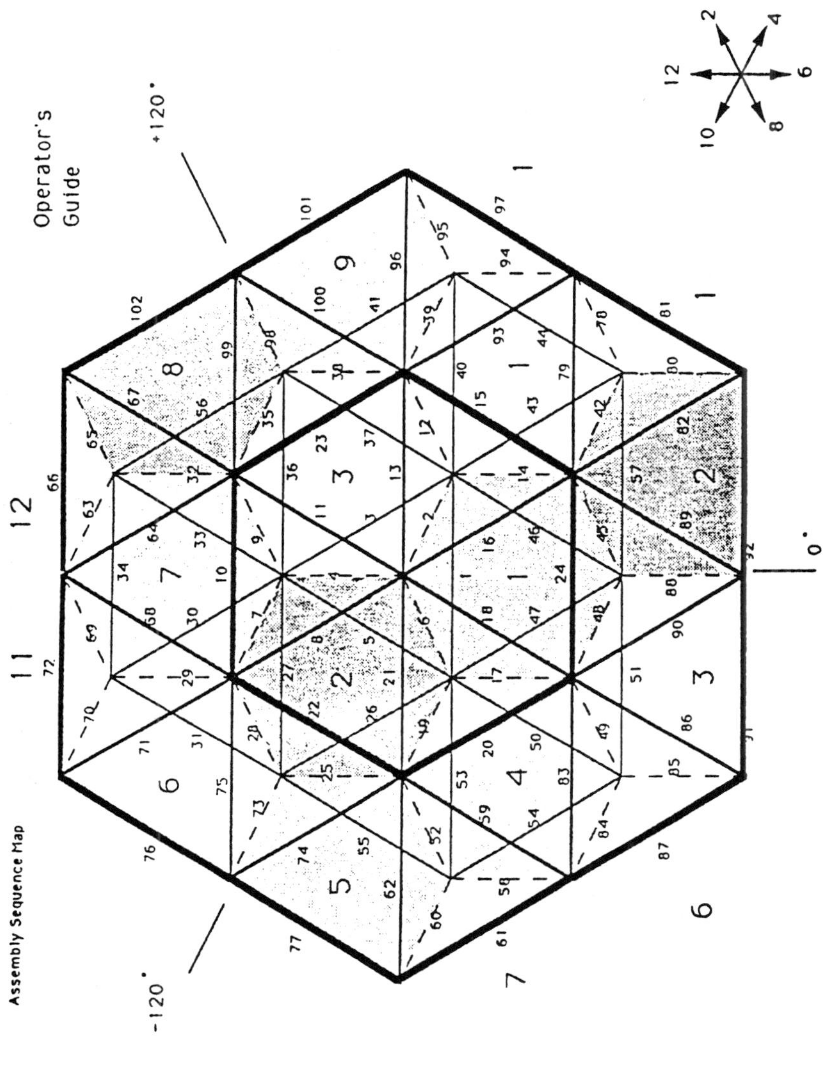


**Figure 2.24: Tetrahedral Truss with Cells and Sequence Numbers**

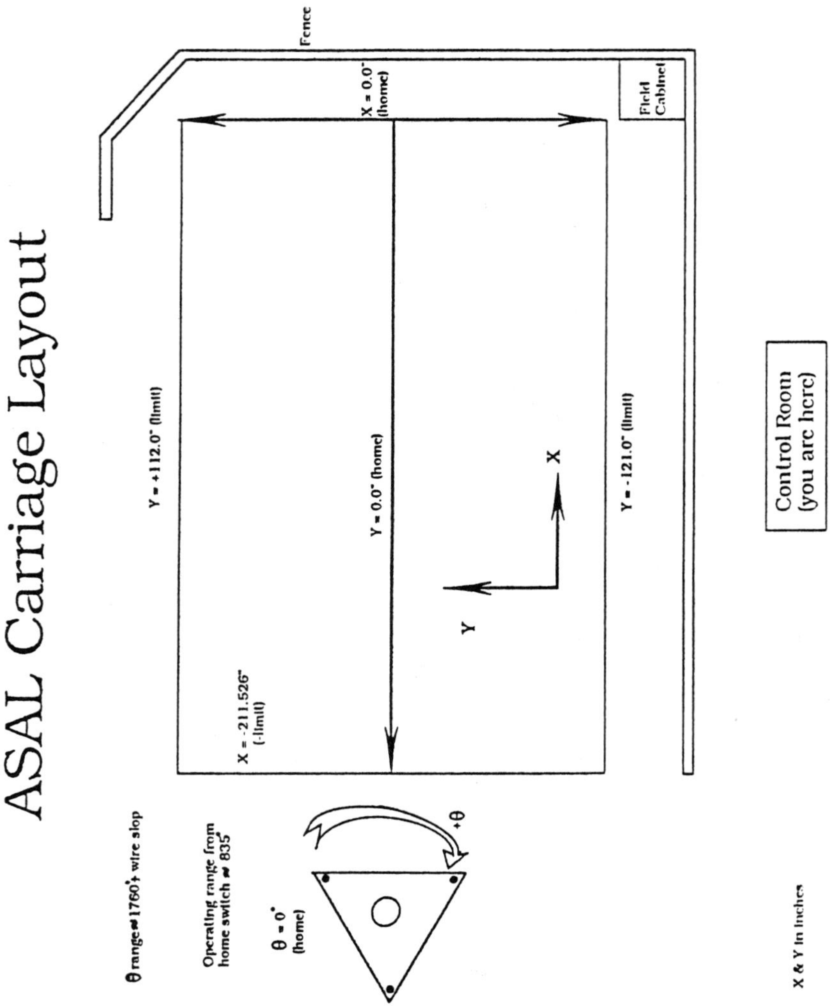


Figure 2.25: Facility Layout Diagram

3. FREE, CAP1, CAP2, CLOSE and FIXED indicate strut capture/release conditions. FREE refers to the attachment of a pre-noded core strut with release of the member as a cantilever beam supported at the root. CAP1 represents a capture of a FREE cantilevered strut receptacle to attach the present strut to. CAP2 means to capture a strut receptacle in the structure at one end, move to and capture the strut receptacle at the other end, then move to the installation point (Condition for 3 struts in first ring only). FIXED represents that present the member is attached to receptacles that are fixed in the structure due to the attachment of several members at both ends. CLOSE is similar to CAP2 condition (two cantilever struts) with capture at only one end. The completion of this installation establishes the FIXED condition noted above.

4. The location and orientation for the core struts with preattached nodes is important. The notation UP (e.g., 12-2/UP) means that the node lies in the top surface of the truss, whereas DN indicates that the node lies in the bottom surface of the truss. The trailing R or F indicates that the taught position is REAL (R-the taught path moves the node from the canister to the correct face of the truss) or FLIPPED (F-requires a 180 degree rotation of the strut to place the node in the correct face of the truss).

5. The motion base positions X/Y/Θ in this sequence are the reference positions from which the robot arm paths are taught. Note that there is a fixed point of reference in one orientation only. In this orientation, the 10_2 strut is always a bottom surface strut and the 8_4 strut is always a top surface strut.

6. The R or L under the Tray/Slot/Node column indicates that the node is on the right or left end of the tray and is picked up on that side of the end-effector. This is associated with the REAL and Flip notation. X NODE means that the location of that strut in the tray is one that normally has a node attached, however, a node is not required for the strut that is currently allocated to this slot. All struts with nodes preattached are either (ROBOT DSN) 12_2, 10_8, or 6_4. There are no 12_10, 8_6, or 4_2 struts (with or without preattached nodes) due to the orientation of the receptacles.

## 2.11 References

[1] Anon, "Report of the advisory committee on the future of the U.S. space program," *Superintendent of Documents, U.S. Printing Office*, December 1990.

[2] W. F. Fisher and C. R. Price, "Space station Freedom external maintenance task team," *Final Report*, July 1990.

[3] M. D. Rhodes, R. W. Will, and M. A. Wise, "A telerobotic system for automated assembly of large space structures," *NASA TM 101518*, March 1989.

[4] K. C. Wu, R. R. Adams, and M. D. Rhodes, "Analytical and photogrammetric characterization of a planar tetrahedral truss," *NASA TM 4231*, December 1990.

[5] J. S. Albus, H. G. McCain, and R. Lumia, "NASA/NBS standard reference model for telerobot control system architecture (NASREM)," *National Bureau of Standards Technical Note 1235*, July 1987.

[6] W. R. Doggett, M. D. Rhodes, M. A. Wise, and M. F. Armstead, "A smart end-effector for assembly of space truss structures," in *Space Operations, Applications and Research Conference*, (Houston, TX), July 1991.

[7] J. W. L. Heard, H. G. Bush, and J. J. Watson, "Space truss construction studies," in *ASCE Space 88 Engineering, Construction, and Operations in Space Conference*, (Albuquerque,NM), August 1988.

[8] S. W. Kenner, M. D. Rhodes, and W. Fichter, "Component count and preliminary assembly considerations for large space truss structures," *NASA TM 102604*, February 1990.

[9] L. S. Homem de Mello and R. Desai, "Assembly planning for large truss structures in space," in *Proceedings of the IEEE International Conference on Systems Engineering*, (Pittsburgh, PA), August 1990.

[10] R. Munger, "Path planning for assembly of strut-based structures," CIRSSE Report 91, Rensselaer Polytechnic Institute, May 1991.

# Chapter 3

# Development of a Truss Joint for Robotic Assembly of Space Structures

by: George F. Parma
Robotics and Mechanical Systems Laboratory
Structures and Mechanics Division
NASA - Lyndon B. Johnson Space Center
Houston, TX 77058

This chapter will present and discuss the results of a study undertaken to quantify the advantages of designing a structural fastener for ease of robotic assembly–a term we will refer to as "robot-friendly". The study was conducted by the Robotics and Mechanical Systems (RAMS) Laboratory in the Structures and Mechanics Division at the NASA-Johnson Space Center (JSC) in 1989-90. Four prototype robot-friendly structural joints were designed, built, and evaluated for use in a full-scale test called the Automated Robotic Assembly of Space Station (ARASS) demonstration. The joints were tested for both ease of robotic assembly as well as for structural properties.

The first section of the chapter includes background material on space-based structural assembly. It discusses Extra Vehicular Activity (EVA) structural assembly, robotic structural assembly, and introduces the ARASS demonstration.

The next section discusses the general field of Design for Assembly (DFA) and outlines the present study. It discusses the design requirements that were developed for these truss joints, then describes the four joints that were designed and built based on these requirements.

Section 3.3 describes the series of tests that the joints were put through. These included tests conducted with the robot, as well as structural testing. The next section discusses the results of these tests and the methodology used to determine the most robot-friendly joint design. An evaluation matrix is presented that was used to rank the joints on six criteria considered important for space-based robotic truss assembly.

Finally, Section 3.5 concludes by bringing out points for consideration when designing for robotic assembly, and discusses some observations about working with real robotic devices (as opposed to computer simulations). We hope that these will be of benefit to others who are learning about, and designing for, robotic assembly.

## 3.1 Space-Based Assembly

Until now, the U.S. space program has relied upon launching space vehicles as completely assembled and integrated systems. As we look to future programs, the sizes of the necessary vehicles become prohibitively large to launch as complete assemblies. The next logical step is to launch the vehicle in pieces and assemble it on-orbit or on another celestial body. Vehicles that are assembled on location can usually be of lighter, more efficient structures because they do not have to withstand launch and landing loads in their as-assembled configurations. Assembly and maintenance of many of these structures will be best accomplished by astronauts conducting EVA or by robotic devices. We will presently consider these two methods.

### 3.1.1 EVA Structural Assembly

The subject of EVA assembly has been under study by NASA for at least 15 years. In that time, numerous evaluations have been performed in neutral buoyancy simulators, like the Weightless Environment Training Facility (WETF) at JSC, and in numerous other water tanks throughout the country. These simulations have used successively higher and higher fidelity mockups to prove the concept of EVA structural assembly and investigate the problems associated with it. Flight experiments to demonstrate on-orbit EVA construction techniques first flew on STS-61B in December 1985–these were the Experimental Assembly of Structures in EVA (EASE) and the Assembly Concept for Construction of Erectable Space Structure (ACCESS). Results from many of these tests are summarized in [1].

Numerous builds of the 5 meter (16.4 ft.) space station truss have been performed in neutral buoyancy simulators since that time. The crew comments have mainly been related to positioning and reach from assembly work platforms, hand fatigue from multiple joint actuations, difficulties in handling and distinguishing between longeron and diagonal struts, glove wear, difficulties in achieving a connection due to joint intolerance to misalignment, and the need for visual cues to confirm the state of the connector [2], [3], [4]. It is a repetitive, labor intensive task, but quite doable.

### 3.1.2 Robotic Structural Assembly

Truss assembly is a rather inefficient use of the valuable flexibility of human presence in space. EVA time for space station assembly missions is limited to 24 man-hours per flight, and there are numerous other more difficult and less monotonous tasks that would be better suited for the astronaut. A better approach would be to relegate such repetitive and well-structured tasks to robotic devices.

A limited number of ground-based robotic structural assembly demonstrations have been conducted. Most of these evaluations have been conducted with EVA hardware and under teleoperator control. For example, the Beam Assembly Teleoperator (BAT) at the Massachusetts Institute of Technology has been used to assemble the EASE truss, and the Laboratory Teleoperator Manipulator (LTM) was used to assemble ACCESS (in conjunction with a human assistant) at Oak Ridge National Laboratory.

The RAMS lab has also studied telerobotic assembly with the ACCESS Telerobotic Assembly Demonstration (ATAD). This demonstration was performed in our own structures laboratory using WETF ACCESS hardware and a teleoperated hydraulic manipulator. This hardware had been optimized for gloved EVA assembly, and that led to numerous difficulties for teleoperation. The joints were difficult to align because of their tight tolerances, often tripped shut unintentionally (and were impossible to reset), and were difficult to visually align and verify [5].

However, structural assembly is repetitious, well structured, and labor intensive; it is, therefore, well suited for automation. Very few groups have studied automated structural assembly. Apart from our colleagues at NASA's Langley Research Center (whose experimental work was discussed in Chapter 2), no other groups have extensively studied this technique. This was the impetus that led to the inception of the ARASS demonstration.

### 3.1.3 The ARASS Demonstration

The ARASS demonstration was conceived to establish general confidence in the usefulness of robotics as a *tool* that can be used to accomplish *meaningful tasks.* It will do so by demonstrating the assembly of two bays of full scale 5-meter box truss in a totally automated robotic workcell as shown in Figure 3.1. This workcell configuration is representative of one that could be used to build a a 5-meter truss structure out of the Space Shuttle orbiter payload bay in low earth orbit. We have had to design heavy tower structures in order to offset

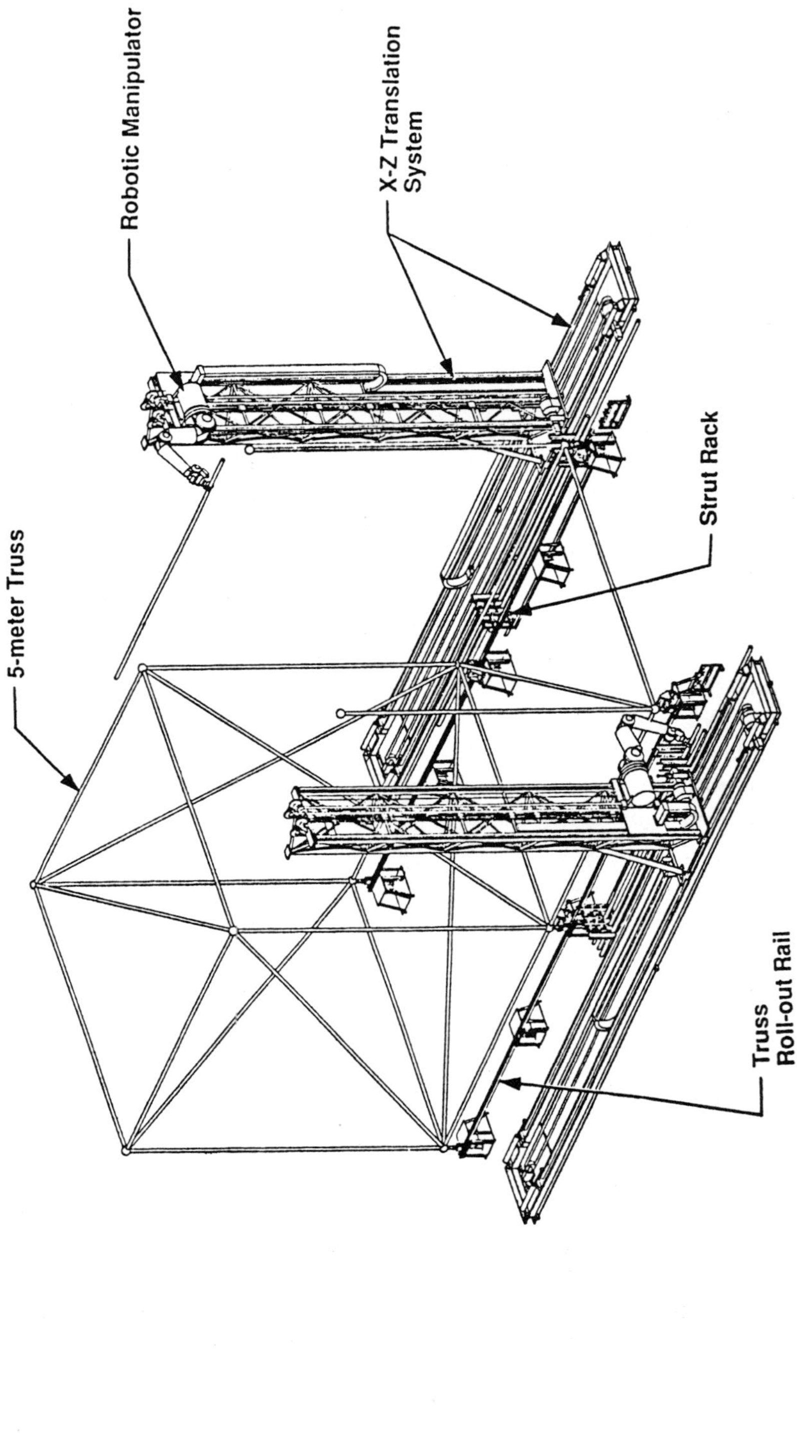


Figure 3.1: ARASS Workcell Configuration

the pull of gravity, but an actual flight system might use something like the shuttle's Remote Manipulator System (RMS) instead.

The ARASS demonstration is a ground-based test that will utilize two 7-degree-of-freedom (DOF) Robotics Research K-1607 manipulators each outfitted with a specialized end-effector, force/torque sensor, and machine vision system. Each robot is mounted on a large 2-DOF platform, and the two of these together are called the Automated Space Structure Assembly Platform (ASSAP). The ASSAP is a general purpose system, which will allow the robots to translate 4.6 meters (15.1 ft.) vertically and 7.8 meters (25.7 ft.) horizontally on either side of the truss. The entire assembly process will take place autonomously, with all the devices being controlled by a central workcell control program called the Workcell Integration System (WIS). WIS is being developed and implemented by the RAMS lab.

## 3.2 The Robot-Friendly Structural Joint Study

The key element to the success of the ARASS demonstration is the use of hardware that is easy for robots to assemble, or "robot-friendly". This means that the hardware incorporates many of the principles generally associated with DFA, e.g., simple insertion operations, generous lead-in chamfers, and self-aligning parts. Discussions on the principles of DFA may be found in [6] and [7]. There are additional advantages to be gained by designing specifically with automation in mind. In fact, in most cases in industry where cost comparisons are made in order to show the supposed benefits of automatic assembly, careful study usually shows that the benefits have been obtained through redesigning the product for assembly rather than through automation [7]. By designing with robotic assembly in mind, we anticipate creating a joint that is easier for both an astronaut *and* a robotic device to put together.

The objective of this study was to start with a clean slate and design a truss joint to accommodate both automated and EVA assembly while taking maximum advantage of the robotic assembly tool. The result would be a truss structure with all the advantages of tooled assembly plus easier EVA assembly.

This was the study procedure:

1. Develop design requirements consistent with the space station truss design while meeting the requirements of the ARASS demonstration.

2. Design and build prototype hardware to meet those requirements.

3. Test prototypes for ease of robotic assembly.

4. Test prototypes for structural properties.

5. Analyze test data.

6. Choose the best overall joint/end-effector system; incorporate the best features of all joints into a final design to be manufactured for the full-scale ARASS demonstration.

### 3.2.1 Designing for Robotic Assembly

The current state-of-the-art truss joints for on-orbit space structural assembly were designed specifically for EVA hand assembly (as discussed in Section 3.1.1). The design of these fasteners focused on the ease with which a gloved astronaut could assemble hundreds of these joints without excessive fatigue or glove wear.

These requirements forced a compromise of operational simplicity at the price of weight, complexity and cost. Each joint has its own internal mechanism *plus* an external mechanism (e.g., a lever or collar) that adapts the connector to the gloved astronaut's hand (one might think of this as incorporating the tool necessary to tighten the joint into every single unit). The designs also assume that whatever motions the astronaut does to operate the joint, a robot could be made to do as well. However, in order to duplicate these operations, the robot must perform extremely difficult motions and/or the gripper must incorporate an *additional* mechanism that allows it to mimic the motions of the gloved astronaut. This duplication results in an overall system that is twice as complex as is necessary.

Another drawback shared by these EVA truss joints is that they are very intolerant of misalignments. They assume that the person performing the assembly will provide the compliance and exhibit the dexterity necessary to overcome such problems. Robots are, in general, not as adaptive to such conditions. This can result in the task becoming difficult or impossible to accomplish. Even if the robot were able to compensate for such deficiencies, it still operates on a perfect model of the world, and assumes its own actions are perfect. There are still many unaccounted errors in the system, such as gripper misalignment, deflection in the manipulator structure, inaccuracies in the robot drive system, and inaccuracies in the jig that holds the truss during assembly. Fasteners that are not designed to accommodate such misalignments are, therefore, unsuitable for automated assembly.

Another advantage of designing a joint for robotic assembly is increased system efficiency. Since we are designing the robotic end-effector and the truss joint as a system, we can optimize the end-effector to joint interface to achieve the most efficiency. We can also place as much of the complication as possible in the end-effector (of which we are building only two), and make the truss joint (of which we may be building hundreds or thousands) much simpler, stiffer, lighter, and less expensive than its EVA counterpart.

### 3.2.2 Design Requirements

Considerable discussion went into the formulation of the design guidelines for this fastener. These requirements (outlined in Table 3.1) were developed based on our best assessment of the space station truss requirements at the time.

A comment should be made here about the "soft capture" requirement (the eighth requirement listed in Table 3.1). A "soft capture" device is a mechanism that loosely connects a joint so that a person can let go of the joint without it coming apart. "Soft capture" has always been a requirement for EVA joints as echoed by [1] and [2]. Videos of the crew assembling truss structures in both the WETF and on-orbit showed the reason: manipulating and actuating the joint was necessarily a two part task in order to prevent accidental triggering of the joint; a "soft capture" allowed the crewman to change his grip location from manipulation to actuation. However, we felt that by using a clever assembly sequence, and the fact that our end-effector did not have to let go of and re-grip the joint, that there was no need for such a feature. This helped simplify our designs even more.

### 3.2.3 Design Descriptions

In the process of discussing robotic truss assembly and its requirements, four joint design concepts were developed. Each of these designs addressed the robotic alignment problem and the structural fastening task in a different way. These joints are pictured in Figures 3.2 through 3.5, and their cross sections are shown in Figure 3.6. For all joints, the half that is mounted to the node ball (the common attachment point at each corner of the truss) is referred to as the "scar", as shown in Figure 3.6-A.

Notice that the Collet/Flex Drive and Bolted/Flex Drive joints have guides built onto them. This would appear to be a violation of one of the design requirements. However, we felt that if the joint needed an external guide, it should be built onto the prototype so that four different end-effectors would not be needed for these evaluations. Ultimately, the guide would need to be incorporated onto the end-effector.

An important note about one common feature of all these designs: in some way, each incorporates belleville spring washers. In all cases, this stack of bellevilles is deflected at the end of the joint tightening process and, thus, transmits a preload to the joint interface.

There is a very good reason for this. Observe that the connection is being made along the axis of a tube. When tightening these connectors, the moment arm available (i.e., the radius of the strut) for straightening out the tube is extremely short (one inch, maximum) compared to that available to any sideways resistance forces along the length of the tube. Viz., in the laboratory, we bolted and preloaded (without any bellevilles) a vertically oriented 5-meter truss tube

| Requirement | Comments |
|---|---|
| Design loads: ± 1140 lb. axial<br>± 2620 in.-lb. bending | Based on early space station docking and crew tether loads. |
| Preload after tightening: 3000 lb | Good preloaded joint design practice. |
| Capture envelope: ± 0.10" linear<br>± 3.0° angular | Based on conservative estimate of accuracy of vision system. |
| Sideways insertion. | Required for strut replacement. |
| One arm fastening operation. | Only one arm available. |
| Joint must be clocked* & torsion countered via a positive mechanism (not friction alone). | Certain struts require a specific orientation and that orientation must be maintained throughout the life of the station. |
| Alignment guide should be provided as part of the end effector if possible. | Minimizes joint complexity, weight. |
| A "soft capture" device is unnecessary. | This reduces complexity of the joint. The robot never needs to let go of the strut so long as the end effector provides the necessary degrees of freedom to accomplish fastening. |
| Must make provision for vision system fiducials (reference markings) on joint & scar. | Vision system must have an alignment target which is visible from directly over the insertion point. |
| EVA backup operation. | A tool will probably be required. |

* "Clocking" refers to the orientation of the joint about the strut axis

Table 3.1: Robot-Friendly Structural Joint Design Requirements for ARASS

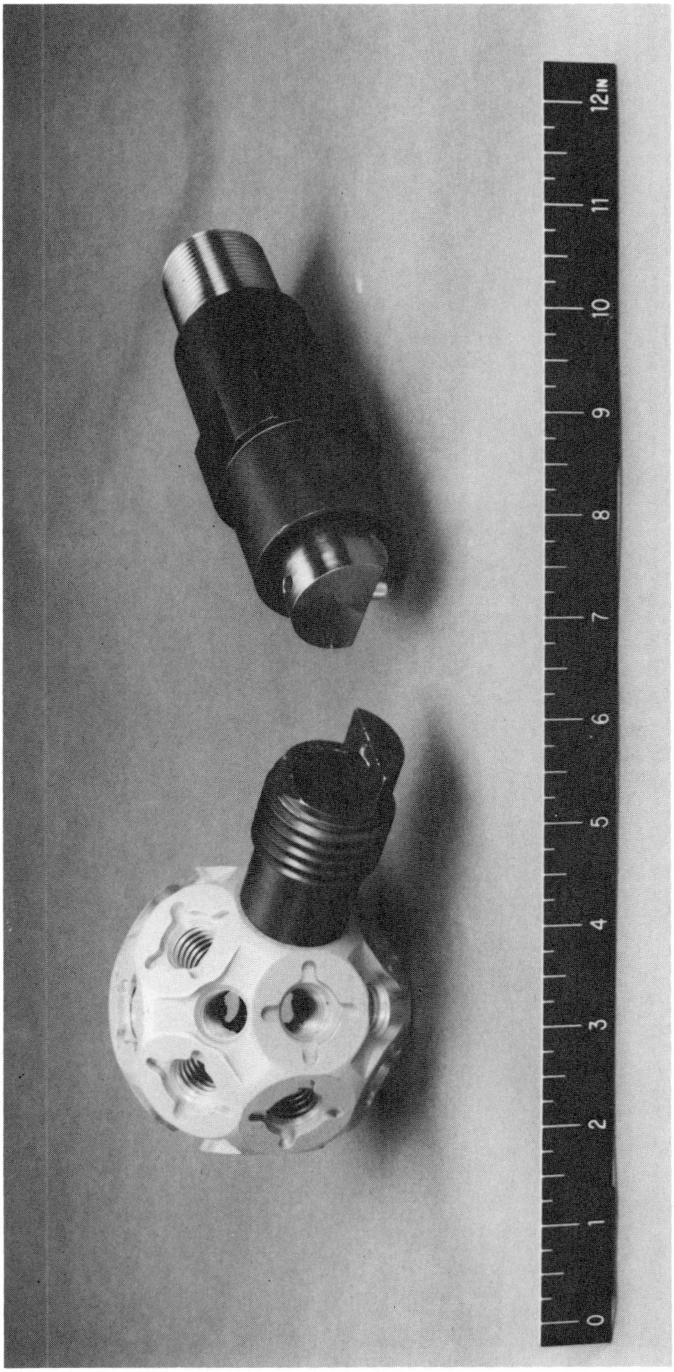


Figure 3.2: Threaded Collar Joint

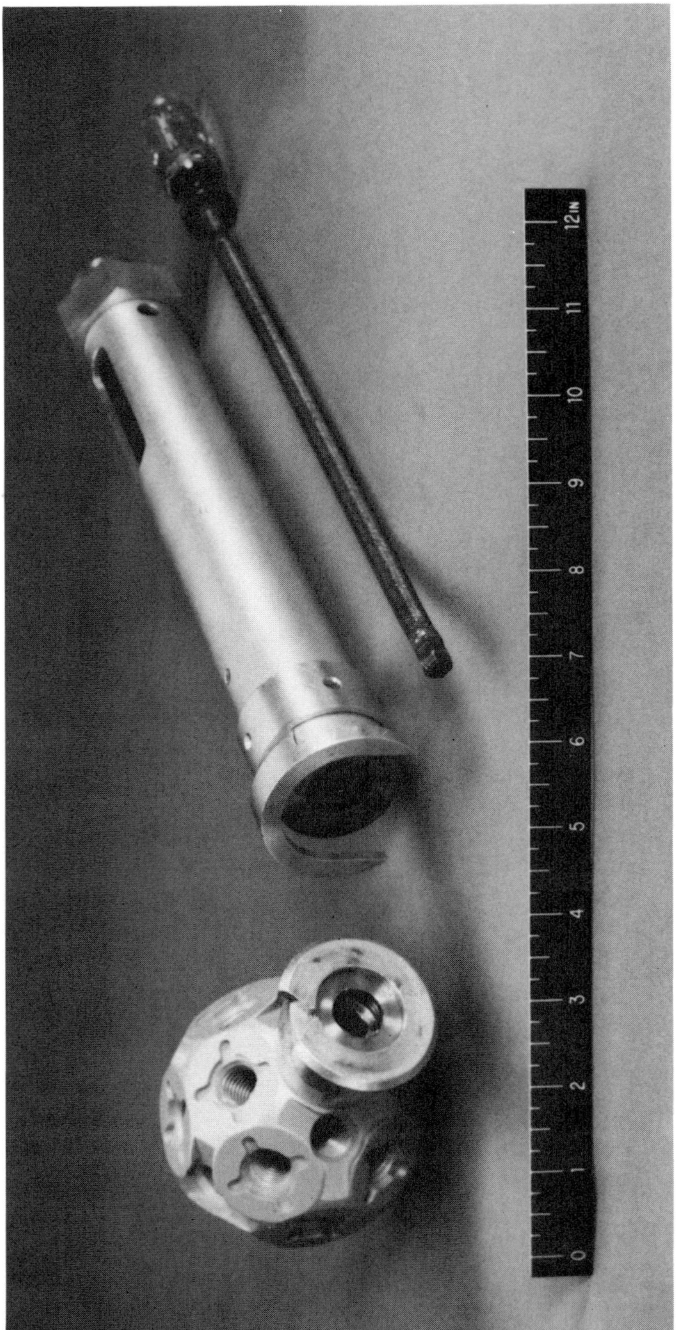

Figure 3.3: Collet/Flex Drive Joint

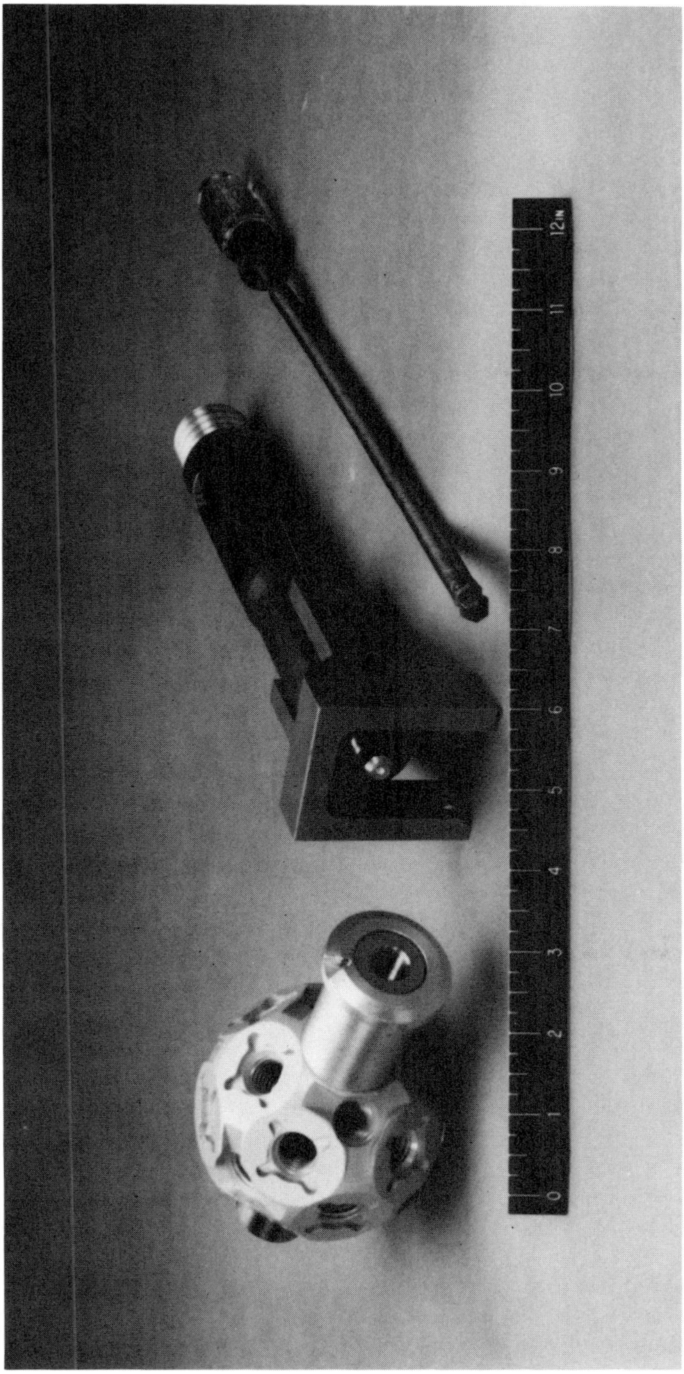

Figure 3.4: Bolted/Flex Drive Joint

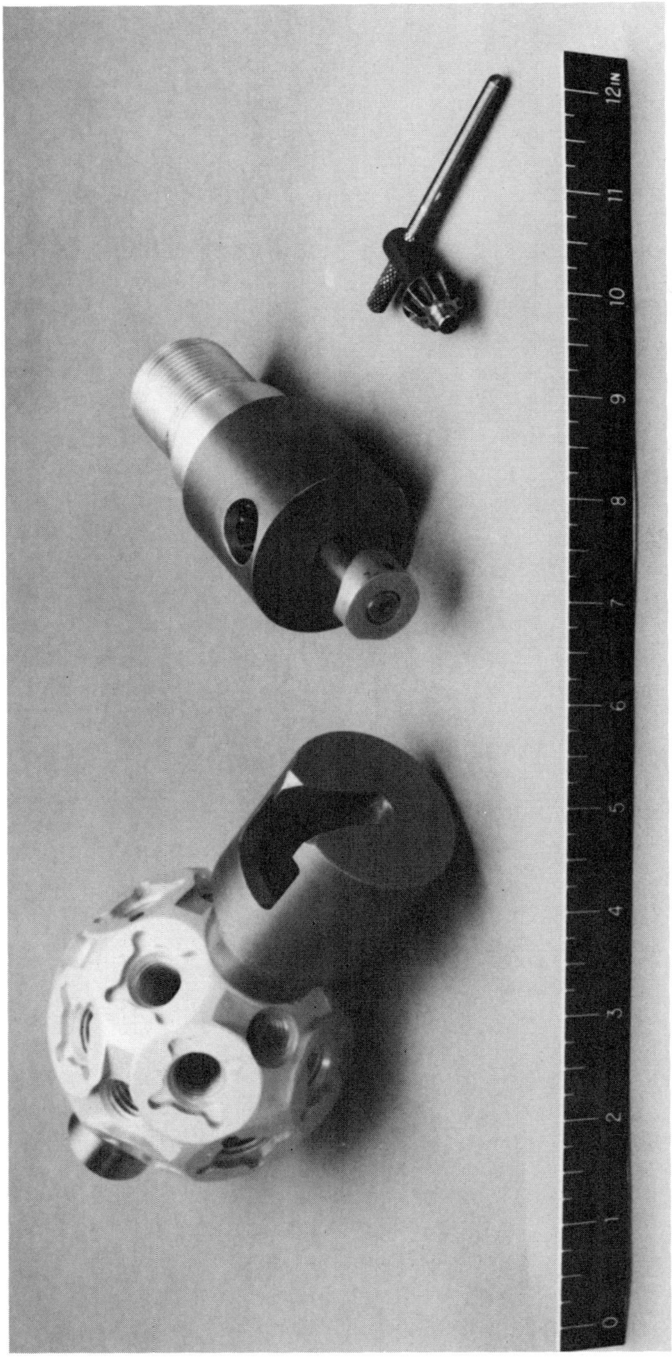


Figure 3.5: Hammer-Head Joint

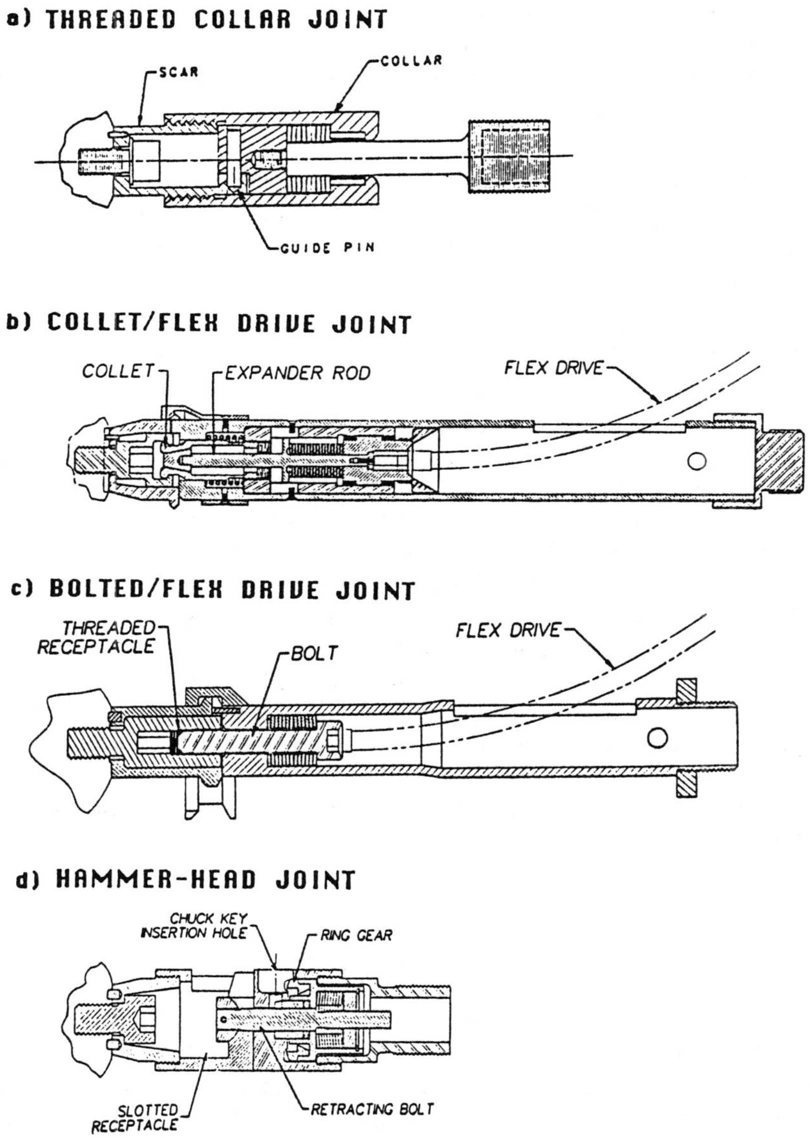


Figure 3.6: Robot Friendly Structural Fasteners Evaluated by the Robotics and Mechanical Systems Laboratory

at one end, then let go of it. The tube righted itself, making the joint completely loose! The best way to prevent this phenomenon from occurring was to include belleville washers, which build up their preload over many thousandths of an inch of deflection. Thus, when the truss tube righted itself after being released, the "relaxation" of a few thousandths of an inch would only minimally reduce the preload.

A noteworthy side benefit of this belleville arrangement is that the motor applying the preload torque now ramps up to the full preload torque value over a longer time period. This is much better for the life of the joint tightener's drive train.

### Threaded Collar Joint

The Threaded Collar Joint is shown in Figures 3.2 and 3.6-A. The principle feature of this joint is that it reduces the insertion problem to that of the well-studied pin-in-hole task. The pin and the scar's guide pin hole both have chamfers to facilitate the insertion process. The scar has an external thread, and the collar has a matching internal thread. Ultimately, the collar would also have an externally cut gear profile, which would be engaged by a worm gear on the end-effector. This would provide the collar rotation.

Initially, the collar is held retracted by a light compression spring force such that the guide pin is fully exposed (as shown in Figure 3.2). As the two joint halves slide together, the guide pin is inserted into the guide pin hole on the scar and accomplishes clocking of the joint. Then the collar is slid to the left and simultaneously rotated by the worm gear to engage the threads. It is theoretically possible for the worm gear alone to accomplish both motions, i.e., the worm's rotation would slide the collar to the left until it met resistance, then force the collar to rotate and thread onto the scar.

### Collet/Flex Drive Joint

The Collet/Flex Drive Joint is shown in Figures 3.3 and 3.6-B. This design is essentially a robot-compatible version of a previously developed EVA design (see "JSC Truss Joints" in [4]). Its principle of operation involves an expander rod, which forces a set of collet fingers to expand outward into a receptacle to achieve connection. End-effector drive engagement is accomplished by using a section of commercially available flexible drive shafting with an Allen-head wrench welded to the end. The hand operated version of this tool is shown to the right of the joint in Figure 3.3.

Before connection, the entire collet assembly is held inside the joint body by a light compression spring force. The joint halves are slid together with the assistance of a two-fingered guide. Clocking is performed by a wedge protruding from the joint face which engages a matching receptacle on the scar as the two halves are slid together (these are visible at the top of the joint interface surfaces in Figure 3.3). Then, the flex drive is inserted into a slot in the joint body. The tip of the Allen-head wrench simply slides along the inside wall of the body, up a conical ramp, and into a matching hex hole in the end of the collet assembly. Continued axial pressure by the flex drive overcomes the compression spring force, and pushes the collet assembly out of the joint body and the collet fingers are extended into the receptacle. The flex drive now begins to rotate, screwing out the expander rod so that it forces the collet fingers outward. Note that this joint needed the lowest input torque to achieve a 3000 lb. preload as a result of the collet's wedging action combined with its threaded actuation.

### Bolted/Flex Drive Joint

The Bolted/Flex Drive Joint is shown in Figures 3.4 and 3.6-C. This joint is probably the simplest possible–save for its capture guide. It is a basic, bolted joint that incorporates the same flex drive system as in the previous design. The bolt is a modified, standard socket head cap screw: its tip has been heavily chamfered, 3/4 of a diameter's worth of threads near the tip have been removed, one diameter's worth of threads retained beyond that, and then the remaining threads from there to the head removed as well. Both the joint body and the receptacle are threaded.

Initially, the bolt is completely retracted into the body. The two halves of the joint are slid together, again with the help of an external guide. Clocking is achieved via a pin on the joint side that slides into a narrowing, wedge-shaped slot on the scar (the slot is visible on the front of the scar in Figure 3.4). The flex drive is inserted, begins to rotate, and engages the bolt. As the bolt advances, its tip brings the two halves into proper alignment before the threaded portion begins to engage the receptacle. By the time the bolt head begins to compress the bellevilles, the threaded portion of the bolt is engaged only in the threads of the receptacle, and the interface can be properly preloaded.

### Hammer-Head Joint

The Hammer-Head Joint is shown in Figures 3.5 and 3.6-D. It operates on the principle of inserting a cylindrical head into a wide, well-chamfered hole, then using the joint's actuation to achieve final alignment as it is tightened. A standard chuck key is inserted through a hole in the side of the joint, and engages a ring gear. When the key is turned, the ring gear turns a nut which then retracts the bolt. It is essentially a bolted joint; however, it has avoided the possibility of cross-threading by turning a nut whose threads are always engaged with the bolt.

Hammer-Head operation starts with the head extended as shown in Figure 3.5. The capture guide is essentially incorporated into both halves of the joint. As the two halves of the joint slide together, clocking is accomplished by flats on either side of the retracting bolt conforming to the flat sides of the slotted receptacle. The chuck key is inserted, turns the ring gear and nut, and the joint is drawn into alignment as the transverse cylindrical head seats itself in the mating receptacle.

## 3.3 Test Descriptions

The four prototypes were manufactured, and each joint put through a series of identical tests. This test battery is illustrated in Figure 3.7. The coordinate system referenced in the remaining discussions is shown in Figure 3.8.

### 3.3.1 Robot-Friendliness Tests

All robot tests were performed in the RAMS lab using a Robotics Research K-1607 7-DOF manipulator. A commercial JR3 force/torque sensor was used to measure the forces and torques at the joint interface. The testing was performed in three phases.

**Phase One.** The first phase of the testing was designed to quantify the misalignment tolerance of each joint. We felt this to be the most important measure of "robot-friendliness". A node ball with the appropriate scar attached was mounted rigidly to the floor in front of the robot. Thirty two misalignment measurements were made.

The first battery of misalignment tests was aimed at determining how well each joint met the design requirements. These first tested each joint at one-half of its requirement ($\pm.05$ in. for linear and $\pm1.5°$ for angular) and then at the full requirement ($\pm.1$ in. and $\pm3°$). These four measurements were made for each of the following five conditions: 1) linear in X; 2) linear in Y; 3) roll about X; 4) roll about Y, or clocking, and 5) roll about Z. In each case, the joint was held by the robot approximately two inches above the insertion point. The misalignment was input in hand coordinates using the teach pendant and measured by an external measuring device. The joint was inserted, tightened, and the maximum forces (induced by those operations) recorded for later analysis. The test was stopped and "no connection possible" declared if the Z-direction insertion force exceeded 20 lb.

The second battery of misalignment tests was meant to determine the maximum misalignment that each design was capable of tolerating. Each of the five conditions above was tested again, but with larger and larger misalignments until the capture guide no longer pulled the joint together. In addition, two Z-direction tests were added: could the joint pull itself together and be tightened 1) with a 20 lb. overload, and 2) when the joint had not been inserted all the way.

**Phase Two.** Vision system compatibility was the second phase of the "robot-friendliness" tests. These tests were intended to determine if there were

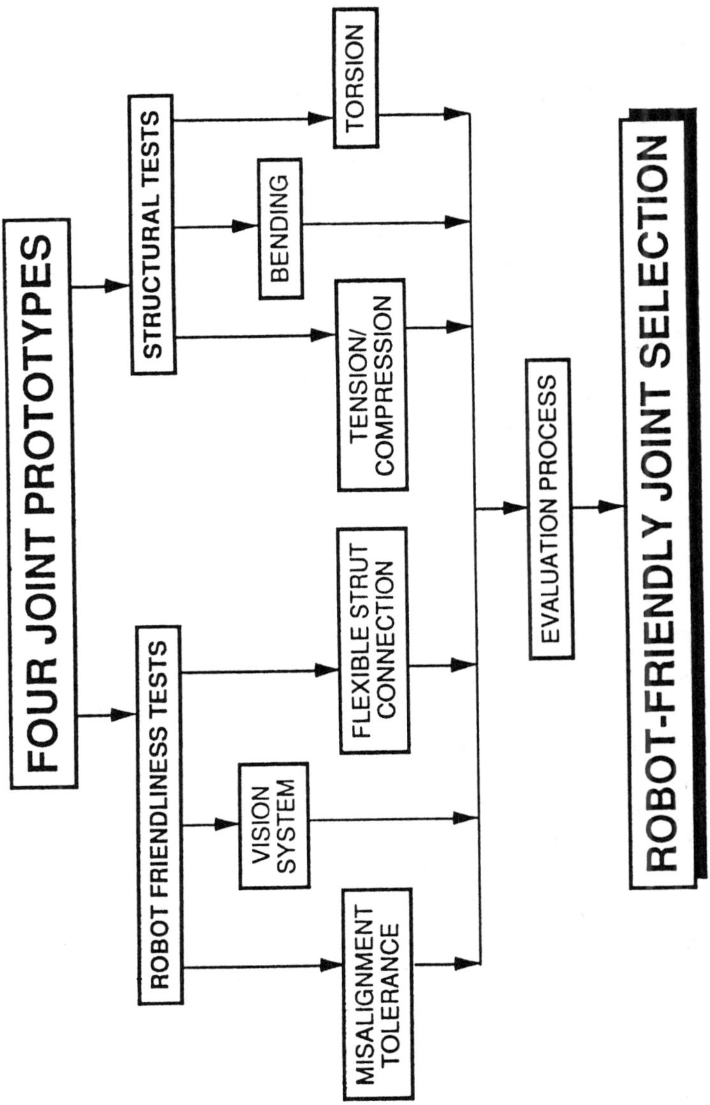


Figure 3.7: Robot-Friendly Joint Selection Process

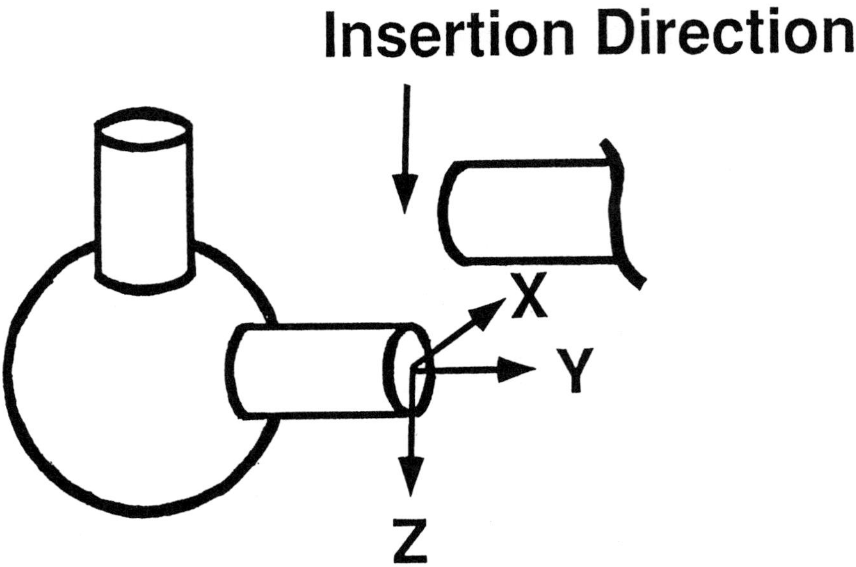


Figure 3.8: Joint Coordinate System

any incompatibilities between the joint and the planned vision system alignment and insertion scheme. These might be in the form of camera view blockage, features on the scar that might confuse the vision system, etc. In addition, we wanted to know if there were any residual misalignments after vision system-controlled insertion that would inhibit proper fastening.

The vision system camera was located on the end-effector just beyond the end of the joint shown in Figure 3.9. A fiducial, or reference mark, was placed on the joint's scar such that when the joint was properly aligned, the camera was centered over the fiducial. At the start of the test, the robot, then, needed only to be positioned so that the fiducial was within the camera's view. The vision system then commanded the robot in X, Y, Z, and roll about Z, until the joint was positioned 2.5 inches directly above the insertion point. The robot then moved forward 2.5 inches in Z to make the insertion.

Five different initial misalignments were used in the testing. These five points are depicted graphically in Figure 3.10. For instance, point 1 was at a location +X, +Y and -Z from what would be the perfect alignment point (2.5 inches above the actual insertion point). Point 2 added a roll misalignment to that, and the remaining points tested other combinations of X, Y, Z, and roll misalignment (including having the joint too close). Residual forces and torques were recorded immediately after insertion and then again after the joint was tightened.

**Phase Three.** We predicted that the most difficult joint connection neces-

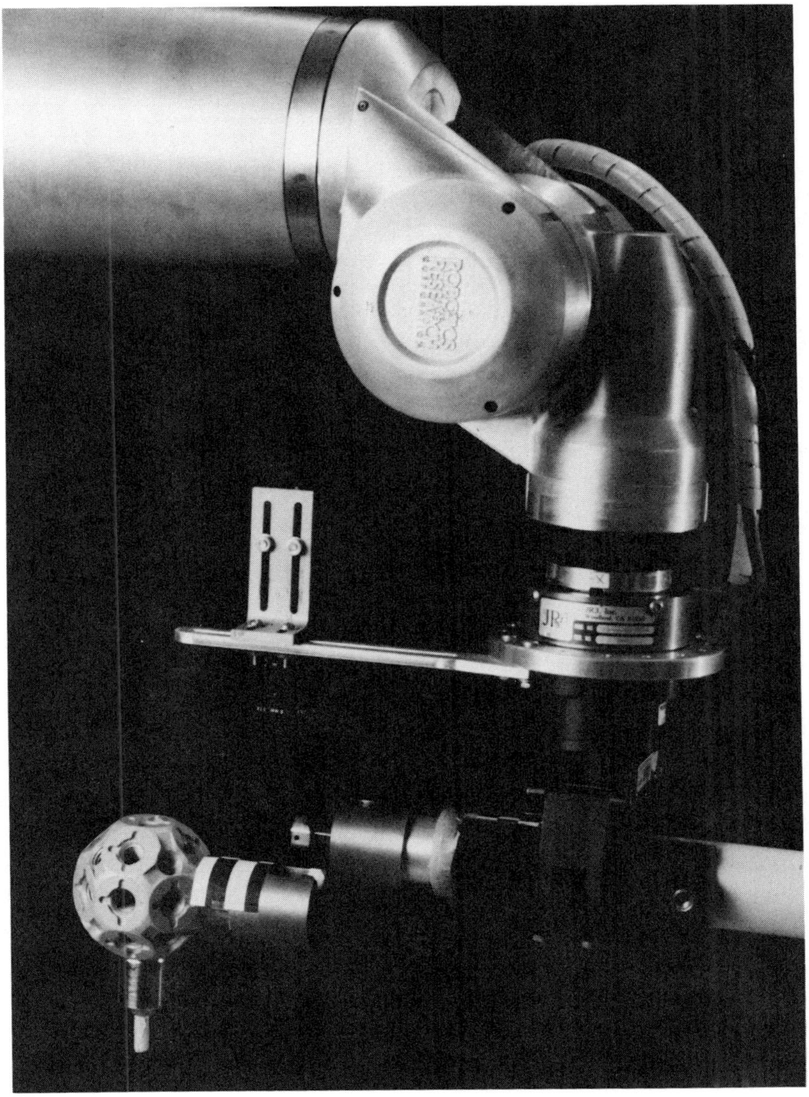

Figure 3.9: View of general test set-up. Visible are the robot, force/torque sensor, vision system camera, truss joint, scar, and node ball atop the flexible node simulator (see text).

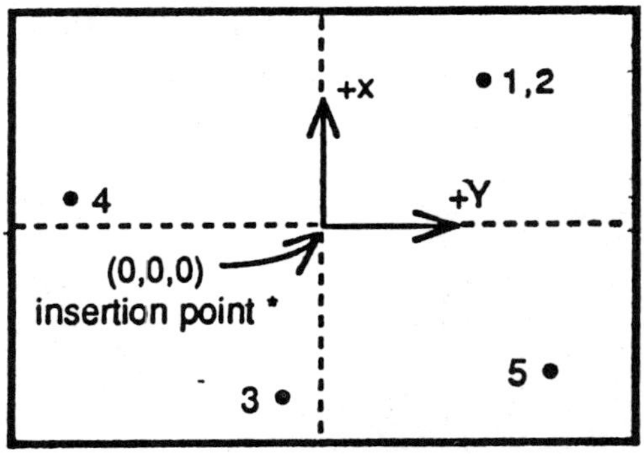


Figure 3.10: Location of Vision System Test Points (as seen from the end-effector camera)

sary in assembling the 5-meter truss was going to be that to a node which was at the top of a vertical truss member which was anchored only at its bottom end (this situation is depicted for the robot on the right in Figure 3.1). A "simulation" of this particular task was constructed by attaching a node ball with some additional mass to the end of a steel rod which was then anchored to the floor in front of the robot as shown in Figure 3.9 (Note: additional mass is not shown). The stiffness of the rod and the additional masses were chosen such that the simulator mimicked the stiffness and frequency of vibration of an actual 5-meter strut with a node ball attached. The third phase of "robot-friendliness" tests were performed with this flexible node simulator. Both of the previously described maximum misalignment and vision system tests were repeated.

### 3.3.2 Structural Tests

The purpose of the structural test program was to gather data to compare the joints in structural stiffness, linearity and repeatability. The tests were conducted in the Structures Test Laboratory at JSC. Exact details of the test may be found in the structural test plan [8].

Axial stiffness tests were run first. Each joint was mounted on an Instron test machine with one Linear Voltage Differential Transformer (LVDT) (used to measure linear displacement) attached at every 120° on its periphery. Together, the three LVDT's would measure the relative displacements between the planes of each joint half and therefore any out-of-plane bending induced by the tensile or compressive load. Their average would constitute the total joint deflection. Each joint was preloaded to 3000 lb. before testing. The loads applied were ±100, ±1000, and ±2000 lb. Loads were cycled three times for each case, and data continuously recorded by a data acquisition computer.

Bending tests were run next. A pure couple was applied at the right end (as viewed in Figure 3.6) of each joint using a pair of opposed hydraulic actuators. Again, LVDT's were used to record the actual deflections along each of the three axes. The joints were again cycled three times to ±500, and ±2000 in-lbs. These load cases were repeated in four different joint orientations (every 45° about the Y-axis) to look for any variances in bending dependent upon the joint's orientation. Again, data were recorded continuously.

The third set of structural tests examined torsional stiffness. Since relatively little torsional loading was expected for this space station truss configuration, the torsional data were only collected for completeness. Continuous data collection was not deemed necessary. Dead weights and pulleys were used to apply the load in five increments to a maximum of 240 in-lb. The loads were cycled three times–first clockwise, then counter-clockwise. The torsional deflection was measured by a string extensometer and recorded by hand at each load increment.

## 3.4 Test Results and Evaluation

All the joints met or exceeded the misalignment design requirements. Vision system testing revealed no major incompatibilities, nor were there any difficulties in making attachments to the flexible node simulator. No joint failures occurred during structural testing.

### 3.4.1 Robotic Testing Results

All the joints were able to capture under all the misalignment requirements. However, no joint could be *tightened* under every misalignment. Very high Z-forces developed as the capture guides forced the joints together during insertion. Tighter alignment guides (i.e., those which provided a closer fit with their scars) produced higher capture forces because they tried to correct the misalignment as the robot pushed the joint halves together. Higher actuation torques were usually required during the tightening process as the misalignments became more severe. Angular misalignments created the highest forces and moments (with roll about the Z-axis producing the highest) and every instance of "no connection possible" was caused by these misalignments. The two threaded joints were very intolerant of angular misalignments, but, otherwise had no cross-threading problems.

The vision system performed very well. Our fiducials were a pair of cross-hairs placed one inch apart on the scar. We were able to compensate for up to approximately eight inches in depth and ±7° of roll. Tightening a joint that had been inserted under vision control never produced more than 10 lb. of residual Z-force.

There were no problems in connecting to the flexible node simulator. In fact, the compliance inherent in the simulator usually *helped* the joint to slide together under severe misalignments.

The following are some specific comments on each joint:

> *Threaded Collar*: Although the guide pin worked well to align the joint for lateral misalignments, it tended to jam under severe angular misalignments. This could have been alleviated by greater chamfering of the guide pin hole and a hard surface finish on the pin (the prototype was made of unanodized 6061 aluminum). This joint was felt to have the best visual verification, both for tightening and for loosening, because of its highly visible external collar.

> *Collet/Flex Drive*: The Collet/Flex Drive's capture guide gave it a very large X-direction misalignment tolerance. However, under some angular misalignment conditions, it appeared that the capture guide had pulled the joint together, but the collet assembly could not be advanced into the receptacle. Also, sometimes the assembly would stick in its advanced position, making joint removal difficult. There were also some minor problems with vision system compatibility: the capture guide partially blocked the camera view; and, since the scar had a tapered section, the use of a special fiducial mark was required (because the cross-hair became distorted when taped to the surface of the conical section).

> *Bolted/Flex Drive*: Like the previous joint, this one also had very large X-direction misalignment tolerance. Even though the capture guide pulled the joint together well, and the tip of the bolt would start into the scar, it always jammed under Roll-X and Roll-Z misalignments. This joint also produced the highest moments about the Y-axis during roll about Y misalignments, indicating that it had the best clocking correction device (it is also worth noting that the joint always went together under this misalignment with only a slightly higher than normal tightening torque). The tangs of the alignment guide stuck well down below the joint (see Figure 3.6-C), and these partially blocked the corners of the camera view during vision system testing.

> *Hammer-Head*: The Hammer-Head turned in the best performance in misalignment tolerance, jamming only once on the most severe roll about Z misalignment. However, its clocking mechanism was virtually useless–the retracting bolt was free to rotate about the Y-axis enough, that it was unable to clock the joint. Its negative Roll about X tolerance was outstanding; (27°) due to the cylindrical shape of its head (about the X-axis).

This joint also had the lowest capture forces, and this is worth noting. Its roomy capture guide permitted it to become aligned very easily. The retraction of the head then removed all remaining misalignments and pulled it together nicely. As for vision system compatibility, the Hammer-Head itself just barely stuck into the camera view. The vision system was fooled once by the discontinuity where the two halves of the scar are screwed together (see Figure 3.6-D). This was because the discontinuity appeared as a line to the camera (when viewed from the correct angle), and the vision system then mistook it for part of a cross-hair.

### 3.4.2 Structural Testing Results

The structural test data for the axial and bending tests were processed with specially written FORTRAN computer code. The data were then downloaded to Macintosh computers for plotting. A least-squares curve-fitting algorithm was applied to the data to extract stiffnesses.

The amount of out-of-plane bending in the axial tests turned out to be insignificant–never more than 0.3°. The 100 lb. load tests' data were very noisy and inconsistent. The best linearity and repeatability was achieved in the 2000 lb. load range. The values listed in the evaluation matrix, Table 3.2 (to be discussed in the next section), are for axial stiffness (K) in this load range. The stiffest was the Threaded Collar joint, which may be due in part to its large collar area and the fact that its narrow shaft was constructed of a very stiff material, MP35N. The least stiff was the Collet/Flex Drive, which showed indications of having lost some of its preload.

In the bending stiffness tests, the higher load case again gave more consistent values. The Threaded Collar and the Hammer-Head exhibited consistent stiffness in all bending planes, whereas the Bolted/Flex Drive and the Collet/Flex Drive showed considerable variance. However, there appears to be no relationship between these variances. The Hammer-Head had the highest stiffness (114,310 in-lb./rad) probably due to the fact that it had the largest diameter (2 inches) contact area. The Threaded Collar had the lowest stiffness (51,403 in-lb./rad), evidently due to its narrow section.

Although the figure of merit for axial loading–axial stiffness–was rather obvious, this was not so for bending. It seems that the most critical failure mode for the truss is column buckling. Therefore, the figure of merit for bending is end fixity. This is because a joint with a lower end fixity would require a thicker-walled, heavier truss tube in order to accommodate the same buckling load as a joint with a higher end fixity. Therefore, from a systems standpoint, end fixity becomes more important than bending stiffness [9]. Values for column-end fixity (C) for both longeron and diagonal truss members were computed using the joint bending stiffnesses. The value of C was less for longerons because of their thinner wall. It is these values that we chose to use for comparison in the matrix (to be discussed in the next section). The values of C shown for each joint in Table 3.2 were averaged over the four bending planes.

The torsional stiffness test data were very inconclusive. The deflections were in the noise range. However, they did reveal how much clocking "slop" or "play" each joint had when the load direction was reversed. The Hammer-Head displayed the highest "slop", almost 11°, while the next highest was only 1.2°.

### 3.4.3 Evaluation

At the conclusion of the testing, the joint design and evaluation team met to discuss a methodology for determining the best overall fastener. We decided to use a weighted matrix evaluation scheme. The final matrix is given in Table 3.2. Six criteria were deemed important to our structural assembly task:

1. *Reliability of fastening*: The figure of merit here was the number of times the joint could be connected out of the 20 misalignment cases in the first battery of misalignment tests.

2. *End-effector simplicity and reliability*: The fewer the degrees-of-freedom, the fewer complex motions, and the lower the technology risk of the end-effector design, the better the overall system from a reliability standpoint. The end-effector also scored higher if it did not need to provide an external capture guide for the joint. This score was assigned after group consensus.

3. *Structural performance*: There was a considerable discussion as to whether axial stiffness or end fixity was the more important attribute. We eventually compromised and weighted them equally. All the joints were ranked as a percentage of the best K (for axial) and C (for bending); these two scores were then added, averaged, and normalized to result in the scores shown.

4. *Low cost to manufacture*: Cost is always an important consideration, especially when many hundreds of these joints might be manufactured. For these evaluations, the cost to build each prototype was used as the figure of merit.

5. *Visual verification*: This criterion was best evaluated by answers to some key questions: could you see if the joint was tight, or verify that it was completely disengaged when trying to take it apart? Could the joint be visually inspected as part of regular space station maintenance to verify that it was fully tightened? This score was assigned after group consensus.

6. *Low weight*: Minimal weight is always important in space flight systems. It was assumed that the prototype weights would be indicative of the relative weights of each concept even after weight optimization. The weight of each prototype, which includes its scar, was used as the figure of merit.

| Criteria / Joint | WT | Bolted/Flex Drive | SC | WT SC | Collet/Flex Drive | SC | WT SC | Threaded Collar | SC | WT SC | Hammer Head | SC | WT SC |
|---|---|---|---|---|---|---|---|---|---|---|---|---|---|
| Reliability of Fastening | 9.2 | 12 of 20 connections | 6 | 55.2 | 13 of 20 connections | 7 | 64.4 | 13 of 20 connections | 7 | 64.4 | 19 of 20 connections | 10 | 92.0 |
| EE Simplicity/ Reliability | 6.2 | Complex cap. guide Flex drive actuator | 3 | 18.6 | Simple cap. guide Flex drive actuator | 4 | 24.8 | No capture guide Untested worm gear drive | 6 | 37.2 | No capture guide Simple drive | 10 | 62.0 |
| Structural Performance* | 4.8 | K=4.82 x 10E5 lb/in C=2.01 | 8 | 38.4 | K=3.96 x 10E5 lb/in C=1.87 | 8 | 38.4 | K=8.49 x 10E5 lb/in C=1.65 | 10 | 48.0 | K=6.21 x 10E5 lb/in C=2.18 | 10 | 48.0 |
| Low cost to Manufacture | 4.2 | Prototype cost: $4,800 | 9 | 37.8 | Prototype cost: $9,200 | 5 | 21.0 | Prototype cost: $4,200** | 10 | 42.0 | Prototype cost: $5,600 | 8 | 33.6 |
| Visual Verification | 3.0 | Poor: Guide is in way, Cannot tell if bolt is retracted | 2 | 6.0 | Fair: Guide is not large, but still hard to see collet | 6 | 18.0 | Best: Easy to see how far collar has advanced | 10 | 30.0 | Good: Easy to see head | 8 | 24.0 |
| Low weight | 2.6 | 1.94 lb | 9 | 23.4 | 3.11 lb | 5 | 13.0 | 1.99 lb | 9 | 23.4 | 1.72lb | 10 | 26.0 |
| Total weighted score | 30 | | | 179.4 | | | 179.6 | | | 245.0 | | | 285.6 |

* K = axial stiffness, C = column end fixity

** Evaluation team estimate

Table 3.2: Robot-Friendly Structural Joint Evaluation Matrix

The team voted on what weighting to assign each of the six criteria with a requirement that the weights must add up to 30 (5 for each criterion). The team then considered each criterion and assigned scores for each joint. The best joint for each criterion was always assigned a score of "10" and the others ranked relative to it. The score, times that criterion's weighting resulted in a weighted score, and the weighted scores were then added for the final total. Note that in this type of evaluation, the top score must be at least 10% higher than the runner up. As can be seen, the highest scoring joint was the Hammer-Head joint, scoring 285.6 out of a possible 300 points.

## 3.5 Conclusions

The RAMS laboratory completed a detailed study of mechanical, structural fasteners which were designed for ease of robotic assembly while considering structural properties and overall system design. The Hammer-Head joint was chosen as the best overall fastener. Among its merits were its high reliability of fastening under misalignment, highest end fixity, simplest end-effector, integral capture guide, good visual verification, and lightest weight.

### 3.5.1 DFA Observations

We learned a lot about designing for robotic assembly in the course of this study. First of all, designing so that chamfers slide on chamfers always produces smoother assembly than corners working against chamfers. Corners are much more likely to dig into the chamfer and jam. Second, chamfers sliding on a cylindrical surface works amazingly well. This was demonstrated by the Collet/Flex Drive's and Bolted/Flex Drive's capture guides against their large cylindrical scars. The Hammer-Head also had many of these types of surfaces–two cylinders on the head and one on the retracting bolt shaft–which worked well against both chamfers *and* corners. Furthermore, sliding surfaces must have a hard surface finish or they will gall upon each other. Some ways to obtain good sliding surfaces include using hard anodizing aluminum, using smoother surface finishes (e.g., 63 RMS or finer), or considering dry film lubrication of one or more surfaces.

We found the best clocking alignment device to be the pin-in-slot, which was used in the Bolted/Flex Drive design. This is illustrated in Figure 3.11. Notice that this embodies the concept of cylinder-on-chamfer as well.

Finally, it is a good idea not to try complete final alignment while inserting, as was demonstrated by the Hammer-Head. Tighter alignment guides try to do too much at once, and thus jamming the connector before the actuator has a chance to tighten it. The Hammer-Head, for instance, does not try to seat the joint until the motor drive is retracting the head, giving 3000 lb. of alignment force to pull the head into its seat.

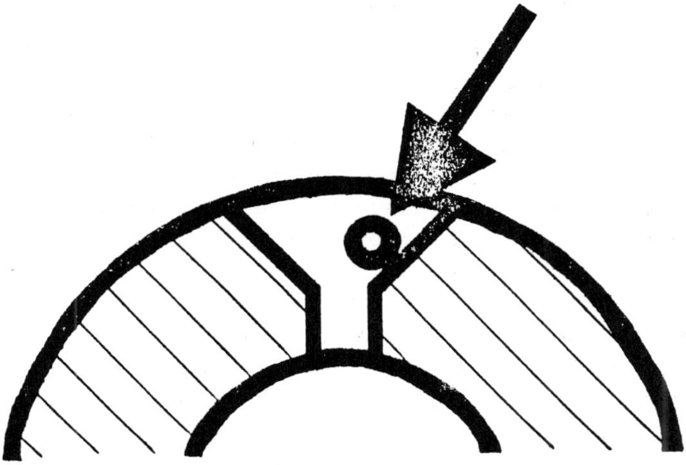

Figure 3.11: Pin-in-Slot Clocking Device

### 3.5.2 Real World Robotic Observations

There are many problems that will never be identified by studying the perfect world of computer graphic models. These can only be flushed out by tests with real hardware. The following are some observations made during the course of this testing:

*Be careful how you interpret force/torque sensor data.* The force/torque sensor data from the first battery of robotic misalignment tests had been plotted using bar graphs to better analyze any overall trends. What we noticed after looking at the plots was that if you put the linear misalignment in X data plots next to the roll misalignment about Y plots, it was difficult to determine which was which. The same held true for linear misalignment in Y and roll misalignment about X. Evidently, the reaction of the capture guide to an X-misalignment, is a force that induces a torque about the Y-axis. Thus, the reaction looks like a roll about Y. Similarly, the guide's reaction to a Y-misalignment is an induced torque about the X-axis, and that is why these two plots look so similar. The significance of this result is that given *one* set of data from a force/torque sensor, you cannot absolutely determine with which type of misalignment condition you are currently faced. Some supplemental data from other sensors, or perhaps even some diagnostic test movement, must be applied in order to make the assembly operation more robust. This diagnostic testing could even be an application for artificial intelligence or a neural network.

*Real robots are not rigid.* For instance, our robot complies 0.1 inches under a 15 lb. load (in one direction). Furthermore, this compliance is dependent upon the configuration of the arm and the direction in which the load is applied.

*Real robots sag under gravity.* Never believe the position and orientation quoted by your robot control computer. Our measurements showed that a supposedly level gripper was actually angled by about 1°. This is why we always tried to use external measurements when setting up our offsets for misalignment testing. This same sag also makes it difficult to confirm a tool offset, because the robot's sag changes as the manipulator's configuration changes. The manipulator *wants* to rotate about the correct toolpoint, but it ***physically cannot.*** The robot's sag under gravity also caused the linear in X misalignment data to be biased toward the robot, i.e., the robot was more compliant in the +X direction because it was sagging in the -X direction. Note that this phenomenon might not occur in the micro-gravity environment of low earth orbit, but larger, crane-type manipulators handling large payloads might see a similar effect due to orbital mechanics.

*Real robots are machines–they experience hysteresis, thermal expansion, and friction.* We saw sluggish response when reversing direction, and mechanical drift when making orthogonal changes in directions. We only noticed these effects because we were taking measurements to the thousandth of an inch, and these effects were on the order of 20 to 30 thousandths. For this same reason, we had to "exercise" the robot every morning for about 45 minutes before testing. The robot, being all aluminum, would grow some 20 or so thousandths of an inch as it warmed up. We also noted that this warm-up changed the friction at the joints, and that affected the drift and hysteresis of the system. Again, all of these effects were dependent on the arm configuration.

### 3.5.3 Concluding Remarks

A second version of the Hammer-Head joint has been designed that incorporates lessons from all the joint designs and test data. Recently, much emphasis has been put on space station design for robotic assembly and servicing. Hopefully, this study will serve as a good example of the application of DFA to hardware design for the space station program including robotic and EVA servicing.

## Acknowledgments

As is alluded to in most of this chapter, this project was a team effort, and the author is, in part, the messenger of all their good work. Particular thanks belongs to Erik Evenson, Steve Ruiz, Lebarian Stokes, John Thompson, Mark Vandeberghe, and all the members of the RAMS lab, who made significant contributions to the success of this endeavor. Finally, my thanks to Dr. Alan Desrochers of CIRSSE for allowing us to contribute a chapter on our work to this volume.

## 3.6 References

[1] W. L. Heard, Jr., H. G. Bush, J. J. Watson, S. C. Spring, and J. L. Ross, "Astronaut/EVA construction of space station," in *AIAA SDM Issues of International Space Station Conference*, AIAA Paper 88-2459, (Williamsburg, VA), April 21-22, 1988.

[2] L. C. Thomas, "Space station truss assembly simulation and fixture evaluation (second series) final test report," Report 32057, JSC, February 9, 1988.

[3] L. C. Thomas and L. P. Matranga, "Space station LaRC fixture truss assembly simulation (third series) final test report phase 1 - fixture evaluation," Report 32070, JSC, May 17, 1988.

[4] L. C. Thomas and L. P. Matranga, "Space station utility system integration phase II - integrated truss and component evaluation final test report," Report 32038, JSC, June 7, 1989.

[5] P. Sheridan, "Telerobotic truss assembly," in *Proceedings from the First Annual Workshop on Space Operations Automation and Robotics (SOAR '87)*, (NASA CP 2491), pp. 487–491, 1987.

[6] A. J. Scarr, D. H. Jackson, and R. S. McMaster, "Product design for robotic and automated assembly," in *Proceedings of the 1986 IEEE International Conference on Robotics and Automation*, vol. 2, (San Francisco, CA), pp. 796–802, April 7-10, 1986.

[7] G. Boothroyd, "Effects of assembly automation on product design," *CIRP Annals: Manufacturing Technology*, vol. 32, no. 2, p. 511, 1983.

[8] G. F. Parma, "Space station truss robot-friendly structural joints structural test plan," *Internal Document*, March 26, 1990.

[9] J. E. Keever, "Final report, erectable joint trade study," McDonnell Douglas Report 2ITA409, McDonnell Douglas, May 14, 1990.

Copies of JSC publications may be obtained by writing to: Freedom of Information Office, Mail Code AP4, NASA-Johnson Space Center, Houston, Texas 77058.

# Chapter 4

# Hierarchical Planning for Space-Truss Assembly

by: Rajive K. Mathur
GTE Spacenet Corporation, McLean, VA 22102
Rolf Münger
Buchholzstrasse 151, 8053 Zurich, Switzerland
Arthur C. Sanderson
Electrical, Computer, and Systems Engineering Department
Center for Intelligent Robotic Systems for Space Exploration
Rensselaer Polytechnic Institute
Troy, New York 12180-3590

Construction, repair, and maintenance of space-based structures will require extensive planning of operations in order to effectively carry out these tasks. This chapter describes a hierarchy of planning algorithms focusing on space-truss assembly as a specific application domain. This planning paradigm generates a sequence of feasible assembly operations required to build a space-truss structure, and provides the motion plans for a robot arm to carry out these operations. In this chapter, the two planning levels: task planning of assembly sequences, and path planning of robot motions, are described in detail. These planning algorithms have been implemented and evaluated using existing space-truss designs, and are being integrated into the RPI-CIRSSE Testbed environment.

## 4.1 Introduction

Continued progress in the exploration of space will include orbital, extra-orbital, and planetary missions involving more extensive and complex vehicles, structures and systems for scientific exploration, life support, and instrumentation. These future missions will require extensive planning and logistics support for both manned and unmanned portions of their activities. This chapter describes a set of planning paradigms that address tasks needed for construction, maintenance, and repair in many space-based operations. This approach is described specifically for the domain of space-truss assembly, but generalizes to many related problem areas.

The space-truss assembly problem is broken down into two major levels. First, the sequence of operations required for assembly is planned to guarantee that steps are feasible and efficient for the human or automated system. Second, the path is generated to move each strut into its final position (with the associated motion of a robot arm) and to provide a smooth assembly motion without collision of the strut or robot. The first of these problems is a *task planning* problem and is described in Part I of this chapter. The second problem is a *path planning* problem, and it is described in Part II.

A variety of approaches to both task and path planning have been described previously. In the approach presented in this work, we emphasize a *hierarchical representation* of the problems, and demonstrate an efficient solution to both problems in a manner which supports integration. The task planner described here is *domain dependent* and relies on a particular hierarchical representation of truss assemblies in order to efficiently find feasible operation sequences. This task planner is based on our prior work [1], [2], [3], [4], [5], [6], on generalized assembly sequence planning from CAD models.

The path planning algorithm described here is a general approach to generating paths that guarantee collision avoidance for a single chain nonredundant or redundant robot. The algorithm uses a graph search of feasible points in position space of the part followed by a local potential field method that guarantees collision avoidance among objects, structures, and the robot arm. This algorithm is novel in its computation of goal attractive potential fields in Cartesian space, and computation of obstacle repulsive fields in robot joint space. These effects are combined to generate robot motion. Computation is efficiently implemented through the computation of the robot arm Jacobian and not the full inverse arm kinematics.

The algorithms described here are being implemented in the CIRSSE Testbed environment. The examples described here have been used to generate an assembly sequence plan for a space-truss that has been designed as a general structural component of several NASA projects, and has been used as the focus for research and development of structure assembly methods at NASA Langley, as described in Chapter 2 of this book. The output of the assembly sequence planner is a geometric description of the goal positions of successive struts, and provides the input to the path planner. Our current implementation of the path planner is for one 9-degree-of-freedom (DOF) redundant robot arm from the CIRSSE Testbed. In this environment, the sequence of robot joint positions is input to the motion control system for execution. In related work, we are exploring the integration of this path planning algorithm to enhance the NASA Langley Assembly Testbed capability, as well as, the extension of our current planning models to incorporate two arms.

## 4.2 Part I: Assembly Sequence Planning

### 4.2.1 Introduction

The purpose of this section is to present a hierarchical task planning paradigm and use it to generate a feasible, realistic and near-optimal assembly sequence for large, 3-D, space-based truss structures consisting of struts and nodes, similar to the ones that have been proposed to build space platforms in the near future. The assembly sequence is part of a high-level task plan that will be used by human, teleoperated, or robotic systems. Additionally, this section demonstrates the extensibility and use of theory previously developed for the assembly planning scenarios in a standard manufacturing environment, [1], [2], [3], [4], [5], [6]. By extending previous theory to the strut-node domain, a hierarchical representation is presented that uses this domain's inherent properties.

### 4.2.2 Background

Much of the research on task planning systems has emphasized *domain independence* in which the representation and search methods are not developed to solve problems that fall in any specific category. These types of planners generally concentrate on abstract problems subject to generalized definitions of actions and ordering constraints. The properties of these domain independent planning algorithms have been studied to evaluate time and space complexity issues that are directly impacted by the representation and search methods that may be used. Representation and search are key elements in the success of a planner, and have been a recurring theme in planning research. The choice of an adequate knowledge representation scheme and search paradigm is essential to the generation of an efficient and successful solution. The hierarchical planning system for space structures implements an efficient *domain specific* representation of constraints as a basis for planning sequences of assembly operations.

There are two distinct approaches to domain independent planning: hierarchical and non-hierarchical planning. While both types of planners solve subgoals that lead up to the final goal, hierarchical planners also *represent* the plan that is generated in a hierarchical manner. This means that a plan generated from a hierarchical planner consists of more than one level in which the higher levels are abstracted, and the lower levels contain a complete and detailed sequence of the plan's operations [7].

STRIPS [8] is an example of a domain independent planning system. It is a non-hierarchical problem solving system that uses a closed-world and means-end strategy to achieve a goal. However, because of extensive backtracking and the lack of ability to focus on the essential subgoals of a problem, STRIPS may result in a computationally intensive solution or no solution at all. Other important non-hierarchical planners include NOAH [9] and MOLGEN [10]. Alternatively, ABSTRIPS [11], is a noteworthy hierarchical planning system since it plans in a hierarchy of abstraction spaces. An ABSTRIPS state space contains a partial order of operators and objects based on some human-specified preconditions. A criticality value is assigned to these conditions and when ABSTRIPS begins planning, its plan is focused on achieving only those goals that have a high criticality, leaving the ones with low values as details to be solved at some later time. Thus, it plans in the highest abstraction of space and achieves only those goals that are critical to the success of the plan *at that level.* Once it finishes planning at the highest level, ABSTRIPS plans at the next abstraction level, which has subgoals at the next highest criticality level. It continues in this manner until all the preconditions that were initially specified are satisfied and the plan is created for all levels.

It is important to note that ABSTRIPS utilizes an efficient representation as well as employing the effective means-end search method from STRIPS. Since a complete plan is developed at each level before the next level of detail is explored, ABSTRIPS can detect dead-ends earlier than STRIPS. A more modern and more practical example of a domain independent planner is provided by SIPE [12].

PLEIADES, (PLanning Environment for Integrated Assembly system DESign), [3], is a domain dependent system for planning of assembly sequences based on geometric descriptions of parts constraints. PLEIADES consists of a set of software tools that integrates a solid modeling system, an assembly modeling system, and an assembly planning system into one environment. In this environment, a designer can develop solid models of the parts, then use an assembly modeling system to specify the geometric and mechanical relations between the parts. The resulting assembly sequence plan is represented in the form of an AND/OR graph, precedence relations, or a directed graph that can then be used as an input to the assembly planning system [13]. The hierarchical planner that is the focus of this work is also domain dependent and utilizes some of PLEIADES' general features associated with representation and search methods.

The key aspects of PLEIADES that are incorporated into the hierarchical planner for space-truss assemblies are as follows:

- Relational graph representation of an assembly state. This relational graph representation provides the basis for the generation of feasible assembly operations through the use of cutsets. Each feasible cutset may be considered a removal operation.

- AND/OR graph representation of all possible assembly sequences. The AND/OR graph representation provides a compact and complete representation method for all possible states since its subassemblies are generated from the cutsets of the relational graph.

- Two, and no more, subassemblies are joined at any single assembly step. This feasibility criterion governs the decomposition of the assembly to include only certain operations. In effect, each cutset must adhere to this guideline in order for the removal operation to be considered valid.

- Resource dependent feasibility conditions include geometric feasibility (stability), structural feasibility (rigidity), and adjacency conditions.

### 4.2.3 Assembly Sequence Planning in the Space-Truss Domain

A general strut-node structure, *3D(N, S)*, consists of $N$ nodes and $S$ struts. Assembly sequence planning addresses the following problem: what ordering of struts will assemble a 3-D structure, *3D(N, S)*, such that all intermediate states are physically rigid and stable and every strut operation is a feasible one? In general, the plan must begin with a set of individual, unconnected struts and nodes and assemble a sequence of struts that will result in the desired configuration. Homem de Mello and Sanderson [4] showed that the planning of the assembly of one product can be viewed as a path search in the state space of all possible configurations of that set of parts, and that for most assembly problems, the branching factor from the initial state to the goal state is greater than the branching factor from the goal state to the initial state. The backward search is therefore more efficient and corresponds in this case to the problem of disassembling the product using reversible operations. Hence, the approach used in the space-truss planner is one that develops the assembly sequence by disassembling the intact structure strut-by-strut with the assumption that the disassembly operations are *inverses* of the assembly operations and the disassembly sequence itself is an inverse assembly sequence.

In order to remove any strut $S_i$, from a space-truss structure *3D(N, S)*, the strut must first satisfy the *geometric feasibility condition* and the *structural feasibility condition*. Finally, the stability of the resulting structure relative to other possible structures must be assessed. The structure with the most relative stability is considered as the next subassembly in the plan and the removal operation is considered the next step in the disassembly plan.

The following sections briefly describe the geometric feasibility and the structural feasibility criteria of the disassembly sequence planner.

### Geometric Feasibility Condition

*Geometric feasibility* of a strut is defined as any strut that is accessible to the robot arm and end-effector at the points where the next assembly operation is to be executed. The removal of the strut from a given position assumes that the robot gripper location on the strut must be completely accessible and that the strut can be connected in the correct, secure manner. (The problem of trajectory planning for the insertion of the strut is discussed in Part II of this chapter). In order to determine if the strut under consideration has such a property, the *convex hull* [14] of the current subassembly is constructed. The convex hull of a set of vertices is defined as the collection of vertices that lie in the extreme points of the space that would be created if the set were to form a volume (3-D space) [15]. Intuitively, the convex hull can be created by taking a large rubber sheet and placing it around the set of vertices, *V*. The resulting outline formed by the rubber sheet's deformations can be considered the convex hull.

It is assumed that the nodes that are associated with the strut as endpoints are rigidly connected to the struts but do not hinder its removal, i.e., the removal of the strut may or may not imply removal of the nodes, depending on whether there are other struts attached to the node. The computation of the 3-D convex hull was implemented using a tetrahedral decomposition algorithm and will not be discussed here in detail. After computing the convex hull, *CH(N)*, of *3D(N, S)* consisting of *N* nodes and *S* struts, the geometric feasibility of a strut is defined as follows:

1. Determine all struts that lie in the convex hull and place them on the *active* list, $A$—:
2. $\forall S_i \in S$ with endpoints $(N_k,\ N_l)$:

   $(\text{If } N_k \in CH(N) \ \vee \ N_1 \in CH(N) \implies A \leftarrow S_i)$

   $\vee$

   $(\text{If } N_k \in CH(N) \ \wedge \ N_1 \in CH(N) \implies A \leftarrow S_i)$
3. End.

As mentioned previously, for this implementation, an accessible strut is one that has at least one endpoint or node which is on the convex hull. For some purposes it is also useful to consider only those struts with both nodes on the convex hull.

### Structural Feasibility Condition

The *structural feasibility condition* is the second criterion used to determine resulting feasibility for the removal of a strut. If the removal of a strut results in a maximally stable and rigid structure, the 3-D structure is structurally feasible and the strut possesses the property of satisfying the *structural feasibility condition*. All intermediate states of disassembly must be stable.

The inherent hierarchical nature of the space-truss domain allows 3-D space structures to be viewed from two levels:

1. Struts and nodes interconnected to form a complete super-structure.
2. Collections of struts and nodes that form rigid sub-structures that are interconnected to form the complete structure.

For planning purposes, structural feasibility of a disassembly operation is based on this hierarchical representation of the truss structure.

A *Structural Volume Element* (SVE) is a collection of struts and nodes that, when interconnected in a specific manner, forms a structurally rigid and stable state. SVE elements can be pentahedra (or pyramids), tetrahedra, braced cubes, or any other rigid, 3-D structure that can be used as a building block or substructure to build a larger, more expansive 3-D structure. Structural rigidity and stability of an ideal SVE element is indistinguishable from that of a single strut or node and is more stable and rigid than a planar polygon. The relative stability analysis of the structures just mentioned is not presented here, but a static force analysis can be used to assess relative rigidity of these substructures. In the current planning system, the tetrahedron and pentahedron are used as SVE elements. A few examples of SVE elements are given in Figure 4.1.

A *Non-Structural Volume Element* (NSVE) is any collection of struts and nodes that does not form a structural volume element. A *non-structural volume element node* (NSENode) is any node that is not part of any existing SVE element. Such a node must have at least one strut attached to it. In addition, any strut that is attached to a NSENode must not be a member of any existing SVE element. A *non-structural volume strut* (NSEStrut) is any strut that is not a member of an existing SVE element and connects two or more SVE elements. An NSEStrut will always contain endpoints that lie in SVE elements. Figure 4.2 gives some examples of structures with NSVE elements.

### Connections and Stability

SVE and NSVE structures provide the hierarchical representation of the space-truss assembly, and their interrelationships affect the stability of the overall structure. There are three types of SVE interconnection relationships.

1. A *point connection* occurs when two or more SVEs have only one node in common and has order $O(1)$. Note that the SVEs do not have any other struts or nodes in common and a point connection may be a part of a higher order connecting relationship.

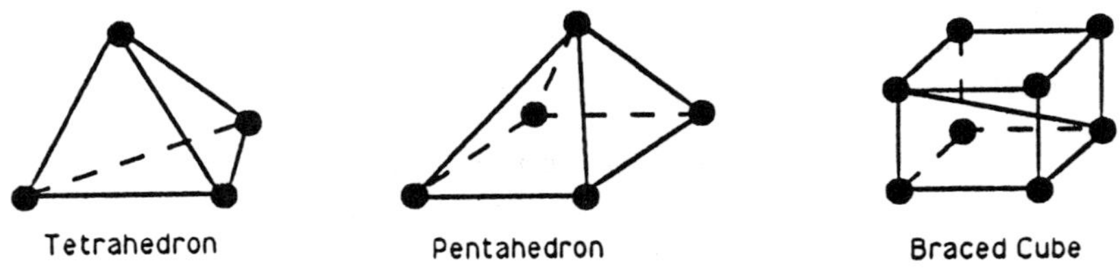


Figure 4.1: Examples of Structural Volume Elements

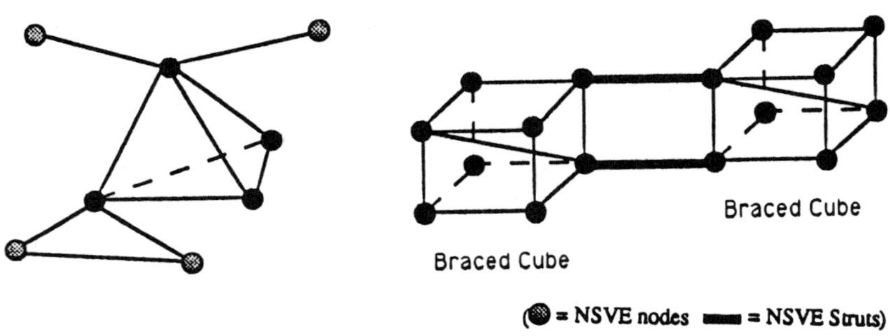


Figure 4.2: Examples of Non-Structural Volume Elements

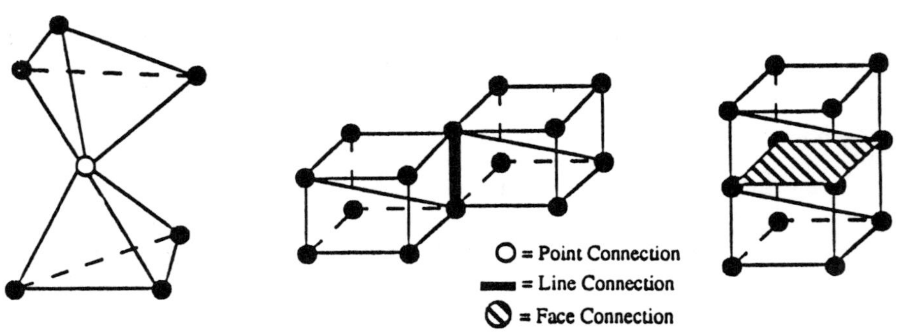


Figure 4.3: Examples of Structured Volume Element Connections

2. A *line connection* (order $O(2)$) exists between two or more SVE elements if the SVEs have a single strut in common.

3. A *face connection* is the highest order of connection possible in the 3-D domain. It occurs when two SVE elements are connected by a set of line connections that form a closed polygon. In the current implementation of the planner, face connections are triangular or a square polygon.

Homem de Mello described relative stability in terms of degrees-of-freedom of a candidate part, [6]. The greater the relative degrees-of-freedom between two subassemblies, the less the degree of stability for the assembly. Using this heuristic, the relative stability (ST) of the three types of SVE connections are, in decreasing order, ST(Face connection) > ST(Line connection) > ST(Point connection).

Examples of point, line and face connections are shown in Figure 4.3.

### Relating the Geometric and Structural Feasibility Criteria to the Hierarchical Components

The notion of geometric and structural feasibility of struts can also be applied to SVE elements. An SVE is geometrically accessible if it is on the convex hull of the 3-D structure and can be removed by removing any member struts that are on the convex hull of the 3-D structure. Note that an SVE can be *uniquely* removed by removing all active member struts which are unique to that SVE element. Any active struts that are members of more than one SVE cannot be removed.

Structural feasibility or rigidity can also be extended to the 3-D structure composed of SVE elements. Using the relative stability of the SVE connections, the goal of achieving maximal structural feasibility is reduced to maximizing the number of face connections and minimizing the effects of the less stable point and line connections.

### State and Plan Representation

The planner uses a hierarchical representation which consists of lower-level components that include struts and nodes and higher-level components which include structural and non-structural volume elements, point, line, and face connections.

A relational graph consisting of the components described above is used to represent a 3-D assembly state. Feasible cutsets of the relational graph represent the feasible assembly operations on the given assembly state. These operations result in a state space of all possible assembly states.

Sanderson and Homem de Mello [1], [2], [3], [4], [5], introduced the AND/OR graph as a means of representing the state space associated with disassembly sequencing. The AND/OR graph representation provides a complete and robust representation of *all* possible configurations of the struts and nodes. Using such a representation, the path from the root node at the leaves can then be generated using a search algorithm with preassigned costs, [6]. Once the state space is determined, the complete paths going from the root node to the leaf nodes represent disassembly sequences.

### The Hierarchical Search Strategy

The search space can be greatly reduced by limiting it to only possibilities that will result in feasible solutions. The hierarchical method of disassembly is used to limit the search tree since it eliminates the paths that are known to be dead-ends based on the constraints of the configuration at hand. More importantly, the hierarchical approach uses the notion of creating subgoals as a method of problem decomposition in order to drastically reduce the state space and resulting search. The hierarchical approach, the low-level details (the struts and nodes) of disassembly are ignored in the first step and only the disassembly of the SVE elements is addressed. The total complexity of the problem is the sum of the complexities of the individual searches, not the product. The hierarchical sequencing with SVE elements as subgoals in this planner results in an optimal sequence. Because of the symmetry of the structures in general, there is always more than one optimal sequence. In the current implementation of this planner, only one optimal sequence is presented, although the implementation can be easily extended to determine other optimal sequences.

The hierarchical method consists of two tiers of disassembly. The first tier sequences the SVE elements using SVE constraints. Once the SVEs are sequenced, the second level of disassembly is invoked in which each SVE in the SVE sequence is taken in turn and disassembled at the node-strut level. As stated previously, a property of independent subgoals is that a near-optimal solution can be reached by ensuring that all subgoals are solved optimally as well. The global solution is just the concatenation of the optimal subsolutions associated with the subgoals [16].

### Level One: SVE Level

Given an arbitrary space-truss structure, $3D(N, S)$, with $N$ nodes and $S$ struts, the following algorithm guarantees an order of disassembly at the SVE level.

**Algorithm** $DISASSEMBLE_{SVE}(3D)$

1. Begin.

2. $SVE$ $\leftarrow$ Determine all SVE elements in 3D.

3. $SVE_{Valid}$ $\leftarrow$ Determine all valid SVE elements.

4. $SVE_{MinFaces}$ $\leftarrow$ Determine SVE elements with minimum number of faces.

5. IF $SVE_{MinFaces} = 1 \Rightarrow SVE_{next} \leftarrow SVE_{MinFaces}$ else $SVE_{ease} \leftarrow$ Determine easiest SVE element to disassemble from $SVE_{MinFaces}$

6. $SVE_{next} \leftarrow SVE_{ease}$

7. $3D\prime \leftarrow REMOVE(SVE_{ease})$

8. $DISASSEMBLE_{SVE}(3D\prime)$

The following criteria are used in the algorithm to develop the SVE sequence:

1. SVE removal should impact a minimum number of face connections. The removal of an SVE must result in a minimum number of face connections being broken.

2. SVE removal must be of maximum ease. The easier it is to disassemble an SVE, the less likely that there will be unstable intermediate states.

3. SVE removal must be resource efficient. Each subsequent SVE must be *adjacent* to the previous one. This approach eliminates excessive movement on the part of the robot.

**Level Two: Strut Level**

The strut level sequencing is done for struts associated with a specific SVE. Since the maximum number of struts in current SVE elements is at most 8, the strut sequencing is done on a beam-first search in which the valid struts are sequenced in such a manner as to ensure the lowest possible total cost. The convex hull is generated for the structure to determine geometrically accessible struts. The heuristic cost function is applied to the struts and nodes of the SVE to determine the optimal strut removal operation. This cost heuristic is based on maximizing the number of struts that may be attached to any one node during an intermediate disassembly state, thus minimizing unconnected nodes. The disassembly sequence for the tetrahedron structure shown in Figure 4.1 is {6, 5, 4, 3, 2, 1}. The disassembly sequence for the pentahedron structure shown in Figure 4.1 is {7, 3, 2, 1, 6, 4, 8, 5}.

### 4.2.4 Case Studies

This section presents two examples of 3-D structures for which assembly sequence plans have been generated using the method outlined in this chapter. These case studies can be considered representative structures that may realistically exist as parts of space-truss assemblies.

**Case Study One:** $4D(10,\ 24)$

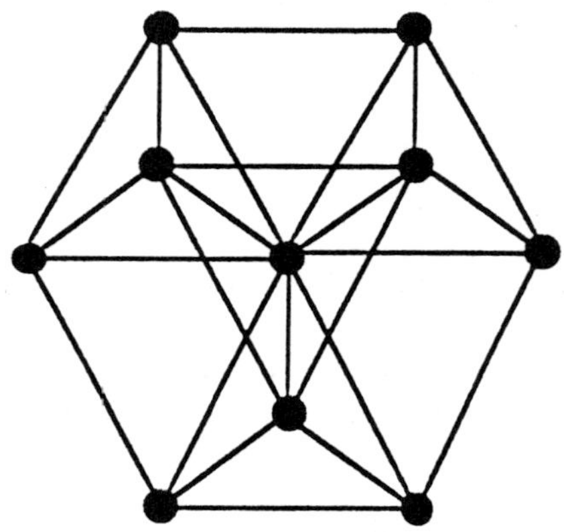

Figure 4.4: Case Study with 10 Nodes and 31 Struts

The first example is the structure shown in Figure 4.4 and consists of 10 nodes and 24 struts. There are seven SVE elements including three pentahedra and four tetrahedron. Figure 4.5 shows six intermediate steps in the assembly plan for this structure. The detailed plan generates the individual strut assembly sequence and the specific geometric position of struts and nodes at each step. This detailed description is sufficient for motion planning of the assembly operation as described in Part II.

**Case Study Two:** $3D(31,\ 102)$

A second example is shown in Figure 4.6 [17]. This structure consists of 31 nodes and 102 struts and contains a unique geometry of SVE elements. Due to space considerations and complexity of the structure, only intermediate states of disassembly are presented in Figure 4.6. As in the previous example, the final detailed plan specifies all the geometric information required for motion planning.

### 4.2.5 Discussion on Assembly Sequence Planning

The results from this task level planner are being integrated into the ongoing research at Rensselaer's Center for Intelligent Robotic Systems for Space Exploration. The overall goal is to build a planner that not only generates a high level task plan, but also develops the low level steps needed to achieve the higher level goal, as well as provide the control mechanisms required to carry

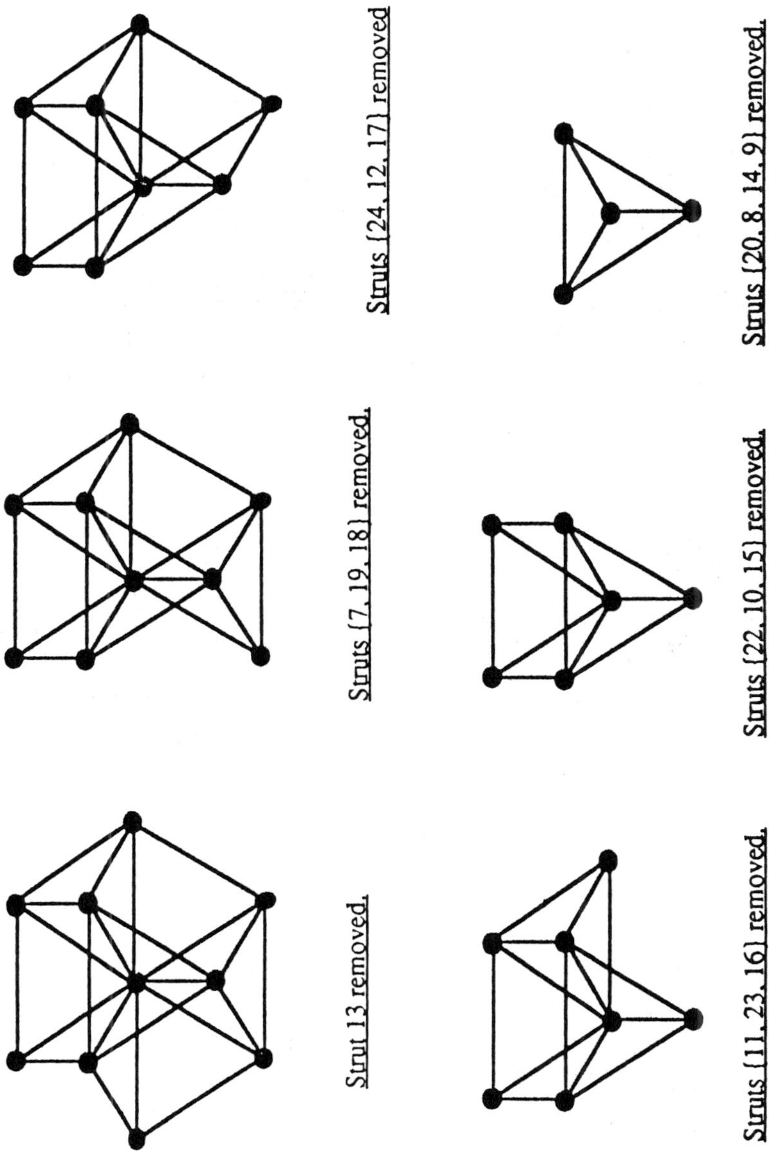


**Figure 4.5: Disassembly Sequence for Structure with 10 Nodes and 24 Struts**

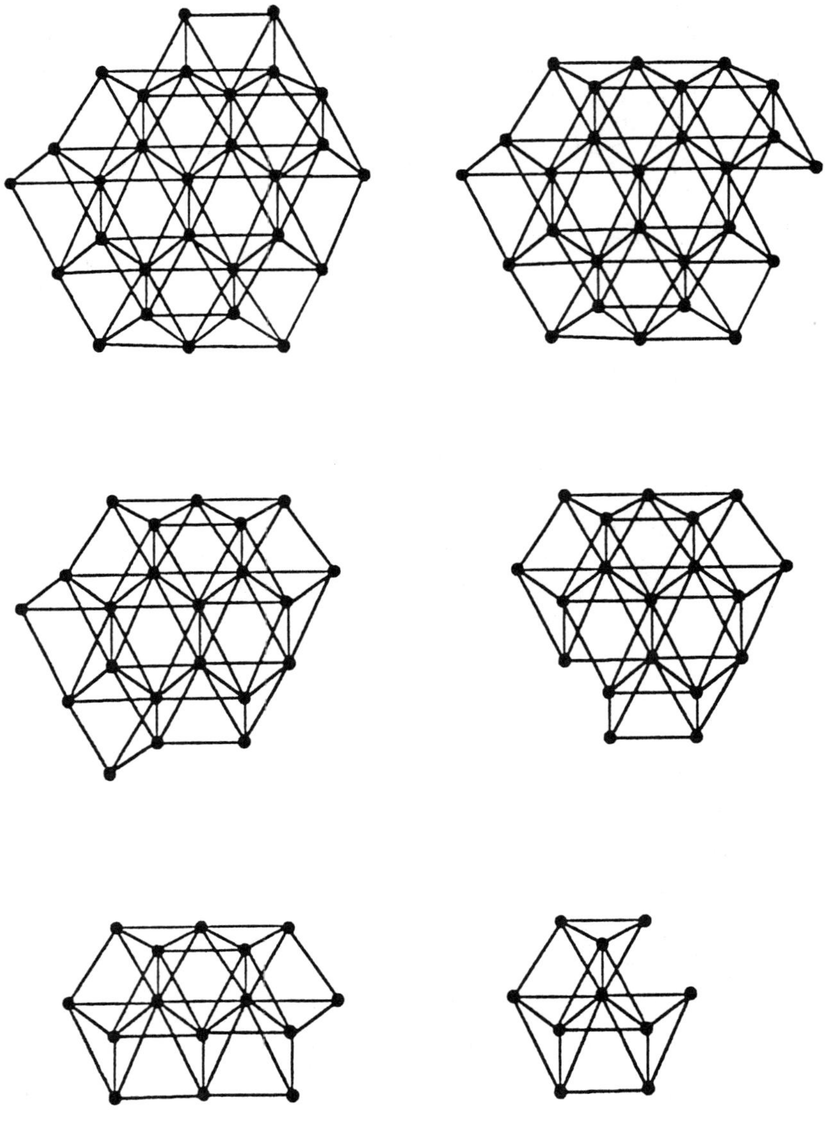

Figure 4.6: Intermediate Disassembly States for Structure with 31 Nodes and 102 Struts

out the physical tasks. The output of this sequence planner provides the input to the path planner discussed in the next section.

While this planner used the tetrahedron and pyramid structural volume elements, additional work needs to be done in order to allow automatic generation of alternative SVE elements based on the geometry and domain at hand. The heuristics could be generated on or off-line, but would be general enough to provide the basic SVE elements in a given configuration. Additionally, further research may include formal structural analyses or simulation of SVE elements that would verify the conclusions of this and future heuristic planners.

Another direction of continuing research is the use of sensor information to determine the geometric feasibility of an operation. In a real world scenario, there may not be enough prior knowledge regarding the geometry, size, and relative positions of the objects and sensory feedback may be used to determine what is and is not accessible.

Finally, in any system, unpredictable occurrences are inevitable and additional research in error recovery will address this issue. The planning system, at both the higher and the lower levels, must be robust enough to react to an unpredictable event or human interaction when required. Sensor integration would play an important role in developing this feature and must be complete enough to be able to address this important need.

## 4.3 Part II: Assembly Path Planning

### 4.3.1 Introduction

The execution of a motion task requires the end-effector to move from its initial position to a goal position and orientation given in Cartesian space. This section proposes a method that provides a sequence of joint vectors that describe a feasible path to accomplish the motion task. All joint vectors are guaranteed to be contained in the robot's workspace and collisions among the links and with parts of the environment are avoided. In the subsequent trajectory planner stage the velocity and acceleration information is added and the result is passed to the low-level robot controller stage.

### 4.3.2 Background

Robot path planning algorithms fall into two categories: global and local.

The global methods typically consider the environment as a whole when planning a path. They generally have the advantage of finding a path if one exists, but they are computationally expensive if the environment is complex. There are three subgroups of global path planning algorithms:

- Cell decomposition methods,

- Search algorithms on a visibility graph, and
- Optimization using calculus of variations.

Cell decomposition methods can be further divided into exact and approximate methods whereas exact methods decompose free space into cells complex enough to exactly represent the free space. Approximate methods however split free space into very simple units, for instance cubes, that cannot represent free space exactly. There will be a number of cubes that contain both free and occupied space. These cells are further subdivided into smaller cells, until the desired accuracy is obtained. After the cell model is established, graph search techniques are used to generate the shortest path that includes only cells representing free space [18], [19].

Instead of splitting the space into cells, one can also find a sufficiently large set of feasible intermediate goals, usually located near corners of obstacles, and establish a graph based on visibility between pairs of these goals. Both of these methods rely on an efficient graph search algorithm. A suitable algorithm is A*, a best first search algorithm that considers a lower bound estimate of the remaining distance in order to expand the most promising nodes first [20], [21].

A different approach that does not involve graph search is the optimization of a cost function that typically includes path length and clearance from obstacles. This approach uses methods from calculus of variations [22].

The oldest local path planning method is the "hypothesize and test" method. The robot proceeds a step in the desired direction and checks the feasibility of the path that leads to the new position. If the path is not feasible, another direction is chosen according to some heuristic. This approach has not been very successful.

A more recent method is based on potential fields [23], [24], [25], [26]. The gradient of a potential field guides the robot towards the goal and keeps it at a safe distance from obstacles. This potential field has very large values near obstacles in order to produce a repulsive force between the robot and the obstacles. The potential assumes its lowest value at the goal position in order to attract the robot to the goal. The drawback of all potential field methods is the fact that they do not guarantee finding a collision-free path even if one exists. If the potential field function is not carefully chosen, it is likely to contain local minima that stop the path planning algorithm. However, even if a path can be found, it is not guaranteed to be optimal.

Variations of the potential field methods include information about the dynamics of the robot and its payload. These methods have the advantage that they generally produce paths which can be executed at a comparatively high speed [27], [28].

If the robot is not assumed to be a point or a sphere, then its shape must be included in the path planning process. A particular path might be feasible depending on the orientation of the robot as a function of its position. In order to account for the robot's orientation, the path planning problem can be solved in configuration space [29], [30], [31], [32]. The rotational degrees-of-freedom of the robot are added as additional dimensions to the Cartesian space. If we consider a three dimensional space and a robot with another 3-DOFs for rotation, the associated configuration space is six dimensional. In configuration space, the robot is shrunk to a point and the obstacles are enlarged according to the robot's orientation. This reduces the problem to planning a path for a point, however, in a higher dimensional space.

### 4.3.3 Geometric Modeling

Every path planning algorithm needs a geometric description of the objects with which it works. If a very simple model is used to represent the objects, it is in most cases necessary to choose the model's volume considerably larger than the real object's volume, since the object's volume must be completely enclosed in the volume of its model. This restricts the path planning process and might lead to a failure even though a path exists. On the other hand, distance computations on simple models are typically performed very efficiently. More involved models allow for a better approximation to the real shape of the objects, but they are computationally more expensive.

In this work, cylinders with spherical caps have been chosen as the basic unit for geometric modeling. An example is shown in Figure 4.7. It can also be viewed as the volume that is swept by a sphere with radius $r$ moving between the two points $p_1$ and $p_2$ on a straight line. This volume can be described as a set $S$. First, define $s = p_2 - p_1$, then

$$S(p_1, p_2, r) = \{p \mid p = p_1 + ts + q \quad \text{with } t \in [0 \ldots 1], \ |q| \leq r\} \tag{4.1}$$

This model is particularly well suited for the objects of interest in strut–based structure assembly, namely struts and robot links, but any kind of object can be approximated by using a union of several basic cylinders. The algorithm herein is independent of the choice of the model, so any other model (e.g., a polyhedral model) can be used, provided that a distance computation algorithm is available.

In the following, a short overview of the distance computation algorithm for the cylindrical model is presented (Figure 4.8). In order to find the distance $d$ between two objects, we find the distance $d_s$ between the twc line segments in the center of the cylinders and subtract the radii of the cylinders $r_p$ and $r_q$.

$$d = d_s - r_p - r_q \tag{4.2}$$

The line segment in the center of a cylinder $S(p_1, p_2, r)$ can be described as $S(p_1, p_2, 0)$.

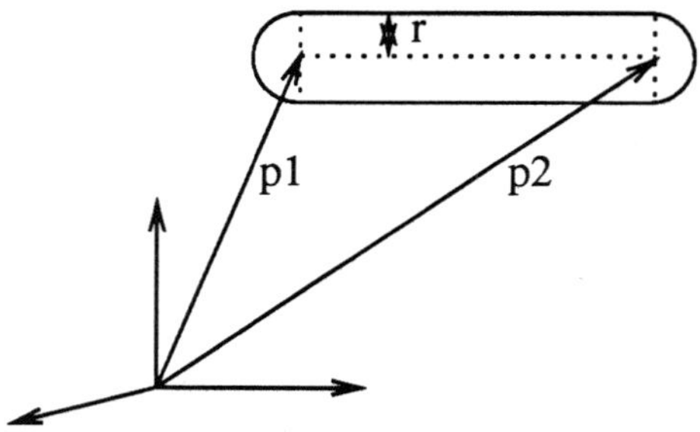


Figure 4.7: Basic Geometric Model

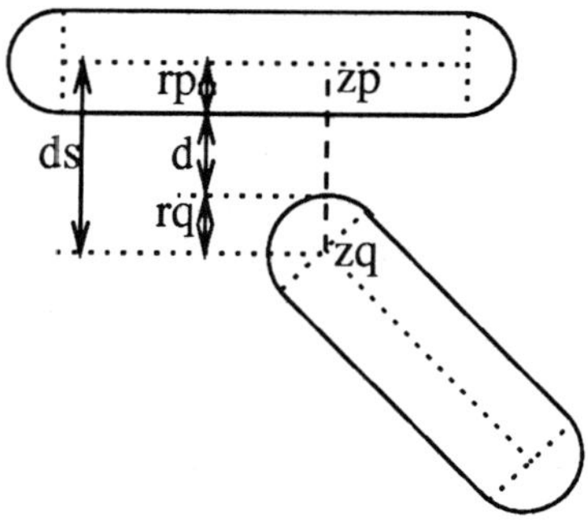


Figure 4.8: Shortest Distance Between Two Swept Sphere Cylinders

The computation of $d_s$ is based on methods from vector geometry. It involves checking for special cases of parallel line segments. The algorithm provides the two points $z_p$ and $z_q$ on the two line segments. Their distance is $d_s$, the minimal distance between the line segments.

## 4.3.4 Robot Path Planning

In this work, a single robot that holds a cylindrical object at its center is considered. The method presented can be used for all single-chain robots with at least 6-DOFs. All joints can be either revolute or prismatic.

### Modeling the Robot's Geometry

In the process of path planning, we must consider collisions between different struts, collisions between a strut and a part of the robot, and collisions among parts of the robot. In order to simplify distance computations, it makes sense to use the same model for robot links and other parts. The cylindrical model shown in Figure 4.7 can also provide sufficient modeling accuracy for the robot links when performing free motion planning. Distance computation is done as described in Section 4.3.3.

### Potential Field Method in the Robot's Joint Space

Potential field methods are defined in Cartesian space. All gradients of the potential fields are vectors in normal 3-D space. They define the direction of motion of the robot's end-effector, so the new location is first given in Cartesian space. This means that an inverse kinematics routine is required to find the new joint vector. Since this must be done at every step of the path, the algorithm becomes very computationally expensive, since in general, the inverse kinematics problem can only be solved iteratively. In this work, however, an alternate strategy is taken whereby the robot's new location is first obtained in joint space, rather than in Cartesian space. To find the location in Cartesian space, a simple forward kinematics routine is required.

Thus, the goal of the algorithm is to find a joint vector increment $\Delta q$ that is added to the current joint vector $q$ to obtain the next step of the path. $\Delta q$ is the sum of several components each performing a different task:

$$\Delta q = \Delta q_a + \Delta q_r \tag{4.3}$$

$\Delta q_a$ is the joint vector increment that moves the payload closer to its goal position and orientation (attractive forces), while $\Delta q_r$ is the sum of all joint vector increments that keep the robot from colliding or running out of joint range (repulsive forces).

$$\Delta q_r = \Delta q_{rc} + \Delta q_{rr} \tag{4.4}$$

where $\Delta q_{rc}$ is doing the collision avoidance and $\Delta q_{rr}$ keeps the robot from running out of joint range.

Note that all fields are defined as gradient fields and not as potential fields. This often leads to a more natural definition and also saves the derivation of the field's gradient.

### Attraction Field

The attractive field is the only field that must be defined in Cartesian space. First the definition of the Cartesian vector $x_a$ is given, then the transformation into joint space is shown.

Let $x_a$ be a $6 \times 1$ vector. The first three elements ($x_{at}$) represent the translational part (force) while the last three elements ($x_{ar}$) represent the rotational part (torque), or

$$x_a = \begin{bmatrix} x_{at} \\ x_{ar} \end{bmatrix} \tag{4.5}$$

If $p_c$ is the Cartesian position of the center of the strut in the robot's gripper (current position) and $g_c$ the Cartesian position of the center of the strut in its goal position, then

$$x_{at} = C_{at} \frac{g_c - p_c}{|g_c - p_c|} \tag{4.6}$$

$C_{at}$ is the field constant of the translational attraction field. The rotational part $x_{ar}$ is more difficult to obtain. Let $p_1$ and $p_2$ be the endpoints of the object in the robot's gripper and let $g_1$ and $g_2$ be the endpoints of the same object in its goal position. Furthermore, let $a_p$ be the approach vector of the end-effector in its current orientation and let $a_g$ be the approach vector of the end-effector when it reaches its goal orientation. The approach vector represents the direction the end-effector must move to properly approach an object before grasping it. Both $a_p$ and $a_g$ are unit vectors. We also need two unit vectors that point along the axes of the struts. These are given by

$$p_p = \frac{p_2 - p_1}{|p_2 - p_1|}$$

$$p_g = \frac{g_2 - g_1}{|g_2 - g_1|}$$

The approach vectors are perpendicular to the axes of the struts:

$$a_p \perp p_p \tag{4.7}$$

$$a_g \perp p_g \tag{4.8}$$

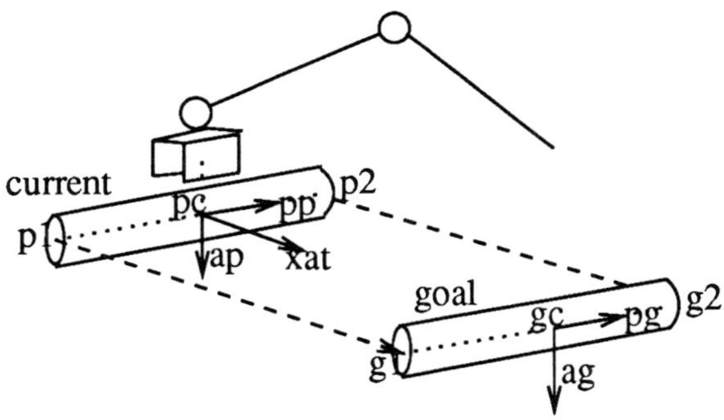


Figure 4.9: Overview

All these vectors are illustrated in Figure 4.9. The current orientation and the goal orientation can be represented by rotation matrices that give the respective orientations with respect to the world coordinate frame as

$$T_p = [\ p_p \quad (a_p \times p_p) \quad a_p\ ] \tag{4.9}$$

$$T_g = [\ p_g \quad (a_g \times p_g) \quad a_g\ ] \tag{4.10}$$

Since rotation matrices are orthonormal, their inverse is equal to their transpose. Thus, the rotation matrix that rotates the current gripper into the gripper in its goal orientation is defined as

$$T = T_g T_p^T \tag{4.11}$$

This is the rotation we are interested in, but we need the axis of rotation and not the rotation matrix. In order to find the axis of rotation, an eigenvalue problem must be solved. Every rotation matrix has an eigenvalue of 1 and a corresponding eigenvector $a$. The axis of the rotation performed by the rotation matrix and eigenvector $a$ have the same orientation.

The vector $a$ represents the axis of rotation, but it does not give any information about the sense of rotation, since $-a$ is a valid eigenvector if $a$ is an eigenvector. We can of course rotate both ways to reach the goal orientation but in most cases it is preferable to rotate in the sense that involves a smaller angle of rotation. For the method used to find the correct sense of rotation or the correct sign of $a$ consider Figure 4.10.

It is assumed that $p$ can be rotated into $g$ by using $a$ as the axis of rotation. The angle of rotation is less than or equal to $180^o$ if the following condition holds:

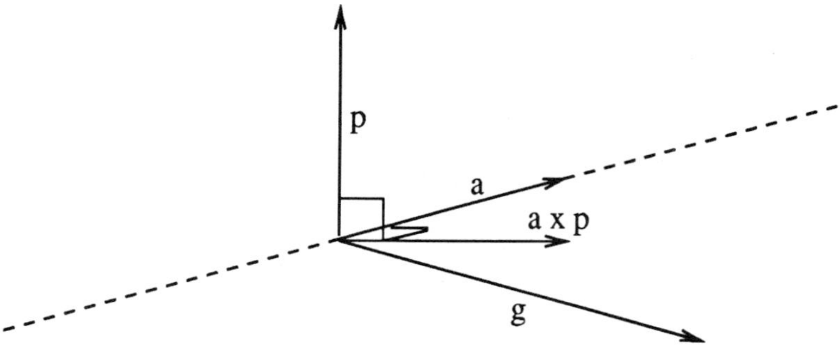


Figure 4.10: Sense of Rotation

$$g \cdot (a \times p) \geq 0 \tag{4.12}$$

If the condition does not hold, we just change the sign of $a$ to get the correct sense of rotation. Then the rotational part from equation (4.5) is:

$$x_{ar} = C_{ar} a \tag{4.13}$$

$C_{ar}$ is again a field constant. All this yields $x_a$, a direction vector in Cartesian space. In order to avoid inverse kinematics, a direction vector in joint space is needed. The direction vectors in either space can be considered velocity vectors, so the manipulator Jacobian, $J$, can be used to transform the direction vector from Cartesian to joint space. The derivation of $J$ is discussed below, but let us assume for now that it is available to us. In order to obtain $\Delta q_a$ from equation (4.3) we need to solve the following linear system:

$$J \Delta q_a = x_a \tag{4.14}$$

The Jacobian, $J$, is a $6 \times n$ matrix, where $n$ is the number of degrees-of-freedom of the manipulator. If $n > 6$ then the linear system is underdetermined, so there is an infinite number of solutions for $\Delta q_a$. In this case the term $\Delta q_a^T Q \Delta q_a$ will be minimized, where $Q$ is a weight matrix that will normally have nonzero terms on the diagonal only. This weight matrix is important for robots with both revolute and prismatic joints, since vector $q_a$ has mixed units in this case. This constrained minimization problem is solved by introducing a Lagrangian vector to transform it to a system with 6 equations in 6 unknowns. In our implementation the system is solved using Householder transformations [33].

This method is a variation of iterative inverse kinematics algorithms that have been developed for redundant robots or for nonredundant robots with kinematic properties for which no closed-form inverse kinematics solution exists, [34], [35], [36], [37], [38], [39], [40].

### Obstacle Avoidance

The goal of this part of the algorithm is to keep the moving objects (the links and the payload of the robot) from colliding with items in the environment and also to avoid collisions among the moving parts. These goals are achieved by two separate algorithms that both contribute a joint vector to $\Delta q_{rc}$ from equation (4.4).

$$\Delta q_{rc} = \Delta q_{env} + \Delta q_{self} \tag{4.15}$$

Before beginning the discussion of the two methods, it is worth mentioning that the moving parts are the manipulator links *and* the object held by the manipulator. The payload is treated exactly like a manipulator link.

The idea behind both methods for collision avoidance is to go through all pairs of items that may collide and for each pair go through all joints that may change the distance between the items of the pair. The joints that have an influence on the distance of two objects are located between the objects in the kinematic chain. Adding an increment to these joint values may increase or decrease the distance or it may have no effect on the distance at all. Figure 4.11 shows two items modeled using the cylindrical model presented earlier. The joint in Figure 4.11 is a revolute joint; prismatic joints are discussed later.

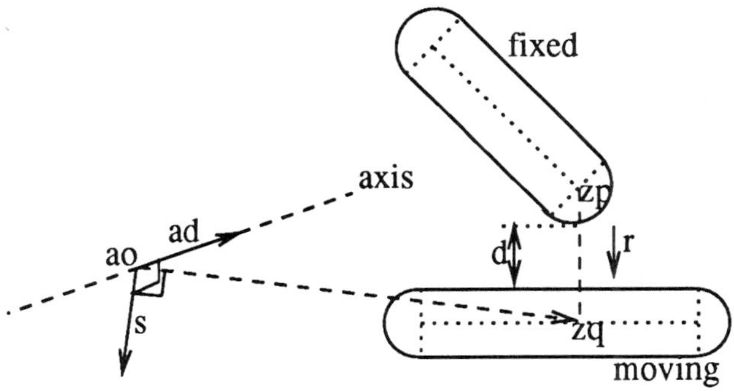


Figure 4.11: Basic Collision Avoidance Procedure for a Revolute Joint

The points $z_p$ and $z_q$ are the two closest possible points on the line segments of the models, they are found using the algorithm described earlier. The distance $d$ is the minimum distance between the models (see equation (4.2)). The vectors $a_o$ and $a_d$ represent the axis of rotation of the joint under consideration. $a_o$ is the origin of the frame rotated by that joint and $a_d$ is a unit vector representing the direction of the axis of rotation. Now we define a unit vector $r$ pointing from $z_p$ towards $z_q$:

$$r = \frac{z_q - z_p}{|z_q - z_p|} \tag{4.16}$$

We assume that model $S_q$ is moving and model $S_p$ is fixed, then vector $r$ is the best possible direction in which model $S_q$, and point $z_q$ in particular, can be moved in order to avoid a collision between the two models. We also define a vector $s$ as follows:

$$s = a_d \times (z_q - a_o) \tag{4.17}$$

Vector $s$ represents the direction in which point $z_q$ *will* move when the joint is rotated in the positive sense of rotation. Vector $s$ is really this joint's column of the Jacobian matrix for point $z_q$. Note that the absolute value of $s$ is equal to the distance between $z_q$ and the axis of rotation. Now we can define the joint increment $\delta q_r$ that will be added to the component of this joint in the joint vector increment as

$$\delta q_r = C_r \frac{\mathbf{r} \cdot s}{d^2} \tag{4.18}$$

It turns out that a degree of repulsion proportional to $d^{-2}$ leads to good results. The repulsive force grows very fast when the models get close together. When the distance gets larger, the force goes down fast enough, so that there is almost no influence on the path when the distances to obstacles are reasonably large. The force is also proportional to $|s|$, thereby increasing the contribution of joints that have their axis of rotation far off the point of interest. These joints can cause a large increase in distance between the objects with comparatively small angles of rotation. The force is finally proportional to the dot product of $r$ and $s$ which is a degree of matching between the optimal direction to increase the critical distance and the direction that $z_q$ will actually go if the current joint is moved. Note that this dot product can be negative, which means that the joint must be rotated in the negative sense of rotation in order to increase the critical distance.

Figure 4.12 shows the same situation with a prismatic joint.

The variables $z_p$, $z_q$, $r$, $a_o$, $a_d$ and $d$ are defined for the case of a revolute joint. The vector $a_d$ is the direction of motion of this joint, so $\delta q_p$ is defined as

$$\delta q_p = C_p \frac{r}{\cdot} a_d d^2 \tag{4.19}$$

Based on this elementary procedure we can define the joint vector increments $q_{env}$ and $q_{self}$ from equation (4.15). For this purpose we define a function $f_q(\cdot,\cdot,\cdot)$ as follows:

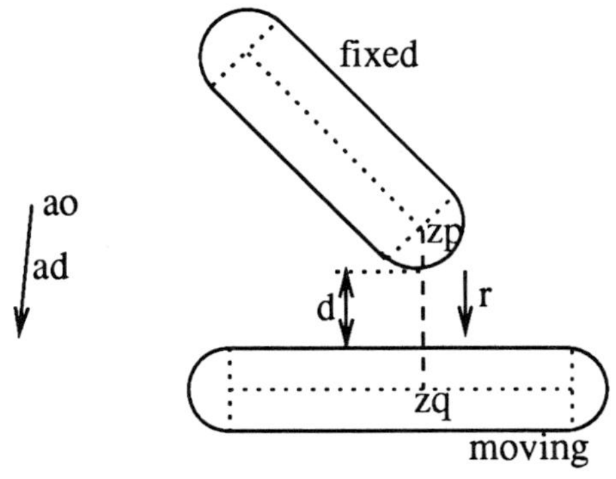


Figure 4.12: Basic Collision Avoidance Procedure for a Prismatic Joint

$$f_q(A, B, i) = [v_1 \dots v_n]^T \qquad \text{where } n \text{ is the number of joints} \tag{4.20}$$

$$v_j = \begin{cases} \delta q_r & \text{for } j = i \text{ and joint } i \text{ revolute} \\ \delta q_p & \text{for } j = i \text{ and joint } i \text{ prismatic} \\ 0 & \text{for } j \neq i \end{cases} \tag{4.21}$$

The points $z_p$ and $z_q$ and the distance $d$ are found by applying the distance computation algorithm described earlier to the models $A$ and $B$.

Furthermore, let $O_i, i \in [1 \dots o]$ be the model of the $i$th obstacle in the environment and $L_j, j \in [1 \dots l]$ model the $j$th link of the robot. Thus, there are $o$ obstacles and $l$ links to be considered. Then $q_{env}$ and $q_{self}$ are defined as

$$q_{env} = \sum_{i=1}^{o} \sum_{j=1}^{l} \sum_{k=1}^{j} f_q(O_i, L_j, k) \tag{4.22}$$

and

$$q_{self} = \sum_{i=2}^{l} \sum_{j=1}^{i-1} \sum_{k=j+1}^{i} f_q(L_i, L_j, k) \tag{4.23}$$

The numbering convention of the joints and links is shown in Figure 4.13. This concludes the derivation of $\mathbf{\Delta q_{rc}}$.

### Derivation of the Manipulator Jacobian

Before proceeding to the joint range problem, let us discuss the derivation of the manipulator Jacobian, $J$. The origins of the link frames $(a_o)_i$ and the joint axes

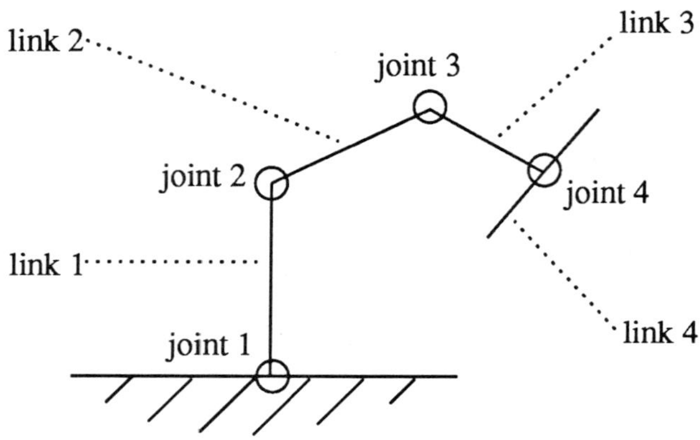


Figure 4.13: Numbering Convention for the Joints and Links of the Robot

$(a_d)_i$ have been introduced in the previous subsection, since they are used for collision avoidance. Once they have been computed, the manipulator Jacobian can be obtained very easily and with relatively little computational effort from,

$$J = \begin{bmatrix} j_1 & j_2 & \dots & j_n \end{bmatrix} \tag{4.24}$$

with $n$ being the number of degrees-of-freedom of the manipulator. Then

$$j_i = \begin{bmatrix} (a_d)_i \times (p - (a_o)_i) \\ (a_d)_i \end{bmatrix} \tag{4.25}$$

if joint $i$ is a revolute joint or

$$j_i = \begin{bmatrix} (a_d)_i \\ 0 \end{bmatrix} \tag{4.26}$$

if joint $i$ is a prismatic joint; vector $p$ is the current end-effector position. The computation of a row corresponding to a revolute joint involves a cross product and a vector subtraction, while the rows corresponding to prismatic joints are already fully computed.

### Joint Range

Besides avoiding collisions, a path planner must also guarantee that every joint vector on the path is in the robot's joint range. This goal can be achieved by monitoring the joint angles and creating a corrective joint value $\delta q_R$ if an angle is close to an end of its range. These values for all joints form the joint vector $\Delta q_{rr}$ from equation (4.4). The value $\delta q_R$ is a function of the respective joint value $q$. The qualitative behavior of this function is shown in Figure 4.14.

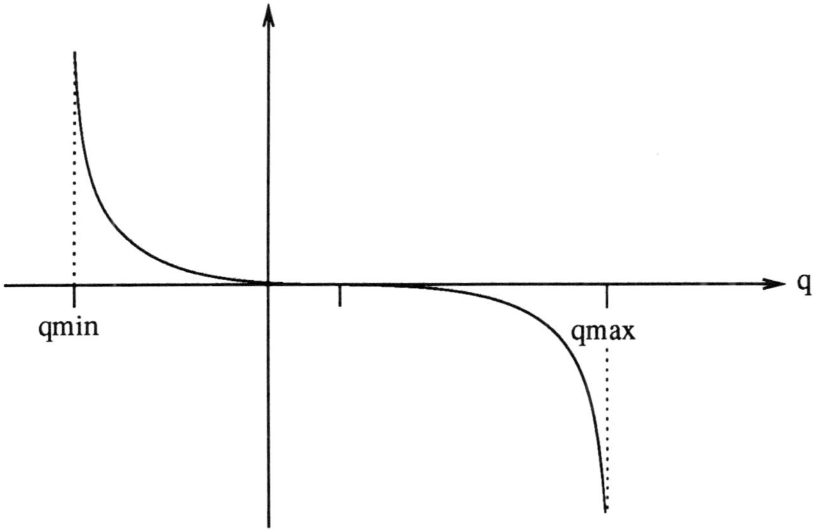


Figure 4.14: $\delta q_r(q)$

In our current implementation, the function $\delta q_R$ is based on the reciprocal function.

**Imposing Restrictions in Cartesian Space**

Recall the attraction field described above. Translation and rotation to the goal position is achieved by finding a direction vector and an axis of rotation that describe the necessary motion (equation (4.5)). Then we solve for the joint vector using the manipulator Jacobian (equation (4.14)). The two algorithms that provide a direction vector and an axis of rotation are independent. Also, the length of the direction vector as well as an angle of rotation are constant for each step. A consequence of this approach is the fact that rotation and translation are generally not completed at the same time. In a given situation it might happen that the payload has reached the correct orientation, but it has not reached the goal yet. In this case we would detect that the correct orientation has been reached and thus, set $x_{ar}$ in equation (4.5) to zero. After solving equation (4.14) we obtain a joint vector increment that will not rotate the end-effector when it is added to the current joint vector of the manipulator, so a rotation free motion is obtained during the remaining part of the path.

A problem arises when the joint vector increments for collision avoidance and joint range protection are added to $\Delta q_a$ to obtain the final joint vector increment $\Delta q$ (equation (4.3)). Since the joint vector increments for obstacle avoidance and joint range protection are computed in joint space, they will in general result in both a translation and a rotation in Cartesian space. This is undesirable, since it will affect the orientation or position, that has been determined to be equal to the goal.

To overcome this problem we add the two joint vector increments to obtain $\Delta q_r$ as done in equation (4.4), and split the result into two orthogonal components $\Delta q_{r0}$ and $\Delta q_{rx}$ such that $\Delta q_{r0}$ is in the null space of the Jacobian, $J$:

$$\Delta q_r = \Delta q_{r0} + \Delta q_{rx} \tag{4.27}$$

$$\Delta q_{r0} \cdot \Delta q_{rx} = 0 \tag{4.28}$$

$$J\Delta q_{r0} = 0 \tag{4.29}$$

If we use $\Delta q_{r0}$ instead of $\Delta q_r$ in equation (4.3), then the collision avoidance will not have any effect on the end-effector position in Cartesian space. Nevertheless, if the robot is redundant, the robot's joint vector will still change and increase the distance to obstacles and joint range boundaries.

As discussed earlier, we would like to stop either translational or rotational motion, once either has reached its goal. If we reduced $\Delta q_r$ to the null space of the full Jacobian, then collision avoidance would have no effect on any Cartesian motion, neither translation nor rotation. But if we assume that the rotation was completed first, then there is no reason to inhibit effects of collision avoidance on the translational motion. However, any effect on the rotational motion must be avoided, since the correct orientation has been reached. This partial effect on Cartesian motion can be obtained by reducing the joint vector to the null space of a part of the Jacobian matrix. As we have seen in equations (4.25) and (4.26), the rows of the Jacobian have a translational and a rotational part. In this sense we can split the complete Jacobian matrix into a translational and a rotational part:

$$J = \begin{bmatrix} J_t \\ J_r \end{bmatrix} \tag{4.30}$$

Both $J_t \times J_r$ are $3 \times n$ matrices, where $n$ is the number of degrees-of-freedom of the manipulator. If we need to inhibit translational motion, we reduce $\Delta q_r$ to the null space of $J_t$ and if we need to inhibit rotation, we reduce it to $J_r$'s null space.

There remains the question of how to reduce a joint vector to the null space of a Jacobian, complete or partial. The derivation follows:

$$J\Delta q_r = J(\Delta q_{r0} + \Delta q_{rx}) \tag{4.31}$$

$$
\begin{aligned}
J\Delta q_r &= J\Delta q_{r0} + J\Delta q_{rx} && (4.32)\\
J\Delta q_{r0} &= 0 && (4.33)\\
\Rightarrow J\Delta q_r &= J\Delta q_{rx} && (4.34)
\end{aligned}
$$

Here we can find the minimum norm solution for $\Delta q_{rx}$. Then

$$
\Delta q_{r0} = \Delta q_r - \Delta q_{rx} \tag{4.35}
$$

provides the null space component of $\Delta q_r$.

### 4.3.5 Combining Global And Local Path Planning Methods

#### Potential Field Methods and Their Problems

Potential fields can be used effectively to produce safe, smooth paths around obstacles, but they have a serious drawback: if a path exists, but the potential field contains local minima, then there is no guarantee that the path will be found. In relatively simple two dimensional cases, it is often possible to choose the potential field such that it can be proven to contain no minima other than the global minimum in the goal. This task becomes very difficult when considering a strut modeled by a line segment moving in an environment of other line segments, and it becomes hopelessly complex when a reasonably accurate model of the manipulator is added.

Thus, it is not possible to rely on a potential field method alone. This chapter describes how a potential field approach can be embedded into a backtracking algorithm that calls the potential field method under various conditions, until a path is found.

#### Different Ways of Performing a Given Task

As an example of a task, consider the insertion of a strut into an existing structure. For the case that a strut is symmetrical, there would be two ways to complete this task. Figure 4.15 shows the two possibilities. The strut is drawn asymmetrically for clarity.

In this example, the choice A is probably more likely to be successful than choice B, since the angle of rotation is smaller. This is of course not more than a heuristic statement and given certain joint angle configurations, it could very well turn out that B is successful and A runs out of joint range. But it is still preferable to try A first, since the probability for success is still higher and the resulting path also looks more natural.

Once we have decided upon an orientation, there are two possible ways to perform the task. We obtained the axis of rotation by finding the eigenvector of the rotation matrix, that corresponds to eigenvalue 1. It was also mentioned

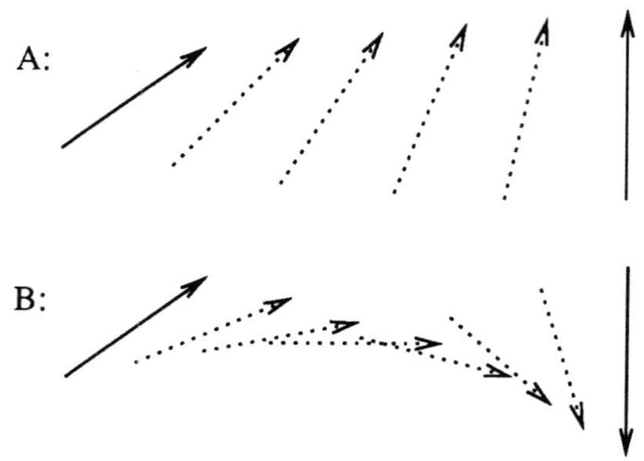


Figure 4.15: Two Ways of Inserting a Strut

that this eigenvector describes the axis of rotation that allows the gripper to reach its goal orientation, but it provides no information about the sense of rotation. In fact, we can reach the goal orientation no matter what sense of rotation we choose, but the angles of rotation will in general be different, since the sum of the two angles is always $360^o$. Here we will first try the rotation involving the smaller angle of rotation, even though this is not always the better choice. Consider Figure 4.16.

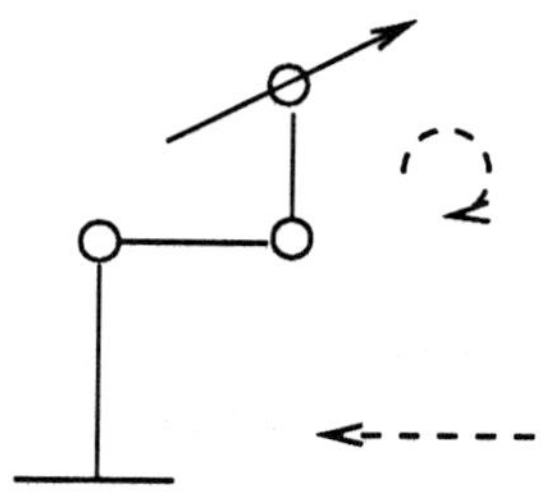

Figure 4.16: Choosing the Correct Sense of Rotation

The indicated sense of rotation is clearly the only possible choice in this particular case, even though the angle of rotation is greater than $180^o$.

This yields four possible ways of performing the same task of moving a strut from a given start to a given goal position.

### Splitting into Subtasks Using Graph Search

If none of the four cases leads to a successful completion of the potential field method, then it is necessary to split the task into subtasks. An example is shown in Figure 4.17.

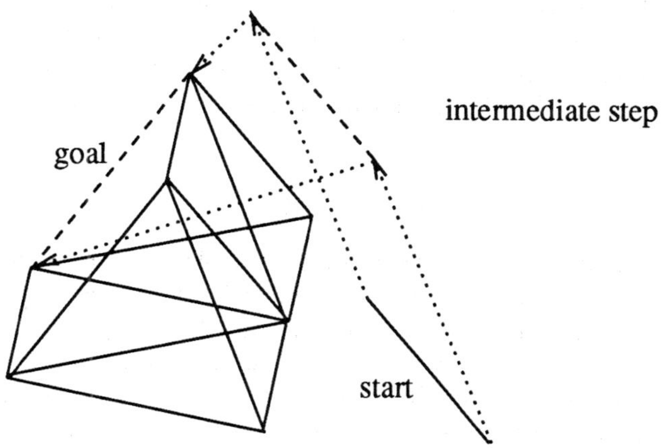


Figure 4.17: Splitting a Task into Two Subtasks

The figure shows only one of many ways the intermediate step could be chosen. The idea is to provide a set of possible intermediate steps and use it to build an undirected graph. The nodes of this graph are the intermediate steps plus the start and the goal position of the strut. The selection of suitable intermediate steps is crucial for the success of the algorithm; it is described in Section 4.3.2. After the nodes are created, edges are inserted such that every node is connected to every other node. In the example of Figure 4.17, we would obtain a graph with three nodes and three edges connecting them.

### Application of the A* algorithm

Now we use the A* algorithm to find the shortest path leading from the start to the goal node. We introduce the notion of distance by assigning a weight to every edge. This weight is the sum of the distance between the centers of the two node struts $p$ and $g$ attached to the edge and a measure for the degree of rotation involved:

$$w(p,g) = \frac{\left|\frac{p_1+p_2}{2} - \frac{g_1+g_2}{2}\right|}{s} + C_w(1-(p_p \cdot p_g)^2) \qquad \text{with} \quad s = |p_1 - p_2| \quad (4.36)$$

Consider Figure 4.9 for the meaning of the vectors: $p_p$ and $p_g$ are unit vectors, $s$ is the strut's length and $C_w$ is a positive constant that determines the weight assigned to rotation differences. If this value is large, more weight is assigned to rotation. A rotation of $90^o$ and a translation along the distance of $sC_w$ have the same weight. Note that identical orientation and exactly opposite orientation are both leading to zero orientation weight, so struts are again assumed to be symmetrical.

The A* algorithm requires an estimate $h$ of the cost of going from the node it currently works on $(p)$ to the goal node. This estimate must be lower than the actual cost of reaching the goal in order to guarantee that A* will find the optimal path. In this application $h$ can be defined using the weight function defined above:

$$h(p) = w(p, \text{goal}) \tag{4.37}$$

The first path found by the A* algorithm is very obvious. It consists of the one edge connecting the start and goal nodes directly. This result is passed to the potential field stage. So before trying any intermediate steps, the algorithm will always give the potential field method a chance to find a path by itself. If it fails (after trying all four possibilities), this edge is eliminated from the graph and A* is applied to the rest of the graph. The result will now include at least two edges; they are both passed to the potential field stage. As soon as one of the edges leads to a failure, this edge is removed from the graph and a new path is generated by A*. This procedure is repeated until the potential field method is successful on all edges of a path or until no path connecting the start and the goal node is left. In the latter case, all resources are exhausted and the program reports a failure to the supervisor level.

### Selecting Subgoals for the Truss Structure Problem

The generation of subgoals is a very important step in every global path planner. There should not be too many intermediate steps, since the number of paths grows exponentially with the number of nodes in the graph. The intermediate steps should also be placed in positions where they can help to guide the potential field algorithm around difficult areas. So examples of good intermediate steps are corners of obstacles.

The approach taken for the specific case of truss structures is simple. Since the tetrahedron is the unit of all larger structures, all tetrahedra are extracted from the list of struts in the environment. When five of the six struts of a tetrahedron are present, then the structure is recognized and treated as a tetrahedron. Then an intermediate step is placed along every edge of every tetrahedron, as shown in Figure 4.18 for one tetrahedron.

This approach could be improved in the following two ways:

- Generate an approximate model of the robot's workspace and omit intermediate steps outside the workspace. If an inverse kinematics procedure

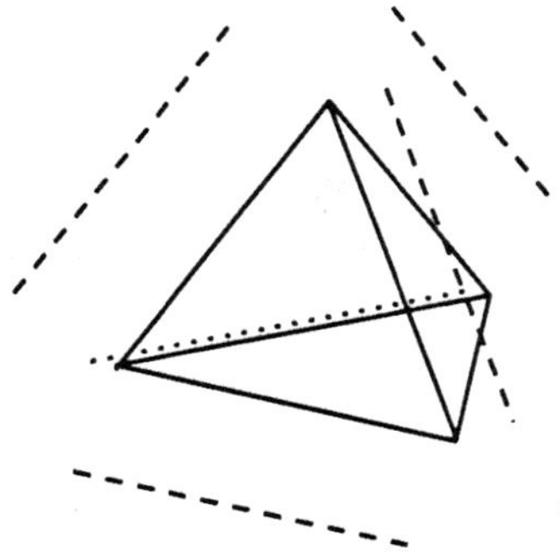

Figure 4.18: Intermediate Steps Placed Around a Tetrahedron

is available, apply it on every intermediate step. If the inverse kinematics has no solution, reject it.

- Find the convex hull of the existing structure and accept only intermediate steps outside the convex hull.

While this approach is described here for specific structural models, analogous methods may be used for many common polyhedral models of objects.

### 4.3.6 Results

The proposed algorithm has been implemented in the C programming language under the UNIX operating system. Its performance has been tested in various simulations.

A model of the robot used for all simulations is shown in Figure 4.19. It is a PUMA arm with 6-DOFs, mounted on a 3-DOF platform. Eight of the nine joints are revolute, the first joint is prismatic. The axes of all joints are shown in the figure.

Figure 4.20 shows the trivial path planning task of moving a strut from an initial to a goal position. The same scene is shown from different points of view in order to help visualization of the three dimensional model. The path is displayed by showing the position of the strut in the robot's gripper at every iteration. In this simulation, the direction of start is along the z axis and the direction of approach is against the direction of the z axis. It can be seen that the program automatically generates an intermediate goal above the true goal position in order to approach the goal position in the specified direction. The seemingly unnecessary rotation during the first few iterations is due to the fact that the program starts rotation towards the goal orientation only after the distance of the end-effector to its goal position has decreased under

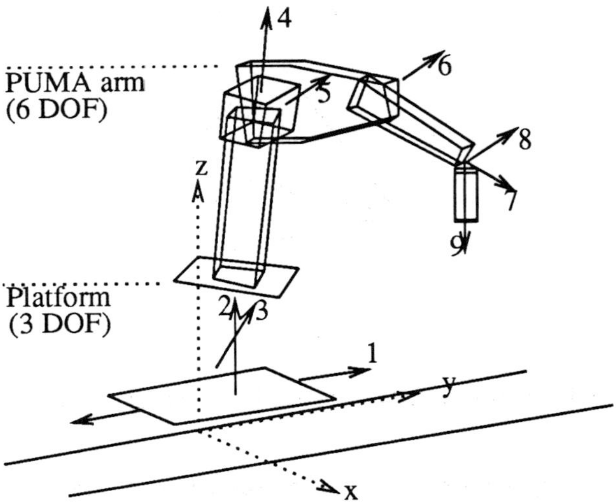


Figure 4.19: A Robot Arm of the CIRSSE Testbed

a critical value. Before this phase, the end-effector orientation is dictated by collision avoidance and joint range requirements. In this case, the purpose of the rotation is to move the joints in the wrist towards the middle of their range.

Figure 4.21 shows the insertion of the last strut into a tetrahedron. This task involves avoiding a collision with the partially completed tetrahedron. Note that the strut is touching the tetrahedron in its goal position. This is possible because all repulsive forces are reduced as the goal position and orientation are approached. The repulsive forces are reduced to zero when

- the strut's orientation is correct,
- the end-effector's distance to its goal position is small enough, and
- the end-effector has reached a position such that a motion along the specified approach direction will lead to the goal position.

Figure 4.22 shows the initial situation of a more difficult problem. The robot must move the strut to the marked goal position approaching it in the x direction.

The result is shown in Figure 4.23.

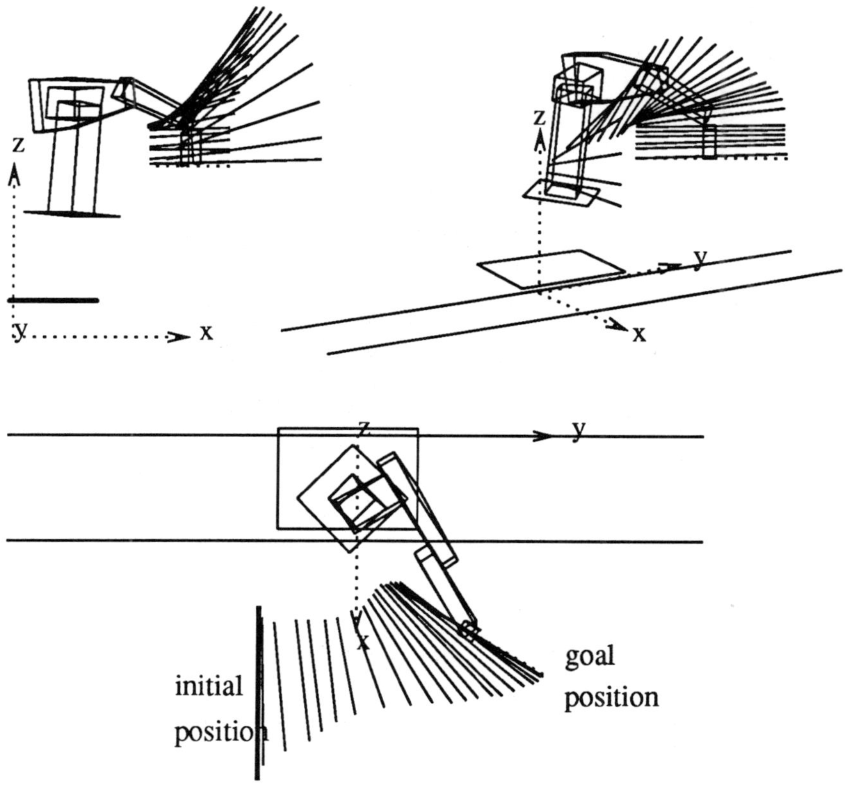


Figure 4.20: A Simple Motion Task

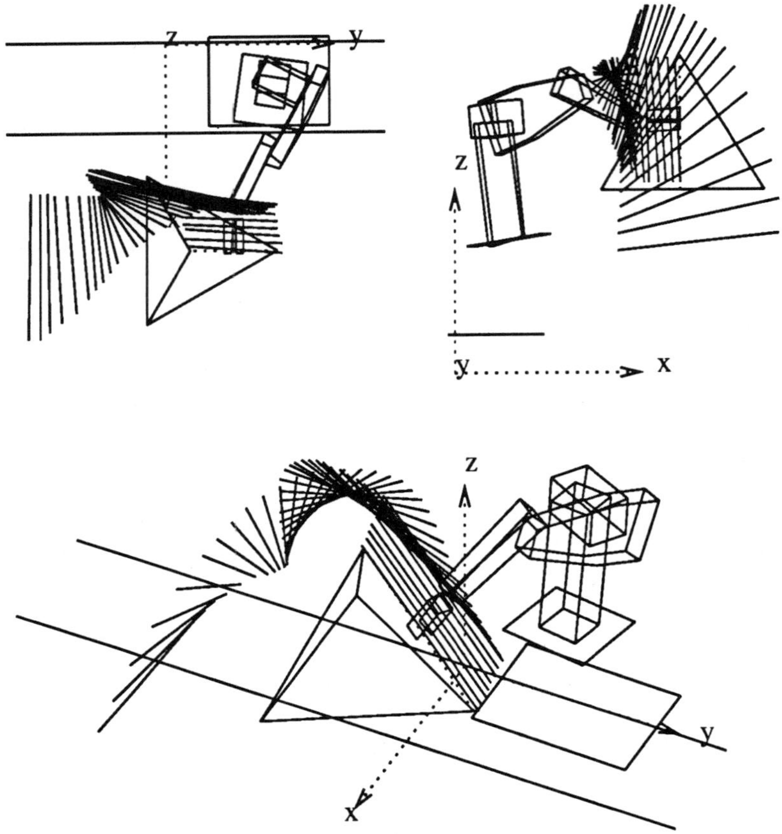


Figure 4.21: Completing a Tetrahedron

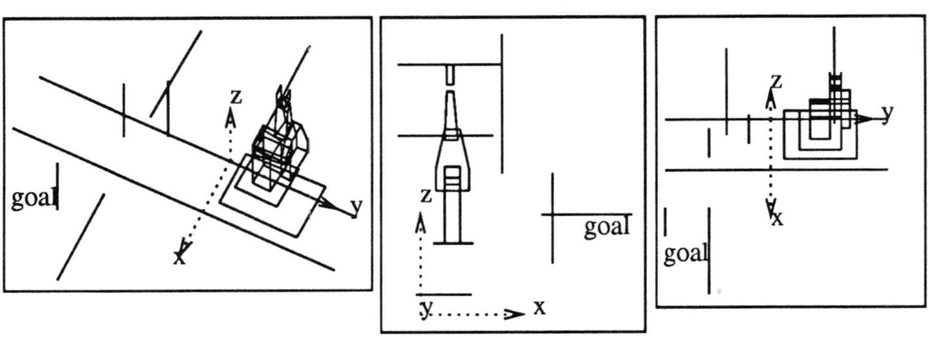


Figure 4.22: Initial State of Obstacle Avoidance Demonstration

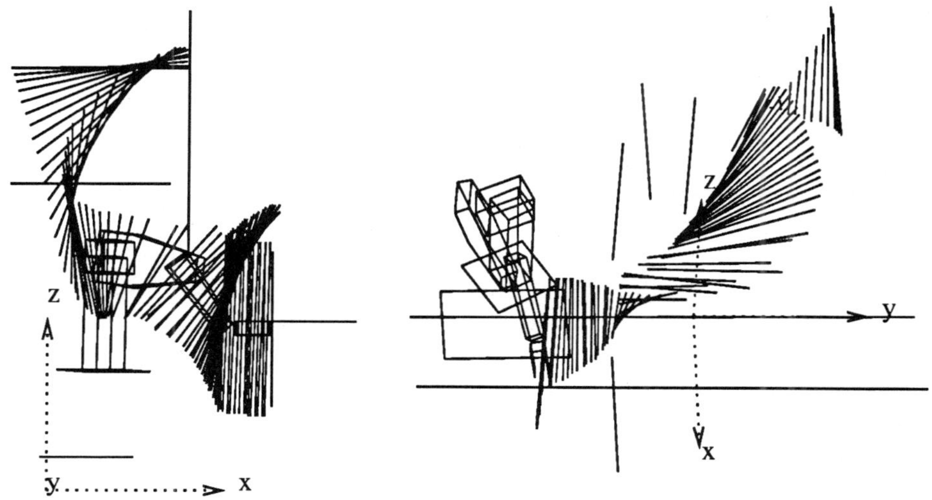


Figure 4.23: Obstacle Avoidance Demonstration

### 4.3.7 Discussion on Path Planning

#### Computational Complexity

In this section, the complexity of the path planning algorithm is discussed. Due to their different structure, the local and the global stage are discussed separately.

The local path planner performs an iteration in three steps. It first finds the attractive joint vector, which involves finding a solution of the Jacobian equation as the limiting algorithm. The Jacobian is a $6 \times n$ matrix, where $n$ is the number of DOFs of the manipulator. The algorithm involves a transformation of this redundant system to a nonredundant system with a $6 \times 6$ system matrix. Thus, the actual elimination algorithm takes a constant computing time per iteration, but the transformation of the nonlinear system involves some matrix multiplications; their complexity is $O(n)$.

The second step in performing an iteration is the computation of the self-collision avoidance vector $q_{self}$. If equation (4.23) is considered and the computing time for $f_q$ is assumed to be constant, then the complexity of $q_{self}$ is $O(n^3)$.

The third step is the computation of the joint vector $q_{env}$. Avoiding collisions between links of the robot arm and obstacles in the environment has a complexity of $O(n^2 m)$, where $m$ is the number of obstacles in the environment. This can be seen from equation (4.22). These three steps are the limiting parts of the algorithm for all cases that have been examined.

Another step is the computation of the joint range vector $\Delta q_{rr}$, which is linear in $n$. However, the computing time for this step is negligible for all $n$.

The $n$-DOFs of the manipulator is not a critical parameter since it will never be very large. A much more critical parameter for the local path planner is the number of obstacles $m$, even though computing time grows only linearly with $m$. A possible solution to this problem is the implementation of a quick distance check between the manipulator and a given obstacle. If this distance was large enough, then this obstacle would be ignored in the collision avoidance procedure. Another method to reduce computing time is the exclusion of pairs of objects from the collision avoidance procedure that cannot possibly collide. This is typically the case for neighboring links and for collisions between very small links and obstacles. The small links can normally be included in the model representing a neighboring link. This method has proven to be very effective and has been implemented in the current software.

The global path planning stage is based on the A* algorithm, which is a very efficient graph search algorithm, but it still doesn't solve the search problem in polynomial time. So care must be taken to keep the number of subgoals as low as possible. However, in all but very large cases the local path planning stage is computationally much more expensive than the graph search.

If the local path planner fails to find a path between two given subgoals, then the corresponding edge will be deleted from the graph and the path planning algorithm will be restarted. If we assume that the number of subgoals is roughly proportional to the number of obstacles, then the number of edges in the graph grows quadratically with the number of obstacles, since every node in the graph is connected to every other node. This leads to a complexity of the backtracking algorithm of $O(m^2)$ in the number of obstacles.

### Generalization of the Geometric Model

The current implementation of the path planner uses the swept-sphere-cylindrical model to represent objects. However, every geometric model can be used, since the path planning algorithm only requires the computation of a distance vector between any two models. It is also required that the volume of the actual object is completely enclosed in the volume of the model representing it.

### Experience with the Potential Field Algorithm

Potential field methods have the advantage that various soft constraints can be incorporated by adding up all components of the potential field. However, every component of the field has a constant factor associated with it and these constants must be carefully balanced in order to obtain good results. This is particularly important for the attractive field and the obstacle avoidance field. If the obstacle avoidance field is too strong compared to the attraction field, then a narrow but feasible path between two obstacles may not be found. If the obstacle avoidance field is chosen too weak, then collision avoidance cannot be guaranteed if the step size is large.

The step size is defined by the norm of $\Delta q$, the incremental joint vector. If this step size is chosen small, then the local path planning process will be slow, since the computation time for one iteration is constant in a given environment. But if the step size is chosen too large, then problems occur when approaching an obstacle. One step might take the robot from a position with sufficient clearance to a position very close to the obstacle or even to a collision. A position very close to an obstacle leads to a strong repulsive force that drives the robot back out of the field's immediate proximity, which leads to oscillations in the path, as shown in Figure 4.24.

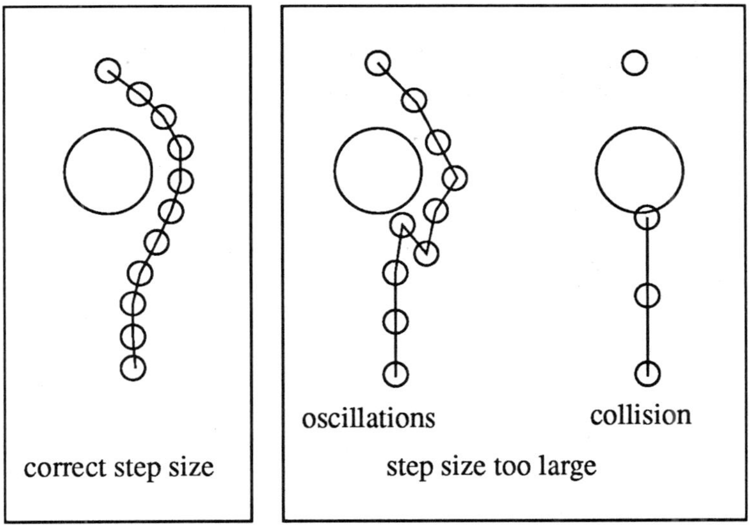


Figure 4.24: Step Size Selection

A related problem is the choice of the range of the repulsive force (equations (4.18) and (4.19)). If the range is very small, then there will be a sudden increase in the repulsive force as the robot approaches an obstacle. Then oscillations may occur even if the step size is small. On the other hand, if the range is chosen too large, then the repulsive field of a cluster of several obstacles accumulates to an excessively strong field, which leads to a path with an unnecessarily large clearance to the cluster. If any part of the path lies in the inside of a truss structure, the repulsive field range must be carefully chosen. In these cases, it may be necessary to adapt the step size and the repulsive field range depending on the proximity of obstacles. This has not been tested in the current implementation.

The last parameter discussed in this context is the function that keeps the joints within their boundaries (see Figure 4.14). If this function permits the joints to run very close to their boundaries and the step size is large, then the joints may run out of range. However, if the function keeps the joints far off their boundaries, then the workspace of the robot is restricted unnecessarily.

### 4.3.8 Configurations and Singularities

Another component to the potential field could be a field that prevents the robot from moving close to singularities. The potential field would be a function of $p = \det^2(\mathbf{J}\mathbf{J}^\mathbf{T})$. If $p = 0$ then the robot is in a singular position. The distance from singularities is related to the dexterity of the manipulator [40] in the sense that the robot is more dexterous if its pose is not close to any singular position. Another advantage of avoiding singular positions is that the Jacobian equation is guaranteed to have a solution. Apart from being very difficult to implement in an efficient manner, singularity avoidance has another drawback; it prevents the robot from changing configuration.

A nonredundant robot normally has several joint angles in which it reaches a given end-effector position and orientation. Every joint angle corresponds to a configuration of the robot. A redundant robot has, in general, infinitely many joint angles for a given end-effector position and orientation, but these joint angles lie on a number of self-motion manifolds in joint space that correspond to the configurations of the nonredundant robot. The robot can only change these configuration manifolds by passing through a singularity. Now assume the robot is in a given initial position that lies in one of the configuration manifolds. It may not be possible to reach a goal position within this manifold due to joint range restrictions. Then the only way to reach the goal is to change the configuration manifold, so it is necessary to pass through a singularity. Our algorithm does not force the robot to explore all configuration manifolds as part of the search process, but it also does not actively prohibit the robot from changing manifolds by driving it away from singularities.

## Acknowledgment

This work was supported by the NASA Center for Intelligent Robotic Systems for Space Exploration at Rensselaer Polytechnic Institute under NASA Grant NAGW-1333. Special thanks go to Dr. Steven Murphy and Michael Eppinger for helpful discussions.

## 4.4 References

[1] L. S. Homem de Mello and A. Sanderson, "An AND/OR graph representation of assembly plans," in *AAAI-86 Proceedings of the Fifth National*

*Conference on Artificial Intelligence*, pp. 1113–1119, 1986.

[2] L. S. Homem de Mello and A. Sanderson, "AND/OR graph representation of assembly plans," *IEEE Trans. on Robotics and Automation*, vol. 6, no. 2, pp. 188–199, 1990.

[3] A. Sanderson, M. Peshkin, L. S. Homem de Mello, and H. Zhang, "Assembly sequence planning," *AI Magazine: Special Issue on Assembly Planning*, vol. 11, no. 1, pp. 62–81, 1990.

[4] L. S. Homem de Mello and A. Sanderson, "A correct and complete algorithm for the generation of mechanical assembly sequences," *IEEE Trans. on Robotics and Automation*, vol. 7, no. 2, pp. 228–240, 1991.

[5] L. S. Homem de Mello and A. C. Sanderson, "Representations of mechanical assembly sequences," *IEEE Trans. on Robotics and Automation*, vol. 7, no. 2, pp. 211–227, 1991.

[6] L. S. Homem de Mello and A. Sanderson, "Evaluation and selection of assembly plans," in *Proc. 1990 International Conference on Robotics and Automation*, pp. 1588–1593, 1990.

[7] R. R. Cohen and E. Feigenbaum, *Planning and Problem Solving*, vol. 3. North Holland, 1982.

[8] R. E. Fikes and N. J. Nilsson, "STRIPS: A new approach to the application of theorem proving to problem solving," *Artificial Intelligence*, vol. 2, pp. 189–208, 1971.

[9] E. D. Sacerdoti, *A Structure for Plans and Behavior.* American Elsevier, 1977.

[10] M. J. Stefik, "Planning with contraints," doctoral dissertation, Stanford Univeristy, 1980.

[11] E. D. Sacerdoti, "Planning in a hierarchy of abstraction spaces," *Artificial Intelligence*, vol. 5, pp. 115–135, 1974.

[12] D. E. Wilkins, *Practical Planning.* San Mateo, CA: Morgan Kaufmann, 1988.

[13] H. Zhang, "Generation of precedence relations for mechanical assemblies," tech. rep., Rensselaer Polytechnic Institute, Troy, New York, August 1989.

[14] F. P. Preparata and S. J. Hong, *Convex Hulls of Finite Sets of Points in Two and Three Dimensions*, vol. 20 of *197.* Communications of the ACM.

[15] G. T. Toussaint, *Computational Geometry.* North Holland, 1985.

[16] R. E. Korf, "Planning as search: A quantitative approach," *Artificial Intelligence*, vol. 33, pp. 65–88, 1987.

[17] M. D. Rhodes, R. W. Will, and M. A. Wise, "A telerobotic system for automated assembly of large space structures," NASA Technical Memorandum 101518, Langley Research Center, Virginia, May 1989.

[18] S. Jun and K. G. Shin, "A probabilistic approach to collision-free robot path planning," in *IEEE 1988 International Conference on Robotics and Automation*, 1988.

[19] B. Paden, A. Mees, and M. Fisher, "Path planning using a Jacobian-based freespace generation algorithm," in *IEEE 1989 International Conference on Robotics and Automation*, vol. 3, pp. 1732–1737, 1989.

[20] E. Rich, *Artificial Intelligence.* Artificial Intelligence, McGraw-Hill, 1983.

[21] R. F. Richbourg, N. C. Rowe, M. J. Zyda, and R. B. McGhee, "Solving global, two-dimensional routing problems using Snell's law and A* search," in *IEEE 1987 International Conference on Robotics and Automation*, vol. 3, pp. 1631–1636, 1987.

[22] Y. Kanayama, "Least cost paths with algebraic cost functions," in *IEEE 1988 International Conference on Robotics and Automation*, vol. 1, pp. 75–80, 1988.

[23] O. Khatib, "Real-time obstacle avoidance for manipulators and mobile robots," *International Journal of Robotics Research*, vol. 5, no. 1, pp. 90–98, 1986.

[24] P. Khosla and R. Volpe, "Superquadric artificial potentials for obstacle avoidance and approach," in *IEEE International Conference on Robotics and Automation*, vol. 3,, pp. 1778–1784, 1988.

[25] R. Volpe and P. Khosla, "Artificial potentials with elliptical isopotential contours for obstacle avoidance," in *IEEE Proceedings of the 26th Conference on Decision and Control*, vol. 1, pp. 180–185, December 1987.

[26] B. Faverjon and P. Fournassoud, "A local-based approach for path planning of manipulators with a high number of degrees-of-freedom," in *IEEE 1987 International Conference on Robotics and Automation*, vol. 2, pp. 1152–1159, 1987.

[27] S. Kheradpir and J. S. Thorp, "Robust real time control of robot manipulators in the presence of obstacles," in *IEEE International Conference on Robotics and Automation*, vol. 2, pp. 1146–1151, 1987.

[28] E. Rimon and D. E. Koditschek, "Exact robot navigation using cost functions: The case of distinct spherical boundaries in $E^n$," in *IEEE 1988 International Conference on Robotics and Automation*, vol. 3, pp. 1791–1796, 1988.

[29] T. L. Perez and M. A. Wesley, "An algorithm for planning collision-free paths among polyhedral obstacles," Tech. Rep. 10, Communications of the ACM, October 1979.

[30] T. L. Perez, "Spacial planning: A configuration space approach," *IEEE Transactions on Computers*, vol. C-32, pp. 108–120, February 1983.

[31] F. Avnaim, J. D. Boissonnat, and B. Faverjon, "A practical exact motion planning algorithm for polygonal objects amidst polygonal obstacles," in *IEEE 1988 International Conference on Robotics and Automation*, vol. 3, pp. 1656–1661, 1988.

[32] W. Meyer and P. Benedict, "Path planning and the geometry of joint space obstacles," in *IEEE 1988 International Conference on Robotics and Automation*, vol. 1, pp. 215–219, 1988.

[33] S. L. Campbell and J. C. D. Meyer, *Generalized Inverses of Linear Transformations*. London, San Francisco, Melbourne: Pitman, 1979.

[34] J. Angeles, "On the numerical solution of the inverse kinematic problem," *International Journal of Robotics Research*, vol. 4, no. 2, pp. 21–37, 1985.

[35] J. Angeles, K. Anderson, X. Cyril, and B. Chen, "The kinematic inversion of robotic manipulators in the presence of singularities," *Transactions of the ASME*, vol. 110, pp. 246–254, September 1988.

[36] K. Anderson and J. Angeles, "Kinematic inversion of robotic manipulators in the presence of redundancies," *International Journal of Robotics Research*, vol. 8, pp. 80–97, December 1989.

[37] M. S. Konstantinov, S. P. Patarinski, V. B. Zamanov, and D. N. Nenchev, "A contribution to the inverse kinematic problem for industrial robots," in *12th International Symposium on Industrial Robots*, pp. 459–465, 1982.

[38] V. J. Lumelsky, "Iterative coordinate transformation procedure for one class of robots," *IEEE Transactions on Systems, Man, and Cybernetics*, vol. SMC-14, pp. 500–507, May/June 1984.

[39] A. A. Goldenberg, B. Benhabib, and R. G. Fenton, "A complete generalized solution to the inverse kinematics of robots," *IEEE Journal of Robotics and Automation*, vol. RA-1, pp. 14–19, March 1985.

[40] C. A. Klein and B. E. Blaho, "Dexterity measures for the design and control of kinematically redundant manipulators," *International Journal of Robotics Research*, vol. 6, pp. 72–83, Summer 1987.

# Chapter 5

# Three-Dimensional Vision

by: J. Russell Noseworthy
Arthur M. Ryan
Lester A. Gerhardt
Electrical, Computer, and Systems Engineering Department
Center for Intelligent Robotic Systems for Space Exploration
Rensselaer Polytechnic Institute
Troy, New York 12180-3590

Three-dimensional vision refers to the process of gathering, processing, and interpreting visual data of a three-dimensional environment. A 3-D vision system may employ either passive or active techniques. A passive technique is one that utilizes available light sources (i.e., those necessary to provide general illumination of the viewed environment), whereas an active technique utilizes the projection of a prestructured light pattern to supply "ground truth" information. This chapter describes the calibration of a fixed camera and a fixed laser scanner. Three-dimensional point estimation methods are then discussed. Finally, a working 3-D vision system used for robotic assembly is described.

## 5.1 Passive Techniques

Using only one camera with passive techniques, it is not possible to determine the 3-D position of a point in space since the projection of the 3-D point onto a 2-D image plane results in a loss of information. However, it is possible to determine the 3-D ray on which the point is constrained to lie. This requires accurate knowledge of the camera's *pose* (position and orientation) relative to some fixed frame of reference (world coordinates). Also required is the transformation the point undergoes when it is projected onto the 2-D image plane of the camera and into 2-D computer pixel coordinates. The process of obtaining the transformation from 3-D world coordinates to 2-D computer pixel coordinates is called *camera calibration*. Camera calibration is essential to 3-D vision and much of this chapter will focus upon this difficult problem (see Section 5.3).

With two calibrated cameras, it becomes possible to estimate a point's 3-D position. Two estimation methods are discussed in this chapter. Both methods pre-suppose knowledge of the 2-D position in each camera's image plane of the 3-D point. In general, the 2-D image plane coordinates (and hence computer pixel coordinates) of the same 3-D point will be different for each camera. The problem of determining the computer pixel coordinates of a particular 3-D point in more than one image is called the *image correspondence problem* or the *point correspondence problem*.

With passive techniques, the image correspondence problem is solved by matching corresponding image events from each camera. This method requires the development of criteria for matching image events from each camera. These criteria are typically based on intensity discontinuities in the images. Compared to active methods, passive methods to obtain image correspondence are generally more computationally expensive and less accurate. However passive methods do offer the ability to generate sparse depth maps of the viewed scene. An excellent review of various 3-D vision techniques is given by Jarvis [1].

## 5.2 Active Techniques

One of the most basic active techniques involves the use of a single calibrated camera and a laser beam. If the laser beam can be calibrated so that the exact position and orientation of its light ray are known with respect to the world coordinate frame of the calibrated camera, the light ray from the laser replaces the ray passively determined from a second calibrated camera. In this case, the image correspondence problem is not an issue since there is only one set of 2-D computer pixel coordinates to be found. These 2-D coordinates correspond to the endpoint of the laser beam (the laser spot) and are easily determined by guaranteeing that the laser spot is the brightest event (or, some event of a unique known brightness) in the camera's field of view (see Section 5.4.5 for a more detailed discussion of this).

The above technique can be extended using *structured lighting*. The simplest structured light techniques use a single calibrated camera and a light source that is capable of producing a known pattern of light such as a line, a circle, or a grid. When an object intersects the light pattern, the pattern is deformed. It is then possible to extract 3-D information from the 2-D image provided by the camera, based on the amount of deformation detected.

With two calibrated cameras (or one calibrated camera moved to two locations), the image correspondence problem can be solved using a laser beam that need not be calibrated. By guaranteeing the laser spot is the brightest spot viewed by both cameras, the image correspondence problem is reduced to the simplified problem of finding the 2-D coordinates of the brightest spot in each computer image. Thus, with two calibrated cameras and a laser to provide point correspondence, the 3-D location of any point in the field of view is found by pointing the laser at it.

This type of 3-D vision becomes redundant if the laser beam can be calibrated, since this provides a third ray on which the point is constrained to lie. If the laser beam can provide a structured light pattern, additional redundancy is created by using both cameras to extract 3-D information from deformation of the structured light pattern intersected by an object. The system described in Section 5.6 can use five calibrated cameras and a laser beam capable of projecting a simple spot or a complex structured light pattern to extract 3-D information. Thus, this system possesses redundancies additional to those described above.

Using passive or active techniques it is possible to obtain 3-D information from combinations of the five cameras taken two at a time. It is also possible to obtain 3-D information from combinations of the five cameras taken two or *more* at a time. This assumes that all cameras used have the 3-D point in question in their field of view. Two methods for estimating the location of the point using more than one camera are discussed in Section 5.5.

The ability to integrate multiple cameras is important to minimize inaccuracies from camera calibration and mechanical vibrations. Multiple cameras may provide information that would not otherwise be available to only two cameras. Also, a multiple camera vision system is capable of continued operation in spite of the failure of one or more of the cameras in the system.

## 5.3 Calibration of Fixed Cameras

As mentioned in Section 5.1, camera calibration is the process of determining the transformation from 3-D world coordinates to 2-D computer pixel coordinates. Using a pin-hole camera model, any 3-D point in the field of view passes through the center of projection of the camera and is mapped to a 2-D point on the camera's image plane. This 2-D image plane point is then mapped into the 2-D computer pixel coordinates. Some calibration methods are based on the pin-hole camera model [2], [3]. However, a more sophisticated method developed by Tsai [4], [5], [6], [7], [8], [9], [10], extends the simplified pin-hole model to account for radial lens distortion.

The key to getting accurate results with any particular camera calibration lies in the accurate extraction of calibration data points. With very accurate data points, any reasonable calibration technique should be able to return very accurate estimates of the camera parameters. For example, Tsai's camera calibration is very fast and quite accurate when given very accurate data points.

Conversely, when given noisy data points, the ability of Tsai's calibration technique to accurately estimate the camera parameters degrades quickly. In particular, the camera rotation matrix returned with his method (see equation (5.1)) is not orthonormal.

In certain applications (e.g., metrology and inspection) it is reasonable to expect accurate determination of the 3-D location of calibration data points. However, in the application of interest to the authors (general robotic assembly problems) it is common that the robots for which the 3-D vision is being provided are unable to determine the position of data points with great accuracy (see Section 5.3.4). Fortunately in this application, the calibration is done infrequently so the calibration process need not be extremely fast.

In the remainder of this section, the modifications to Tsai's calibration technique to make it more robust to noisy data points are described. Several similar techniques will be presented. By their nature, these techniques can be numerically sensitive, and the choice of the "optimal" technique is further complicated by the amount and type of noise in the data points (which is often unknown). Consequently, the techniques are presented with the intent that the reader experiment to find the technique most suitable for his or her application. For the reader's convenience, Tsai's calibration technique will be summarized here.

### 5.3.1 Summary of Tsai's Calibration Technique for Fixed Cameras

The transformation from 3-D world coordinates to 2-D computer pixel coordinates is divided into four steps. The first step involves the *extrinsic parameters* of the camera. Extrinsic parameters are those parameters that only change when the camera is physically moved. They describe the transformation from some arbitrary 3-D world coordinate frame to the 3-D camera coordinate frame at the lens center of the camera, with the $z$ axis coincident with the optical axis.

From Figure 5.1 we have

$$\begin{bmatrix} x \\ y \\ z \end{bmatrix} = \begin{bmatrix} r_1 & r_2 & r_3 \\ r_4 & r_5 & r_6 \\ r_7 & r_8 & r_9 \end{bmatrix} \begin{bmatrix} x_w \\ y_w \\ z_w \end{bmatrix} + \begin{bmatrix} t_x \\ t_y \\ t_z \end{bmatrix} = R \begin{bmatrix} x_w \\ y_w \\ z_w \end{bmatrix} + T \qquad (5.1)$$

where $(x, y, z)$ are the 3-D camera coordinates, $(x_w, y_w, z_w)$ are the 3-D world coordinates, T is a $3 \times 1$ translation vector, and R is a $3 \times 3$ orthonormal rotation matrix of the first kind (i.e., the determinant is positive one, the magnitude of any row or column is one, and all rows and all columns are orthogonal, $RR^T = R^T R = I$).

The next three steps define the camera's *intrinsic parameters*. The intrinsic parameters of the camera (and lens) do not change when the camera is moved. Step two is the transformation from the 3-D camera coordinates to the 2-D

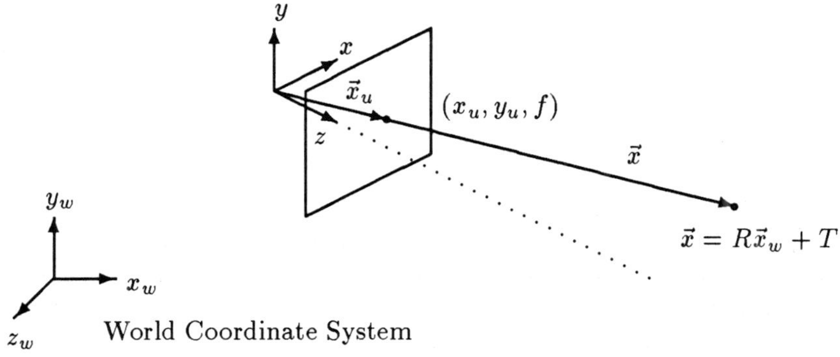


Figure 5.1: Camera Calibration Coordinate Frame Definitions

camera image plane coordinates assuming a pin-hole camera model (i.e., without considering radial lens distortion) as derived in equation (5.119).

$$x_u = \frac{f}{z}x \tag{5.2}$$

$$y_u = \frac{f}{z}y \tag{5.3}$$

where $(x, y, z)$ are the 3-D camera coordinates, $(x_u, y_u)$ are the undistorted image plane coordinates, and $f$ is the effective focal length of the camera's lens.

The third step accounts for the radial distortion of the lens. The formulation given below is taken from Tsai's earlier work [4], [5], [6]. In his later work [7], [8], Tsai presents a different expression for the radial lens distortion. The alternate formulation will be discussed in Section 5.3.3. The undistorted (ideal) image plane coordinates $(x_u, y_u)$ are given by

$$x_d(r_d^2\kappa + 1) = x_u \tag{5.4}$$

$$y_d(r_d^2\kappa + 1) = y_u \tag{5.5}$$

where $(x_d, y_d)$ are the distorted (actual) image plane coordinates, $\kappa$ is the radial lens distortion coefficient, and $r_d$ is given as

$$r_d = \sqrt{x_d^2 + y_d^2} \tag{5.6}$$

The fourth step accounts for the translation from the actual image plane coordinates $(x_d, y_d)$ to the computer pixel coordinates $(x_f, y_f)$ (i.e., pixel row and column numbers),

$$x_f = C_x + \frac{s_x}{d_x}\frac{N_{fx}}{N_{cx}}x_d \tag{5.7}$$

$$y_f = C_y + \frac{1}{d_y}y_d \tag{5.8}$$

where $(C_x, C_y)$ are the computer pixel coordinates for the origin of the image plane, $d_x$ is the center-to-center distance between adjacent sensor elements in the $x$ (scan line) direction, $d_y$ is the center-to-center distance between adjacent sensor elements in the $y$ direction, $N_{cx}$ is the number of sensor elements in the $x$ direction, $N_{fx}$ is the number of pixels in a line as sampled by the computer, and $s_x$ is the uncertainty image scale factor. Note that $N_{cx}$, $N_{fx}$, $d_x$, and $d_y$ are obtained from manufacturer's specifications of the camera and the image frame memory.

Given the above relationships, and assuming $(C_x, C_y)$ to be known, the parameters which need to be calibrated are: R, T, $f$, $\kappa$, and $s_x$. The non-coplanar calibration method requires that many more than 7 non-coplanar data points be gathered. These data points consist of the world coordinates of the points $(x_{w_i}, y_{w_i}, z_{w_i})$ associated with the corresponding computer pixel coordinates $(x_{f_i}, y_{f_i})$.

### Stage 1 of Tsai's Calibration Technique

The first stage of the calibration calculates the rotation matrix, the first two components of the translation vector, and the image scale factor (but assumes $(C_x, C_y)$ is known).

Once the calibration points have been collected, the distorted image coordinates of every calibration point (see equation (5.4) and (5.5)) must be found from

$$s_x x_{d_i} = (x_{f_i} - C_x)d_x\frac{N_{cx}}{N_{fx}} \tag{5.9}$$

$$y_{d_i} = (y_{f_i} - C_y)d_y \tag{5.10}$$

Note that the uncertainty image scale factor, $s_x$, in (5.9) is not yet known, so $x_{d_i}$ cannot be calculated at this point. Tsai showed that $y_d x = x_d y$ and assumed $s_x$ in (5.9) to be 1; however, the authors have removed this assumption using (5.13).

Now, the following overdetermined system of linear equations must be solved for the $n \gg 7$ data points

$$\begin{bmatrix} y_{d_1}x_{w_1} & y_{d_1}y_{w_1} & y_{d_1}z_{w_1} & y_{d_1} & -s_x x_{d_1}x_{w_1} & -s_x x_{d_1}y_{w_1} & -s_x x_{d_1}z_{w_1} \\ y_{d_2}x_{w_2} & y_{d_2}y_{w_2} & y_{d_2}z_{w_2} & y_{d_2} & -s_x x_{d_2}x_{w_2} & -s_x x_{d_2}y_{w_2} & -s_x x_{d_2}z_{w_2} \\ \vdots & \vdots & \vdots & \vdots & \vdots & \vdots & \vdots \\ y_{d_n}x_{w_n} & y_{d_n}y_{w_n} & y_{d_n}z_{w_n} & y_{d_n} & -s_x x_{d_n}x_{w_n} & -s_x x_{d_n}y_{w_n} & -s_x x_{d_n}z_{w_n} \end{bmatrix}.$$

$$\begin{bmatrix} \frac{s_x r_1}{t_y} \\ \frac{s_x r_2}{t_y} \\ \frac{s_x r_3}{t_y} \\ \frac{s_x t_x}{t_y} \\ \frac{r_4}{t_y} \\ \frac{r_5}{t_y} \\ \frac{r_6}{t_y} \end{bmatrix} = \begin{bmatrix} s_x x_{d_1} \\ s_x x_{d_2} \\ \vdots \\ s_x x_{d_n} \end{bmatrix} \quad (5.11)$$

which is based on

$$\frac{s_x y_d}{t_y}(x_w r_1 + y_w r_2 + z_w r_3 + t_x) = s_x x_d(1 + r_4\frac{x_w}{t_y} + r_5\frac{y_w}{t_y} + r_6\frac{z_w}{t_y}), \qquad t_y \neq 0 \quad (5.12)$$

or simply

$$s_x y_d x = s_x x_d y \quad (5.13)$$

The standard approach for solving an overdetermined system of linear equations is to use the technique of least squared error (LSE). LSE solves a system of linear equations of the form $Ax = b$ (where $A$ is $m \times n$, $x$ is $n \times 1$, $b$ is $m \times 1$, and $m > n$) by determining an $\hat{x}$ such that the quantity $(A\hat{x} - b)^T(A\hat{x} - b)$ is a minimum. This leads to the solution $\hat{x} = (A^TA)^{-1}A^Tb$, where $(A^TA)^{-1}$ is known as the pseudo-inverse of $A$. Note that $(A^TA)^{-1}$ only exists when the columns of $A$ are linearly independent, and furthermore, the computation of the pseudo-inverse quite often gives poor results due to numerical problems.

Consequently, the LSE problem is solved using *singular value decomposition* (SVD), which is the "method of choice for solving most linear least squares problems", [11]. SVD is based on the fact that $A = U \cdot W \cdot V^T$ for any $m \times n$ matrix $A$ ($m \geq n$), where $U$ is $m \times n$ and column-orthogonal, $W$ is a positive semi-definite $n \times n$ diagonal matrix, and $V$ is an $n \times n$ orthogonal matrix. Given this fact, the LSE solution of $Ax = b$ is given by

$$\hat{x} = V \cdot W^{-1} \cdot U^T \cdot b. \quad (5.14)$$

Note that the calculation of $W^{-1}$ is trivial and only degenerates when one (or more) of the diagonal elements (termed *singular values*) is zero. In fact, the calculation runs into numerical problems when one or more of the singular values is "small" compared to the largest singular value. The ratio of the largest to smallest singular value is called the *condition number* of the matrix, and when a matrix has a large condition number, it is termed *ill-conditioned.* SVD is able to determine the condition number of a matrix, and can often give a "useful numerical answer" in cases where the matrix is severely ill-conditioned. For greater insight into SVD, the reader is referred to [11].

Once (5.11) is solved, the magnitude of $t_y$ is given by

$$|t_y| = \frac{1}{\sqrt{(\frac{r_4}{t_y})^2 + (\frac{r_5}{t_y})^2 + (\frac{r_6}{t_y})^2}} \tag{5.15}$$

Similarly, $s_x$ may be found from the relation

$$s_x = |t_y| \sqrt{\left(\frac{s_x r_1}{t_y}\right)^2 + \left(\frac{s_x r_2}{t_y}\right)^2 + \left(\frac{s_x r_3}{t_y}\right)^2} \tag{5.16}$$

Once $s_x$ is determined from (5.16), $x_{d_i}$ may be calculated from

$$x_{d_i} = (x_{f_i} - C_x) \frac{d_x}{s_x} \frac{N_{cx}}{N_{fx}} \tag{5.17}$$

To determine the sign of $t_y$, a calibration point with world coordinates $(x_{w_i}, y_{w_i}, z_{w_i})$ is selected so that its pixel coordinates, $(x_{f_i}, y_{f_i})$, are far from the image center $(C_x, C_y)$. Then $t_y$ is assumed to be positive and the following calculations are made:

$$r_1 = \left(\frac{s_x r_1}{t_y}\right)\left(\frac{t_y}{s_x}\right) \tag{5.18}$$

$$r_2 = \left(\frac{s_x r_2}{t_y}\right)\left(\frac{t_y}{s_x}\right) \tag{5.19}$$

$$r_3 = \left(\frac{s_x r_3}{t_y}\right)\left(\frac{t_y}{s_x}\right) \tag{5.20}$$

$$r_4 = \left(\frac{r_4}{t_y}\right)(t_y) \tag{5.21}$$

$$r_5 = \left(\frac{r_5}{t_y}\right)(t_y) \tag{5.22}$$

$$r_6 = \left(\frac{r_6}{t_y}\right)(t_y) \tag{5.23}$$

$$t_x = \left(\frac{s_x t_x}{t_y}\right)\left(\frac{t_y}{s_x}\right) \tag{5.24}$$

$$x_i = r_1 x_{w_i} + r_2 y_{w_i} + r_3 z_{w_i} + t_x \tag{5.25}$$

$$y_i = r_4 x_{w_i} + r_5 y_{w_i} + r_6 z_{w_i} + t_y \tag{5.26}$$

If $x_i$ and $(x_{f_i} - C_x)$ have the same sign and $y_i$ and $(y_{f_i} - C_y)$ have the same sign, then $t_y$ is positive. If not, then $t_y$ is negative and the signs of $t_y$, $t_x$, $r_1$, $r_2$, $r_3$, $r_4$, $r_5$, and $r_6$ must be reversed.

The third row of $R$ is calculated from the cross product of the first two rows, to give

$$r_7 = r_2r_6 - r_3r_5 \tag{5.27}$$

$$r_8 = r_3r_4 - r_1r_6 \tag{5.28}$$

$$r_9 = r_1r_5 - r_2r_4 \tag{5.29}$$

## 5.3.2 Refining Stage 1 of Tsai's Calibration Technique

In (5.11), Tsai sets up an overdetermined system of linear equations to compute the first two rows of the rotation matrix (totalling six parameters). However, a rotation matrix, $R$, is fully determined by three Euler angles (i.e., $R$ represents three and not nine parameters) and so the six parameters calculated by Tsai in Stage 1 of the calibration process are *not independent*, but are instead subject to constraints.

If the calibration data are perfect, then these constraints are automatically satisfied. However, if the data were perfect then it would not be necessary to set up an *overdetermined* system of equations. Since the calibration data will always contain inaccuracies, an overdetermined system of equations must be used to solve (5.11). These inaccuracies give rise to errors in the calculation of the first two rows of the rotation matrix, since (5.11) does not constrain the rotation parameters. The result is that the first two rows do not precisely exhibit the properties of an orthonormal rotation matrix. This error then propagates through the calibration process as other parameters are calculated.

To rectify this, we may state the constraints on the first two rows of the rotation matrix defined in (5.1) as, [12],

$$r_1^2 + r_2^2 + r_3^2 = 1 \tag{5.30}$$

$$r_4^2 + r_5^2 + r_6^2 = 1 \tag{5.31}$$

$$r_1r_4 + r_2r_5 + r_3r_6 = 0 \tag{5.32}$$

Using the above constraint information, (5.11) may be reformulated as a constrained optimization problem. Let

$$
\begin{aligned}
A &= \left[\begin{array}{ccccccc}
y_{d_1}x_{w_1} & y_{d_1}y_{w_1} & y_{d_1}z_{w_1} & y_{d_1} & -s_x x_{d_1}x_{w_1} & -s_x x_{d_1}y_{w_1} & -s_x x_{d_1}z_{w_1} \\
y_{d_2}x_{w_2} & y_{d_2}y_{w_2} & y_{d_2}z_{w_2} & y_{d_2} & -s_x x_{d_2}x_{w_2} & -s_x x_{d_2}y_{w_2} & -s_x x_{d_2}z_{w_2} \\
\vdots & \vdots & \vdots & \vdots & \vdots & \vdots & \vdots \\
y_{d_n}x_{w_n} & y_{d_n}y_{w_n} & y_{d_n}z_{w_n} & y_{d_n} & -s_x x_{d_n}x_{w_n} & -s_x x_{d_n}y_{w_n} & -s_x x_{d_n}z_{w_n}
\end{array}\right] \\
x &= \left[\begin{array}{ccccccc} \frac{s_x r_1}{t_y} & \frac{s_x r_2}{t_y} & \frac{s_x r_3}{t_y} & \frac{s_x t_x}{t_y} & \frac{r_4}{t_y} & \frac{r_5}{t_y} & \frac{r_6}{t_y} \end{array}\right]^T \\
b' &= \left[\begin{array}{cccc} s_x x_{d_1} & s_x x_{d_2} & \cdots & s_x x_{d_n} \end{array}\right]^T
\end{aligned}
$$

where $A$ is $n \times 7$, $b'$ is $n \times 1$, and $n$ is the number of calibration data points $\gg 7$.

The problem now becomes:

$$
\begin{aligned}
\text{Find } x \text{ to minimize:} \quad & (Ax - b')^T(Ax - b') \\
\text{Subject to:} \quad & x_1^2 + x_2^2 + x_3^2 + c_1 = 0 \\
& x_5^2 + x_6^2 + x_7^2 + c_2 = 0 \\
& x_1x_5 + x_2x_6 + x_3x_7 = 0
\end{aligned} \tag{5.33}
$$

where $c_1 = -\left(\frac{s_x}{t_y}\right)^2$ and $c_2 = -\left(\frac{1}{t_y}\right)^2$.

Using the method of Lagrange multipliers, we form

$$
\begin{aligned}
L = {} & (Ax - b')^T(Ax - b') + \lambda_1(x_1^2 + x_2^2 + x_3^2 + c_1) + \lambda_2(x_5^2 + x_6^2 + x_7^2 + c_2) \\
& + \lambda_3(x_1x_5 + x_2x_6 + x_3x_7)
\end{aligned} \tag{5.34}
$$

$$
\nabla_x L = 2A^T(Ax - b') + \begin{bmatrix}
2\lambda_1 & 0 & 0 & 0 & \lambda_3 & 0 & 0 \\
0 & 2\lambda_1 & 0 & 0 & 0 & \lambda_3 & 0 \\
0 & 0 & 2\lambda_1 & 0 & 0 & 0 & \lambda_3 \\
0 & 0 & 0 & 0 & 0 & 0 & 0 \\
\lambda_3 & 0 & 0 & 0 & 2\lambda_2 & 0 & 0 \\
0 & \lambda_3 & 0 & 0 & 0 & 2\lambda_2 & 0 \\
0 & 0 & \lambda_3 & 0 & 0 & 0 & 2\lambda_2
\end{bmatrix} x = 0 \tag{5.35}
$$

Denoting $2A^T A$ as $[a_{ij}]$ $i, j \in [1, 7]$ and $-2A^T b'$ as $[b_i]$, $i \in [1, 7]$, we have:

$$\nabla_{\mathrm{x}}\mathrm{L} = 0 = \begin{bmatrix} b_1 \\ b_2 \\ b_3 \\ b_4 \\ b_5 \\ b_6 \\ b_7 \end{bmatrix} + \left[\begin{matrix} a_{11}+2\lambda_1 & a_{12} & a_{13} \\ a_{21} & a_{22}+2\lambda_1 & a_{23} \\ a_{31} & a_{32} & a_{33}+2\lambda_1 \\ a_{41} & a_{42} & a_{43} \\ a_{51}+\lambda_3 & a_{52} & a_{53} \\ a_{61} & a_{62}+\lambda_3 & a_{63} \\ a_{71} & a_{72} & a_{73}+\lambda_3 \end{matrix}\right.$$
$$\left.\begin{matrix} a_{14} & a_{15}+\lambda_3 & a_{16} & a_{17} \\ a_{24} & a_{25} & a_{26}+\lambda_3 & a_{27} \\ a_{34} & a_{35} & a_{36} & a_{37}+\lambda_3 \\ a_{44} & a_{45} & a_{46} & a_{47} \\ a_{54} & a_{55}+2\lambda_2 & a_{56} & a_{57} \\ a_{64} & a_{65} & a_{66}+2\lambda_2 & a_{67} \\ a_{74} & a_{75} & a_{76} & a_{77}+2\lambda_2 \end{matrix}\right] x \quad (5.36)$$

This has not been solved in closed-form.

### Stage 1 Solved Numerically as an Unconstrained Multi-Dimensional Nonlinear Optimization Problem

If the solution to (5.36) were known, then the optimal solution (in the least squares sense) would also be known for the parameters calculated in Stage 1 ($R$, $t_x$, $t_y$, and $s_x$) subject to the constraint that $R$ be an orthonormal rotation matrix of the first kind. It should be stressed that solving (5.11) and then *normalizing* the result to meet the constraints of the problem is *not* equivalent to solving for the stage 1 parameters *subject* to the constraints (the unconstrained problem can produce a result very different from the constrained problem).

Since the closed-form solution to (5.36) is not known, we are forced to use a numerical technique to perform Stage 1 of the calibration process. Numerical techniques to solve unconstrained nonlinear (and linear) optimization problems are well known, so (5.33) must be reformulated as an unconstrained nonlinear (or linear) optimization problem.

Attempting to reformulate (5.33), one might try expressing (5.13) (which leads to (5.33)) with a unit quaternion instead of with a rotation matrix. A quaternion is a method of representing a rotation matrix as a rotation of angle $\theta$ about a unit vector, [13] and is expressed by

$$\mathbf{q} = \left[\cos\left(\frac{\theta}{2}\right), \sin\left(\frac{\theta}{2}\right)\hat{V}\right] = [s, V_x, V_y, V_z] \quad (5.37)$$
$$\text{where} \quad s^2 + V_x^2 + V_y^2 + V_z^2 = 1 \quad (5.38)$$

The first two rows of the corresponding rotation matrix are given by

$$r_1 = 1 - 2V_y^2 - 2V_z^2 \quad (5.39)$$

$$
\begin{aligned}
r_2 &= 2(V_x V_y - sV_z) && (5.40)\\
r_3 &= 2(V_x V_z + sV_y) && (5.41)\\
r_4 &= 2(V_x V_y + sV_z) && (5.42)\\
r_5 &= 1 - 2V_x^2 - 2V_z^2 && (5.43)\\
r_6 &= 2(V_y V_z - sV_x) && (5.44)
\end{aligned}
$$

Using (5.39) through (5.44) to "simplify" (5.11) does not lead to an improved form since there is still a constraint (equation (5.38)) that must be satisfied. Therefore, the unknowns of the overdetermined system of linear equations are not independent. Thus, the quaternion formulation suffers from the same problem of overparameterization as the rotation matrix formulation.

Expressing the parameters of the rotation matrix in (5.11) in terms of three Euler angles casts (5.11) as an unconstrained multi-dimensional nonlinear optimization problem where the resulting rotation matrix is guaranteed to be orthonormal of the first kind. There are several Euler angle representations. We select the Pitch, Yaw, and Roll formulation corresponding to the following rotations in sequence [14]:

1. Pitch: A rotation of $\psi$ about the World $x$ axis
2. Yaw: A rotation of $\theta$ about the World $y$ axis
3. Roll: A rotation of $\phi$ about the World $z$ axis

where the names of the rotations have been shuffled to agree with the camera coordinate frame shown in Figure 5.1. The corresponding rotation matrix is given by

$$
R = \begin{bmatrix} \cos\phi\cos\theta & \cos\phi\sin\theta\sin\psi - \sin\phi\cos\psi & \cos\phi\sin\theta\cos\psi + \sin\phi\sin\psi \\ \sin\phi\cos\theta & \sin\phi\sin\theta\sin\psi + \cos\phi\cos\psi & \sin\phi\sin\theta\cos\psi - \cos\phi\sin\psi \\ -\sin\theta & \cos\theta\sin\psi & \cos\theta\cos\psi \end{bmatrix} \tag{5.45}
$$

It must be noted that the use of Euler angles to eliminate the overparameterization associated with other methods of representing rotations is not without a pitfall. The difficulty is in extracting the Euler angles (in this case Pitch-Yaw-Roll) from a known rotation matrix.

It is not difficult to see that $\theta$, $\phi$, and $\psi$ can be extracted from (5.45) in the following manner:

$$
\begin{aligned}
\theta &= \mathrm{Atan2}(-r_7, \sqrt{r_1^2 + r_4^2})\\
\phi &= \mathrm{Atan2}(r_4, r_1)\\
\psi &= \mathrm{Atan2}(r_8, r_9)
\end{aligned} \tag{5.46}
$$

However, as Craig [15] shows, when $\theta = \pm 90°$ (or equivalently when $r_7 = \mp 1$), (5.46) degenerates (since Atan2(0,0) is undefined) and only the sum or difference of $\phi$ and $\psi$ is unique. Craig recommends adopting the following convention in these cases:

$$\begin{array}{ll} \theta = +90^\circ & \theta = -90^\circ \\ \phi = 0^\circ & \phi = 0^\circ \\ \psi = +\text{Atan2}(r_2, r_5) & \psi = -\text{Atan2}(r_2, r_5) \end{array}$$

While theoretically the degeneracy only occurs when $\theta = \pm 90^\circ$, in practice as $\theta \rightarrow \pm 90^\circ$ (5.46) becomes increasingly sensitive to small variations in $\theta$.

One possible solution to this problem is to realize that while all Euler angle representations suffer from this type of problem, they do not all occur in the same orientation. Consequently, when $\theta$ is "near" $\pm 90^\circ$ a different Euler angle representation could be used, e.g., one whose singularities occur at orientations "far" from the current orientation. The definitions of "near" and "far" are not clear, but it normally should be possible to select from all Euler angle representations that representation whose singularities are farthest from the current orientation.

However, in order to "avoid" the problem, the authors propose the following. Since the orientation of the world coordinate frame is completely arbitrary, it is possible to select the world coordinate frame so that there is no rotation (or a very small rotation) between it and the camera coordinate frame. When the rotation is very small, the extraction of the Euler angles is quite accurate. The camera parameters can be determined with respect to this "aligned" world coordinate frame after which the extrinsic parameters can be adjusted to locate the camera with respect to any convenient world coordinate frame (e.g., the base of a robot).

The world coordinate frame should be selected so that there is no rotation between it and the camera coordinate frame. To fully understand this, a brief digression relating to the selection of the origin of the world coordinate frame is in order.

Equation (5.11) is the most significant contribution of Tsai's original calibration method, because it allows the calculation of the camera parameters to be performed in two stages. It is based on the fact that, if the lens distortion is assumed to be purely radial, the line normal to the optical axis passing through a point in space is parallel to the line in the image plane normal to the optical axis passing through the projection of the point.

However, (5.11) is very sensitive to the location of the world origin. In theory, $t_y$ cannot equal zero or (5.11) is undefined. In practice, $t_y$ cannot even be "close" to zero or the matrix in (5.11) becomes very ill-conditioned. Theoretically, the world origin can be located anywhere (except where $t_y = 0$) without affecting the solution to (5.11), so it might seem reasonable to place the origin some arbitrarily large distance away from the camera, thereby ensuring that $t_y$ is not "small". Unfortunately, this is unsatisfactory because round-off errors (from multiplication and addition operations) become substantial when the world origin is "too far away".

In order to systematically find a "good" location for the world origin and to align the world coordinate frame with the camera coordinate frame, the authors have developed the following technique. To try to minimize round-off error, the world origin is translated to the centroid of the data points. This makes all the data points as comparable in magnitude as possible.

Now the process of aligning the world coordinate frame with the camera coordinate frame can begin. First equation (5.11) is solved, the stage 1 camera parameters are determined (as described in Section 5.3.1), and the residual sum-squared-error of (5.13) is calculated. To ensure that we do not use a solution to (5.11) where $t_y$ is near zero, this first step is done a total of four times; once with the world origin at the centroid of the data, and once with the world origin translated an arbitrary distance (e.g., 1 m) in the direction of each of its axes. The stage 1 parameters that result in the smallest residual sum-squared-error is then used in the next step. Note that merely checking to see if the $t_y$ calculated is "near" zero is a poor test. When $t_y$ is actually "near" zero (or even exactly zero) a numerical solution to (5.11) will still be found, but frequently the $t_y$ calculated will not be "near" zero.

The world coordinate frame can now be rotated by the rotation matrix determined from the stage 1 parameters selected above. However, that rotation matrix will (in all likelihood) not be orthonormal, so the Euler angles are extracted from it using (5.46) and these angles are used to construct an orthonormal rotation matrix with which to rotate the world coordinate frame. The process is now repeated until the extracted Euler angles become "very small" (as defined by some predetermined criteria).

In the event that the camera rotation matrix is originally near or at the Euler angle extraction singularity, two of the Euler angles extracted during the first iteration may not be correct, but the world coordinate frame is still rotated by $\theta$ which is determined correctly. The result is that in the next iteration the world coordinate frame has rotated away from the singularity and the Euler angles can be reliably determined. In actual experimentation the authors have never seen this entire process require more than two iterations (even when $\theta = \pm 90°$).

Once the world and camera coordinate systems are aligned, the world origin can be easily placed to ensure that $t_y$ is not "small". The first step is to place the aligned world coordinate system back at the centroid of the data points (in the event it had shifted position during the aligning process). Then (5.11) is solved, the stage 1 camera parameters are determined, and the residual sum-squared-error of (5.13) is calculated. This error is compared with the error obtained after translating the origin in the positive and negative $y$ direction an arbitrary (but not "too large") distance (e.g., 1 m). If the error calculated with the origin at the centroid of the data is the smallest, then the origin need not be translated in the $y$ direction. Otherwise, the origin can be translated in the direction that gives the smallest error until the error either increases or ceases to decrease "significantly". After translating the origin in the $y$ direction, the origin can be translated in the $z$ and $x$ directions in an analogous manner.

It should be stressed that the above technique is based on experiments performed by the authors. It is only necessary because of the inaccurate nature of numerical computations with a digital computer. In theory, the residual sum-squared-error obtained with the parameters calculated from (5.11) will be the same global minimum for any pose of the world origin where $t_y \neq 0$. Unfortunately, limited calculation precision results in the residual sum-squared-error being a function of the pose of the world coordinate frame. An alternate solution might be to extend the precision of the calculations performed to where this was no longer a problem. However, such a tactic would, in all likelihood, *drastically* increase the required computation time.

Once the pose of the world origin has been selected in the manner described above and (5.11) has been solved, the stage 1 camera parameters determined will still (in all likelihood) not meet the constraints of the problem (i.e., $R$ will not be orthonormal) due to noise in the data points. To solve this, after extracting the Euler angles from the rotation matrix, these angles and the other parameters determined in stage 1 ($t_x$, $t_y$, and $s_x$) can serve as an initial guess at the minimum of the following cost function:

$$\begin{aligned} J &= \sum_{i=1}^{n} \left\{ \begin{array}{l} \left[d_x \frac{N_{cx}}{N_{fx}}(x_{f_i} - C_x)\right] [(\sin\phi\cos\theta)x_{w_i} + (\sin\phi\sin\theta\sin\psi \\ + \cos\phi\cos\psi)y_{w_i} + (\sin\phi\sin\theta\cos\psi - \cos\phi\sin\psi)z_{w_i} + t_y] \\ - s_x\,[d_y(y_{f_i} - C_y)]\,[(\cos\phi\cos\theta)x_{w_i} + (\cos\phi\sin\theta\sin\psi \\ - \sin\phi\cos\psi)y_{w_i} + (\cos\phi\sin\theta\cos\psi + \sin\phi\sin\psi)z_{w_i} + t_x] \end{array} \right\}^2 \\ &= \sum_{i=1}^{n} \{s_x x_{d_i} y_i - s_x y_{d_i} x_i\}^2 \end{aligned} \tag{5.47}$$

where $n$ is the number of calibration data points and the unknowns are the Euler angles $(\psi, \theta, \phi)$, the first two components of the translation vector $(t_x, t_y)$, and the image scale factor $(s_x)$. The cost function, $J$, is based on $x_d y - y_d x = 0$, where $x_d$ is given in (5.17), $y_d$ is given in (5.10), and $x$ and $y$ are given in (5.1).

To determine the minimum of (5.47), the Polak-Ribiere conjugate gradient multi-dimensional minimization algorithm is used [11]. This requires an initial guess at the minimum and that the gradient of the objective function be known. Determination of a global minimum of (5.47) cannot be guaranteed, but with a "good" initial guess at the minimum (which we have obtained from the solution of (5.11)) a "good" minimum is likely. Note that before minimizing (5.47), the world origin should be translated back to the centroid of the data points as (5.47) does not require $t_y$ to be nonzero. This requires the initial estimates of $t_x$ and $t_y$ to be adjusted accordingly. The required partial derivatives are given below. Note that the common factor of 2 has been omitted.

$$\frac{\partial J}{\partial \psi} = \sum_{i=1}^{n} (s_x x_{d_i} y_i - s_x y_{d_i} x_i) \left\{ \begin{array}{l} s_x x_{d_i}(r_6 y_{w_i} - r_5 z_{w_i}) \\ - s_x y_{d_i}(r_3 y_{w_i} - r_2 z_{w_i}) \end{array} \right\} \tag{5.48}$$

$$\frac{\partial J}{\partial \theta} = \sum_{i=1}^{n} (s_x x_{d_i} y_i - s_x y_{d_i} x_i) \left\{ \begin{array}{l} [-\sin\theta x_{w_i} + \cos\theta(\sin\psi y_{w_i} \\ + \cos\psi z_{w_i})](s_x x_{d_i} \sin\phi \\ - s_x y_{d_i} \cos\phi) \end{array} \right\} \tag{5.49}$$

$$\frac{\partial J}{\partial \phi} = \sum_{i=1}^{n} (s_x x_{d_i} y_i - s_x y_{d_i} x_i) \{ s_x x_{d_i}(x_i - t_x) + s_x y_{d_i}(y_i - t_y) \} \tag{5.50}$$

$$\frac{\partial J}{\partial t_x} = \sum_{i=1}^{n} (s_x x_{d_i} y_i - s_x y_{d_i} x_i) \{ -s_x y_{d_i} \} \tag{5.51}$$

$$\frac{\partial J}{\partial t_y} = \sum_{i=1}^{n} (s_x x_{d_i} y_i - s_x y_{d_i} x_i)\, s_x x_{d_i} \tag{5.52}$$

$$\frac{\partial J}{\partial s_x} = \sum_{i=1}^{n} (s_x x_{d_i} y_i - s_x y_{d_i} x_i) \{ -y_d x_i \} \tag{5.53}$$

Another limitation of (5.11) is that it assumes $C_x$ and $C_y$ are known (which is often not true). Later work [9], [10], shows the accurate determination of the image center is important and suggests methods to determine the image center. Since we have already decided to perform nonlinear optimization to calculate the rotation matrix, the image center can also be calculated at little extra cost. Adding the pixel coordinates of the image center $(C_x, C_y)$ to the list of unknown parameters to be found with (5.47), we increase the dimension of the optimization by two and require the following partial derivatives (omitting the common factor of 2):

$$\frac{\partial J}{\partial C_x} = \sum_{i=1}^{n} (s_x x_{d_i} y_i - s_x y_{d_i} x_i) \{ -d_x \frac{N_{cx}}{N_{fx}} y_i \} \tag{5.54}$$

$$\frac{\partial J}{\partial C_y} = \sum_{i=1}^{n} (s_x x_{d_i} y_i - s_x y_{d_i} x_i) \{ d_y s_x x_i \} \tag{5.55}$$

Theoretically, this is sufficient to determine the location of the image center. However, as discussed in [9] and [10], calculation of the image center in stage 1 is only reliable in the presence of lens distortion and *when the errors in the calibration data points used are small.* Very often the errors in the calibration data points used by the authors are too large to reliably determine the image center in stage 1. Consequently, the image center is determined in stage 2 as described in Section 5.3.3.

### 5.3.3 Methods for Determining the Stage 2 Parameters

Once the first stage of the calibration has been completed, the radial lens distortion coefficient $\kappa$, focal length $f$, $z$ translation component $t_z$, and possibly the image center $(C_x, C_y)$, may be determined.

Four methods of determining $\kappa$, $f$, and $t_z$ are presented in this section. In the interest of clarity, discussion of the determination of the image center will be postponed until the end of the section.

#### Stage 2 Solved Using Overdetermined Linear Equations

In his later work, [7], [8], Tsai advances the following formulation of the radial lens distortion approximation:

$$\begin{aligned} x_d\left(1+\sqrt{1-4\kappa(x_u^2+y_u^2)}\right) &= 2x_u \\ y_d\left(1+\sqrt{1-4\kappa(x_u^2+y_u^2)}\right) &= 2y_u \end{aligned} \tag{5.56}$$

which is equivalent to

$$\begin{aligned} x_u(1+\kappa(x_d^2+y_d^2)) &= x_d \\ y_u(1+\kappa(x_d^2+y_d^2)) &= y_d \end{aligned} \tag{5.57}$$

In his earlier papers, [4], [5], [6] Tsai asserts that

$$\begin{aligned} x_d(1+\kappa(x_d^2+y_d^2)) &= x_u \\ y_d(1+\kappa(x_d^2+y_d^2)) &= y_u \end{aligned} \tag{5.58}$$

Clearly, (5.57) and (5.58) are *not* equivalent.

We begin examining this by verifying for the reader that (5.56) is equivalent to (5.57). Substituting (5.57) into the first line of (5.56) we have

$$\begin{aligned} x_d\left(1+\sqrt{1-\frac{4\kappa(x_d^2+y_d^2)}{(1+\kappa(x_d^2+y_d^2))^2}}\right) &= 2\frac{x_d}{(1+\kappa(x_d^2+y_d^2))} \\ (1+\kappa(x_d^2+y_d^2))+\sqrt{(1+\kappa(x_d^2+y_d^2))^2-4\kappa(x_d^2+y_d^2)} &= 2 \\ 1+\kappa r_d^2+\sqrt{(1+\kappa r_d^2)^2-4\kappa r_d^2} &= 2 \\ 1+\kappa r_d^2+\sqrt{1+2\kappa r_d^2+\kappa^2 r_d^4-4\kappa r_d^2} &= 2 \\ 1+\kappa r_d^2+\sqrt{1-2\kappa r_d^2+\kappa^2 r_d^4} &= 2 \\ 1+\kappa r_d^2+\sqrt{(1-\kappa r_d^2)^2} &= 2 \end{aligned}$$

$$1 + \kappa r_d^2 + 1 - \kappa r_d^2 \quad = \quad 2$$

The second line of (5.56) can be verified in an analogous manner. Therefore, (5.56) is indeed equivalent to (5.57).

Let us now examine (5.57) and (5.58) to determine if they differ only by their constant $\kappa$. We rewrite (5.57) as

$$\begin{aligned} x_u &= \frac{x_d}{1+\kappa(x_d^2+y_d^2)} \\ y_u &= \frac{y_d}{1+\kappa(x_d^2+y_d^2)} \end{aligned} \tag{5.59}$$

and denote $\kappa$ in (5.58) as $\kappa'$. Squaring and adding the components of (5.59) and equating them to the squared and summed components of (5.58) we have

$$\begin{aligned} (x_u^2 + y_u^2) = \frac{x_d^2 + y_d^2}{(1 + \kappa(x_d^2 + y_d^2))} &= (1 + \kappa'(x_d^2 + y_d^2))(x_d^2 + y_d^2) \\ \frac{1}{1 + \kappa r_d^2} &= 1 + \kappa' r_d^2 \\ (1 + \kappa' r_d^2)(1 + \kappa r_d^2) &= 1 \\ 1 + \kappa' r_d^2 + \kappa r_d^2 + \kappa' \kappa r_d^4 &= 1 \\ \kappa' + \kappa + \kappa' \kappa r_d^2 &= 0 \\ \kappa' &= -\frac{\kappa}{1 + \kappa r_d^2} \end{aligned}$$

Since $r_d^2$ varies with position in the image plane, $\kappa' \neq \alpha\kappa$ where $\alpha$ is some arbitrary constant. Hence, (5.57) is not equivalent to (5.58) with a different $\kappa$. However, $\kappa' \approx -\kappa$, since $1 + \kappa r_d^2 \approx 1$. Therefore (5.56) and (5.57) are an approximation to (5.58).

From (5.2), (5.3), and (5.57) the following overdetermined system of linear equations may be set up to solve for $\kappa$, $f$, and $t_z$ [7], [8]:

$$\begin{bmatrix} x_1 & x_1 r_{d_1}^2 & -x_{d_1} \\ y_1 & y_1 r_{d_1}^2 & -y_{d_1} \\ x_2 & x_2 r_{d_2}^2 & -x_{d_2} \\ y_2 & y_2 r_{d_2}^2 & -y_{d_2} \\ \vdots & \vdots & \vdots \\ x_n & x_n r_{d_n}^2 & -x_{d_n} \\ y_n & y_n r_{d_n}^2 & -y_{d_n} \end{bmatrix} \begin{bmatrix} f \\ f\kappa \\ t_z \end{bmatrix} = \begin{bmatrix} (x_{w_1} r_7 + y_{w_1} r_8 + z_{w_1} r_9) x_{d_1} \\ (x_{w_1} r_7 + y_{w_1} r_8 + z_{w_1} r_9) y_{d_1} \\ (x_{w_2} r_7 + y_{w_2} r_8 + z_{w_2} r_9) x_{d_2} \\ (x_{w_2} r_7 + y_{w_2} r_8 + z_{w_2} r_9) y_{d_2} \\ \vdots \\ (x_{w_n} r_7 + y_{w_n} r_8 + z_{w_n} r_9) x_{d_n} \\ (x_{w_n} r_7 + y_{w_n} r_8 + z_{w_n} r_9) y_{d_n} \end{bmatrix} \tag{5.60}$$

where $x_d$ is defined in (5.17), $y_d$ is defined in (5.10), $r_d$ is defined in (5.6), and $n$ is the number of data points.

It should be noted that using (5.58) to generate a set of overdetermined system of linear equations instead of using (5.57) (which is an approximation to (5.58)) results in a system of equations where the unknowns are $f$, $\kappa$, $\kappa t_z$, and $t_z$. These unknowns are *not* independent, and so that system of equations is of little value. Consequently, the following three methods of determining $\kappa$, $f$, and $t_z$ are presented.

### Stage 2 Solved Iteratively

The next three methods to solve for $\kappa$, $f$, and $t_z$, require initial estimates for $\kappa$, $f$, and $t_z$ (denoted by $\hat{\kappa}$, $\hat{f}$ and $\hat{t}_z$, respectively). One method for determining these initial estimates could be to set $\hat{\kappa} = 0$ and solve the following overdetermined system of linear equations for $n$ calibration points using SVD to determine $\hat{f}$ and $\hat{t}_z$:

$$\begin{bmatrix} x_1 & -x_{d_1} \\ y_1 & -y_{d_1} \\ x_2 & -x_{d_2} \\ y_2 & -y_{d_2} \\ \vdots & \vdots \\ x_n & -x_{d_n} \\ y_n & -y_{d_n} \end{bmatrix} \begin{bmatrix} \hat{f} \\ \hat{t}_z \end{bmatrix} = \begin{bmatrix} (r_7 x_{w_1} + r_8 y_{w_1} + r_9 z_{w_1}) x_{d_1} \\ (r_7 x_{w_1} + r_8 y_{w_1} + r_9 z_{w_1}) y_{d_1} \\ (r_7 x_{w_2} + r_8 y_{w_2} + r_9 z_{w_2}) x_{d_2} \\ (r_7 x_{w_2} + r_8 y_{w_2} + r_9 z_{w_2}) y_{d_2} \\ \vdots \\ (r_7 x_{w_n} + r_8 y_{w_n} + r_9 z_{w_n}) x_{d_n} \\ (r_7 x_{w_n} + r_8 y_{w_n} + r_9 z_{w_n}) y_{d_n} \end{bmatrix} \tag{5.61}$$

However, often times better initial estimates can be determined from

$$\hat{\kappa} = -\kappa \qquad \hat{f} = f \qquad \hat{t}_z = t_z \tag{5.62}$$

where $\kappa$, $f$, and $t_z$ used above are found from (5.60).

One method to compute $\kappa$, $f$, and $t_z$ (suggested by D. Sood and M. Repko, [16]) uses (5.61) through (5.65) in an iterative procedure as follows:

$$\begin{bmatrix} r_{d_1}^3 (r_7 x_{w_1} + r_8 y_{w_1} + r_9 z_{w_1} + \hat{t}_z) \\ r_{d_2}^3 (r_7 x_{w_2} + r_8 y_{w_2} + r_9 z_{w_2} + \hat{t}_z) \\ \vdots \\ r_{d_n}^3 (r_7 x_{w_n} + r_8 y_{w_n} + r_9 z_{w_n} + \hat{t}_z) \end{bmatrix} [\hat{\kappa}] = \begin{bmatrix} \hat{f}\sqrt{x_1^2 + y_1^2} - r_{d_1}(r_7 x_{w_1} + r_8 y_{w_1} + r_9 z_{w_1} + \hat{t}_z) \\ \hat{f}\sqrt{x_2^2 + y_2^2} - r_{d_2}(r_7 x_{w_2} + r_8 y_{w_2} + r_9 z_{w_2} + \hat{t}_z) \\ \vdots \\ \hat{f}\sqrt{x_n^2 + y_n^2} - r_{d_n}(r_7 x_{w_n} + r_8 y_{w_n} + r_9 z_{w_n} + \hat{t}_z) \end{bmatrix} \tag{5.63}$$

Equation (5.63) is an overdetermined system of linear equations (solved with SVD) that uses $\hat{f}$ and $\hat{t}_z$ to determine $\hat{\kappa}$. Then $\hat{\kappa}$ is used to refine $\hat{f}$ and $\hat{t}_z$ from

$$\begin{bmatrix} -\sqrt{x_1^2 + y_1^2} & r_{d_1}(1 + \hat{\kappa} r_{d_1}^2) \\ -\sqrt{x_2^2 + y_2^2} & r_{d_2}(1 + \hat{\kappa} r_{d_2}^2) \\ \vdots & \vdots \\ -\sqrt{x_n^2 + y_n^2} & r_{d_n}(1 + \hat{\kappa} r_{d_n}^2) \end{bmatrix} \begin{bmatrix} \hat{f} \\ \hat{t}_z \end{bmatrix} =$$

$$\begin{bmatrix} -r_{d_1}(1+\hat{\kappa}r_{d_1}^2)(r_7x_{w_1}+r_8y_{w_1}+r_9z_{w_1}) \\ -r_{d_2}(1+\hat{\kappa}r_{d_2}^2)(r_7x_{w_2}+r_8y_{w_2}+r_9z_{w_2}) \\ \vdots \\ -r_{d_n}(1+\hat{\kappa}r_{d_n}^2)(r_7x_{w_n}+r_8y_{w_n}+r_9z_{w_n}) \end{bmatrix} \tag{5.64}$$

After refining $\hat{f}$ and $\hat{t}_z$ using (5.64), the residual sum-squared-error is calculated using

$$\sum_{i=1}^{n}\left\{r_{d_i}(1+\hat{\kappa}r_{d_i}^2)(r_7x_{w_i}+r_8y_{w_i}+r_9z_{w_i}+\hat{t}_z)-\hat{f}\sqrt{x_i^2+y_i^2}\right\}^2 \tag{5.65}$$

The estimates of $\kappa$, $f$, and $t_z$ are refined by cycling through (5.63) and (5.64) until the residual sum-squared-error increases or ceases to decrease "significantly". Note that although this method is not guaranteed to reach a minimum of (5.65) (or even to converge), it commonly gives "good" results quickly.

#### Stage 2 Solved by Minimizing the Square of the Euclidean Distance from the Calibration Points to Rays Back-Projected from the Camera

A ray back-projected from a camera is described by

$$\vec{r} = s\vec{d} + \vec{O} \tag{5.66}$$

where $\vec{O}$ (defined in (5.135)) is the origin of a lens centered coordinate frame, $\vec{d}$ (defined in (5.137)) is the direction of the ray emanating from the camera, and $s$ is a scalar.

To find the shortest distance between a point $(x_{w_i}, y_{w_i}, z_{w_i})$ and the ray of (5.66) we must find $s_i > 0 \ni \mathsf{distance}(s_i\hat{d}_i + \vec{O}, \vec{x}_{w_i})$ is a minimum. This occurs when $\hat{d}_i \perp (s_i\hat{d}_i + \vec{O} - \vec{x}_{w_i})$, which implies that $s_i = \hat{d}_i \cdot (\vec{x}_{w_i} - \vec{O})$ or

$$s_i = \frac{d_{x_i}}{|d_i|}(x_{w_i} - O_x) + \frac{d_{y_i}}{|d_i|}(y_{w_i} - O_y) + \frac{d_{z_i}}{|d_i|}(z_{w_i} - O_z) \tag{5.67}$$

and the distance squared, $D_i$, is given by

$$D_i = \left(x_{w_i} - O_x - s_i\frac{d_{x_i}}{|d_i|}\right)^2 + \left(y_{w_i} - O_y - s_i\frac{d_{y_i}}{|d_i|}\right)^2 + \left(z_{w_i} - O_z - s_i\frac{d_{z_i}}{|d_i|}\right)^2 \tag{5.68}$$

Summing (5.68) over all all $n$ calibration data points gives a multi-dimensional nonlinear objective function that can be minimized using the same conjugate gradient technique as in Section 5.3.2. The initial guess at the minimum of $\sum_{i=1}^{n} D_i$ is determined from (5.62).

Using the definitions of $\vec{O}$ and $\vec{d}$ given in Section 5.5.2 and the definition of $x_u$ and $y_u$ given in (5.2) and (5.3) respectively, the necessary partial derivatives of (5.68) (omitting the common factor of 2) can be found to be

$$\frac{\partial D}{\partial \kappa} = \sum_{i=1}^{n} \left\{ \begin{aligned} &\left(x_{w_i} - O_x - s_i \tfrac{d_{x_i}}{|d_i|}\right) \Big\{ -r_{d_i}^2 \tfrac{d_{x_i}}{|d_i|^2} [(r_1 x_{d_i} + r_4 y_{d_i})(x_{w_i} - O_x) \\ &+ (r_2 x_{d_i} + r_5 y_{d_i})(y_{w_i} - O_y) + (r_3 x_{d_i} + r_6 y_{d_i})(z_{w_i} - O_z)] \\ &- s_i \tfrac{r_{d_i}^2}{|d_i|}(r_1 x_{d_i} + r_4 y_{d_i}) + 2 s_i r_{d_i}^2 \tfrac{d_{x_i}}{|d_i|^3} [d_{x_i}(r_1 x_{d_i} + r_4 y_{d_i}) \\ &+ d_{y_i}(r_2 x_{d_i} + r_5 y_{d_i}) + d_{z_i}(r_3 x_{d_i} + r_6 y_{d_i})] \Big\} \\ &+ \left(y_{w_i} - O_y - s_i \tfrac{d_{y_i}}{|d_i|}\right) \Big\{ -r_{d_i}^2 \tfrac{d_{y_i}}{|d_i|^2} [(r_1 x_{d_i} + r_4 y_{d_i})(x_{w_i} - O_x) \\ &+ (r_2 x_{d_i} + r_5 y_{d_i})(y_{w_i} - O_y) + (r_3 x_{d_i} + r_6 y_{d_i})(z_{w_i} - O_z)] \\ &- s_i \tfrac{r_{d_i}^2}{|d_i|}(r_2 x_{d_i} + r_5 y_{d_i}) + 2 s_i r_{d_i}^2 \tfrac{d_{y_i}}{|d_i|^3} [d_{x_i}(r_1 x_{d_i} + r_4 y_{d_i}) \\ &+ d_{y_i}(r_2 x_{d_i} + r_5 y_{d_i}) + d_{z_i}(r_3 x_{d_i} + r_6 y_{d_i})] \Big\} \\ &+ \left(z_{w_i} - O_z - s_i \tfrac{d_{z_i}}{|d_i|}\right) \Big\{ -r_{d_i}^2 \tfrac{d_{z_i}}{|d_i|^2} [(r_1 x_{d_i} + r_4 y_{d_i})(x_{w_i} - O_x) \\ &+ (r_2 x_{d_i} + r_5 y_{d_i})(y_{w_i} - O_y) + (r_3 x_{d_i} + r_6 y_{d_i})(z_{w_i} - O_z)] \\ &- s_i \tfrac{r_{d_i}^2}{|d_i|}(r_3 x_{d_i} + r_6 y_{d_i}) + 2 s_i r_{d_i}^2 \tfrac{d_{z_i}}{|d_i|^3} [d_{x_i}(r_1 x_{d_i} + r_4 y_{d_i}) \\ &+ d_{y_i}(r_2 x_{d_i} + r_5 y_{d_i}) + d_{z_i}(r_3 x_{d_i} + r_6 y_{d_i})] \Big\} \end{aligned} \right\} \tag{5.69}$$

$$\frac{\partial D}{\partial f} = \sum_{i=1}^{n} \left\{ \begin{aligned} &\left(x_{w_i} - O_x - s_i \tfrac{d_{x_i}}{|d_i|}\right) \Big( 2 s_i \tfrac{d_{x_i}}{|d_i|^3}(d_{x_i} r_7 + d_{y_i} r_8 + d_{z_i} r_9) - s_i \tfrac{r_7}{|d_i|} \\ &- \tfrac{d_{x_i}}{|d_i|^2}[r_7(x_{w_i} - O_x) + r_8(y_{w_i} - O_y) + r_9(z_{w_i} - O_z)] \Big) \\ &+ \left(y_{w_i} - O_y - s_i \tfrac{d_{y_i}}{|d_i|}\right) \Big( 2 s_i \tfrac{d_{y_i}}{|d_i|^3}(d_{x_i} r_7 + d_{y_i} r_8 + d_{z_i} r_9) - s_i \tfrac{r_8}{|d_i|} \\ &- \tfrac{d_{y_i}}{|d_i|^2}[r_7(x_{w_i} - O_x) + r_8(y_{w_i} - O_y) + r_9(z_{w_i} - O_z)] \Big) \\ &+ \left(z_{w_i} - O_z - s_i \tfrac{d_{z_i}}{|d_i|}\right) \Big( 2 s_i \tfrac{d_{z_i}}{|d_i|^3}(d_{x_i} r_7 + d_{y_i} r_8 + d_{z_i} r_9) - s_i \tfrac{r_9}{|d_i|} \\ &- \tfrac{d_{z_i}}{|d_i|^2}[r_7(x_{w_i} - O_x) + r_8(y_{w_i} - O_y) + r_9(z_{w_i} - O_z)] \Big) \end{aligned} \right\} \tag{5.70}$$

$$\frac{\partial D}{\partial t_z} = \sum_{i=1}^{n} \left\{ \begin{aligned} &\left(x_{w_i} - O_x - s_i \tfrac{d_{x_i}}{|d_i|}\right) \left( r_7 - (d_{x_i} r_7 + d_{y_i} r_8 + d_{z_i} r_9) \tfrac{d_{x_i}}{|d_i|^2} \right) \\ &+ \left(y_{w_i} - O_y - s_i \tfrac{d_{y_i}}{|d_i|}\right) \left( r_8 - (d_{x_i} r_7 + d_{y_i} r_8 + d_{z_i} r_9) \tfrac{d_{y_i}}{|d_i|^2} \right) \\ &+ \left(z_{w_i} - O_z - s_i \tfrac{d_{z_i}}{|d_i|}\right) \left( r_9 - (d_{x_i} r_7 + d_{y_i} r_8 + d_{z_i} r_9) \tfrac{d_{z_i}}{|d_i|^2} \right) \end{aligned} \right\} \tag{5.71}$$

### Stage 2 Solved by Minimizing the Squared Error in the Perspective Equations

A fourth method to determine the values of $\kappa$, $f$, and $t_z$ is based on the perspective equations (5.2) and (5.3). Using these equations as well as (5.4) and (5.5), the following relationships may be formed:

$$x_i f - x_{d_i}(x_{w_i} r_7 + y_{w_i} r_8 + z_{w_i} r_9 + t_z)(r_{d_i}^2 \kappa + 1) = C_{1_i} \tag{5.72}$$

$$y_i f - y_{d_i}(x_{w_i} r_7 + y_{w_i} r_8 + z_{w_i} r_9 + t_z)(r_{d_i}^2 \kappa + 1) = C_{2_i} \tag{5.73}$$

Thus, we may form the following multi-dimensional nonlinear objective function:

$$C = \sum_{i=1}^{n} \{C_{1_i}^2 + C_{2_i}^2\} \tag{5.74}$$

This objective function can be minimized using the same conjugate gradient technique as in Section 5.3.2. The initial guess at the minimum of (5.74) is determined from (5.62). The partial derivatives of (5.74) are given below with the constant factor of 2 omitted:

$$\frac{\partial C}{\partial \kappa} = \sum_{i=1}^{n} -r_{d_i}^2 z_i (x_{d_i} C_{1_i} + y_{d_i} C_{2_i}) \tag{5.75}$$

$$\frac{\partial C}{\partial f} = \sum_{i=1}^{n} (x_i C_{1_i} + y_i C_{2_i}) \tag{5.76}$$

$$\frac{\partial C}{\partial t_z} = \sum_{i=1}^{n} -(r_{d_i}^2 \kappa + 1)(x_{d_i} C_{1_i} + y_{d_i} C_{2_i}) \tag{5.77}$$

A similar nonlinear optimization can be based on (5.65).

### Calculating the Image Center in Stage 2

As mentioned at the end of Section 5.3.2, the image center can be determined in stage 1 using $(C_x, C_y)$ as additional parameters in the minimization of (5.47). Doing so will result in a smaller residual sum-squared-error for stage 1. However, if the error in the calibration points is too large, it may result in a larger final residual sum-squared-error (as calculated by summing (5.68) over all $n$ calibration points or with (5.74)).

To prevent the *mis*calibration of the image center in stage 1 from adversely affecting the calibration of the stage 2 parameters, the image center may be calibrated in stage 2 using any of the four stage 2 calibration methods described above. The image center coordinates become additional parameters in the methods using nonlinear optimization, while the linear and iterative methods require a separate nonlinear optimization be performed to determine the image center. This separate optimization can be based on either (5.68) or (5.74). Since the concept is identical regardless of the method chosen, only the extension to the minimization of (5.74) will be explicitly presented here.

The required partial derivatives, omitting a common factor of 2, are

$$\frac{\partial C}{\partial C_x} = \sum_{i=1}^{n} z_i \frac{d_x}{s_x} \frac{N_{cx}}{N_{fx}} \left\{ \left[(2x_{d_i}^2 + r_{d_i}^2)\kappa + 1\right] C_{1_i} + 2x_{d_i} y_{d_i} \kappa C_{2_i} \right\} \tag{5.78}$$

$$\frac{\partial C}{\partial C_y} = \sum_{i=1}^{n} z_i d_y \left\{ 2x_{d_i} y_{d_i} \kappa C_{1_i} + \left[(2y_{d_i}^2 + r_{d_i}^2)\kappa + 1\right] C_{2_i} \right\} \tag{5.79}$$

If the image center is assumed when calculating the stage 1 parameters and then calculated in stage 2, it is reasonable to start the entire calibration process again with the new image center. Iterating continues until the final residual sum-squared-error increases or ceases to decrease "significantly".

Alternatively, the results of stage 1 and stage 2 can be used as the initial guess at the minimum of a global objective function where all of the camera parameters are variables. This would require an 11 dimensional nonlinear numerical optimization of the global objective function, which could be based on (5.68) or (5.74) or even (5.65).

### 5.3.4 Sources of Camera Calibration Error

The accuracy of the resulting camera calibration is a function of the accuracy and number of points used (as many as 512 have been used by the authors). The method used to gather the calibration points may also vary. A 24-point calibration pattern positioned by a robot was developed by D. Sood and M. Repko for their original implementation of Tsai's algorithm at CIRSSE [16]. However, that pattern was unsuitable for calibration with wide-angle lenses because it was unable to fill the camera's field of view. As a result, the authors have used a small lightsource positioned in the workspace by a robot (see Section 5.6.3).

An obvious source of camera calibration error arises from the spatial quantization of the image. However, many times the image event of interest spans many pixels. It is then possible to calculate the centroid of this event to sub-pixel accuracy, thereby reducing the spatial quantization error. This is the case with the collection of calibration points described in Section 5.6.3.

In this section we show that a circle, when projected onto a camera's image plane, becomes, in general, an ellipse. Practical use of this knowledge can be used in examining the calibration point gathering procedure described in Section 5.6.3. We are interested in determining the error introduced by assuming the center of a circular lightsource is mapped to the centroid of the ellipse formed in the image plane by the perspective transformation of the circular lightsource. However, the primary source of the camera calibration error described in Section 5.6.4 arises from errors in the kinematic model of the robot as explained at the end of this section.

#### Centroid of the Projection Does Not Equal the Projection of the Centroid

We have a circle lying in the $x_u y_u$ plane with center $(x_w, y_w, z_w)$ and radius $r$. Assuming a pin-hole camera model with focal length $f$ and pose given by $R$

and $T$ we get from (5.129) and (5.130)

$$\begin{bmatrix} (r_7x_u - fr_1) & (r_8x_u - fr_2) \\ (r_7y_u - fr_4) & (r_8y_u - fr_5) \end{bmatrix} \begin{bmatrix} x_w \\ y_w \end{bmatrix} = \begin{bmatrix} (ft_x - t_zx_u - z_w(r_9x_u - fr_3)) \\ (ft_y - t_zy_u - z_w(r_9y_u - fr_6)) \end{bmatrix} \tag{5.80}$$

$$\begin{bmatrix} x_w \\ y_w \end{bmatrix} = \frac{\begin{bmatrix} (r_8y_u - fr_5) & -(r_8x_u - fr_2) \\ -(r_7y_u - fr_4) & (r_7x_u - fr_1) \end{bmatrix}\begin{bmatrix} (ft_x - t_zx_u - z_w(r_9x_u - fr_3)) \\ (ft_y - t_zy_u - z_w(r_9y_u - fr_6)) \end{bmatrix}}{(r_7x_u - fr_1)(r_8y_u - fr_5) - (r_7y_u - fr_4)(r_8x_u - fr_2)} \tag{5.81}$$

$$\begin{aligned} x_w &= (r_8y_ut_x + r_8y_uz_wr_3 - fr_5t_x + r_5t_zx_u + r_5z_wr_9x_u - fr_5z_wr_3 \\ &\quad - r_8x_ut_y - r_8x_uz_wr_6 + fr_2t_y - r_2t_zy_u - r_2z_wr_9y_u + fr_2z_wr_6) \\ &\quad /(-r_7x_ur_5 - r_1r_8y_u + fr_1r_5 + r_8x_ur_4 + r_2r_7y_u - fr_2r_4) \qquad (5.82) \\ y_w &= -(r_7y_ut_x + r_7y_uz_wr_3 - fr_4t_x + r_4t_zx_u + r_4z_wr_9x_u - fr_4z_wr_3 \\ &\quad - r_7x_ut_y - r_7x_uz_wr_6 + fr_1t_y - r_1t_zy_u - r_1z_wr_9y_u + fr_1z_wr_6) \\ &\quad /(-r_7x_ur_5 - r_1r_8y_u + fr_1r_5 + r_8x_ur_4 + r_2r_7y_u - fr_2r_4) \qquad (5.83) \end{aligned}$$

We know that $x_w^2 + y_w^2 - r^2 = 0$, so after multiplying through by the denominator we have,

$$\begin{aligned} 0 &= (r_8y_ut_x + r_8y_uz_wr_3 - fr_5t_x + r_5t_zx_u + r_5z_wr_9x_u - fr_5z_wr_3 \\ &\quad - r_8x_ut_y - r_8x_uz_wr_6 + fr_2t_y - r_2t_zy_u - r_2z_wr_9y_u + fr_2z_wr_6)^2 \\ &\quad + (r_7y_ut_x + r_7y_uz_wr_3 - fr_4t_x + r_4t_zx_u + r_4z_wr_9x_u - fr_4z_wr_3 \\ &\quad - r_7x_ut_y - r_7x_uz_wr_6 + fr_1t_y - r_1t_zy_u - r_1z_wr_9y_u + fr_1z_wr_6)^2 \\ &\quad - r^2(-r_7x_ur_5 - r_1r_8y_u + fr_1r_5 + r_8x_ur_4 + r_2r_7y_u - fr_2r_4)^2 \qquad (5.84) \end{aligned}$$

Multiplying through by $-1$ and collecting the terms with $x_u$ and $y_u$, we have the general form of a planar ellipse:

$$Ax_u^2 + Bx_uy_u + Cy_u^2 + Dx_u + Ey_u + F = 0 \tag{5.85}$$

where

$$A = r_3^2r^2 + (r_9t_y - r_6t_z)^2 - t_y^2 - t_z^2 - 2(r_6t_y + r_9t_z)z_w + (r_3^2 - 1)z_w^2 \tag{5.86}$$

$$C = r_6^2r^2 + (r_9t_x - r_3t_z)^2 - t_x^2 - t_z^2 - 2(r_3t_x + r_9t_z)z_w + (r_6^2 - 1)z_w^2 \tag{5.87}$$

$$F = f^2\left[r_9^2r^2 + (r_6t_x - r_3t_y)^2 - t_x^2 - t_y^2 - 2(r_3t_x + r_6t_y)z_w + (r_9^2 - 1)z_w^2\right] \tag{5.88}$$

$$\begin{aligned} B = 2\big[&(r_7r_8(r_1r_5 + r_2r_4) - r_1r_4r_8^2 - r_2r_5r_7^2)r^2 + r_9t_z(r_6t_x + r_3t_y) \\ &- r_3r_6t_z^2 + (1 - r_9^2)t_xt_y + (r_6t_x + r_3t_y)z_w + r_3r_6z_w^2\big] \qquad (5.89) \end{aligned}$$

$$\begin{aligned} D = 2f\big[&(r_4r_5(r_1r_8 + r_2r_7) - r_1r_5^2r_7 - r_2r_4^2r_8)r^2 + r_6t_y(r_9t_x + r_3t_z) \\ &- r_3r_9t_y^2 + (1 - r_6^2)t_xt_z + (r_9t_x + r_3t_z)z_w + r_3r_9z_w^2\big] \qquad (5.90) \end{aligned}$$

$$\begin{aligned} E = 2f\big[&(r_1r_2(r_4r_8 + r_5r_7) - r_1^2r_5r_8 - r_2^2r_4r_7)r^2 + r_3t_x(r_9t_y + r_6t_z) \\ &- r_6r_9t_x^2 + (1 - r_3^2)t_yt_z + (r_9t_y + r_6t_z)z_w + r_6r_9z_w^2\big] \qquad (5.91) \end{aligned}$$

The center of the ellipse, $(\bar{x}, \bar{y})$, is given by the simultaneous solution of the equations, [17],

$$A\bar{x} + \frac{1}{2}B\bar{y} + \frac{1}{2}D = 0 \tag{5.92}$$

$$\frac{1}{2}B\bar{x} + C\bar{y} + \frac{1}{2}E = 0 \tag{5.93}$$

Using (5.92) and (5.93) we can solve for $(\bar{x}, \bar{y})$ in the following manner:

$$\begin{bmatrix} A & \frac{1}{2}B \\ \frac{1}{2}B & C \end{bmatrix} \begin{bmatrix} \bar{x} \\ \bar{y} \end{bmatrix} = \begin{bmatrix} -\frac{1}{2}D \\ -\frac{1}{2}E \end{bmatrix} \tag{5.94}$$

$$\begin{aligned} \begin{bmatrix} \bar{x} \\ \bar{y} \end{bmatrix} &= \frac{1}{AC - \frac{1}{4}B^2} \begin{bmatrix} C & -\frac{1}{2}B \\ -\frac{1}{2}B & A \end{bmatrix} \begin{bmatrix} -\frac{1}{2}D \\ -\frac{1}{2}E \end{bmatrix} \\ &= \frac{1}{AC - \frac{1}{4}B^2} \begin{bmatrix} (\frac{1}{4}BE - \frac{1}{2}CD) \\ (\frac{1}{4}BD - \frac{1}{2}AE) \end{bmatrix} \end{aligned} \tag{5.95}$$

$$\begin{bmatrix} \bar{x} \\ \bar{y} \end{bmatrix} = \frac{1}{B^2 - 4AC} \begin{bmatrix} (2CD - BE) \\ (2AE - BD) \end{bmatrix} \tag{5.96}$$

The centroid of the ellipse formed by projecting a circle onto the image plane of a camera does not, in general, coincide with the projection of the centroid of the circle onto the image plane of the camera.

However, we are primarily interested in the error introduced by this fact into the procedure of collecting calibration points described in Section 5.6.3. In the configuration of the CIRSSE Testbed system used in Section 5.6.4, the error introduced by assuming the center of the circular lightsource (i.e., the point determined by the kinematics of the robot) is the centroid of the ellipse formed by the projection of the disk of light produced by the lightsource, is quite small (on the order of 1/50th of a pixel).

### Robot Kinematic Errors

The calculation of a robot's forward kinematics is relatively straightforward and is given in Fu, Gonzalez, and Lee, [14], and also in Paul, [18]. The general method defines a coordinate frame at each joint of the robot and gives an exact closed-form solution to the forward kinematics based on the homogeneous transformation from one coordinate frame to the next.

Unfortunately, the pose of these coordinate frames is not known perfectly. The problem of measuring robot link lengths and the relative rotations between them, coupled with the simple fact that the joints and links themselves are not machined to perfection, results in a robot calibration problem as complex as the camera calibration problem.

The net effect is that, given all the joint positions, the location of the end-effector of a robot can be calculated with error between this calculated position and the actual position. In addition, this error is by no means constant or even linear. It is this error that is usually responsible for the largest portion of the camera calibration inaccuracies. The problem is that the location of the calibration points is simply not known exactly.

Given "perfect" calibration points, the calibration methods described in Section 5.3 return the exact calibration parameters of the camera viewing the "perfect" points. Given the imperfect calibration points obtained in practice, these calibration methods cannot return the correct calibration parameters. These camera calibration errors force the examination of methods to estimate the location of a point using two "calibrated" cameras. Two such methods are presented in Section 5.5.

### 5.3.5 Summary of Refinements to Calibration for Fixed Cameras

As a convenient reference, the steps used to calibrate a fixed camera with noisy data points are given below. Steps 1–5 align the world coordinate frame with the camera coordinate frame so that the Euler angles of the camera's rotation matrix can be reliably determined. Steps 6–9 place the world origin so that the $t_y$ is not "near" zero. Steps 10 and 11 calculate the stage 1 parameters subject to the constraint that the camera rotation matrix be orthonormal. Step 12 calculates the stage 2 parameters and step 13 ensures that the final extrinsic camera parameters are given with respect to the original world origin.

1. Translate the world origin to the centroid of the calibration data points (to reduce round-off error) and an arbitrary distance along each axis (to ensure that $t_y \neq 0$ at all four locations).

2. Assuming the image center is known, solve (5.11) for each of the four translations performed in step 1.

3. Use (5.46) to find the Euler angles of the camera rotation matrix determined from the solution found in step 2 with the smallest residual sum-squared-error.

4. Rotate the world coordinate frame by the Euler angles found in step 3.

5. Repeat steps 1–4 until the Euler angles found in step 3 are "small".

6. Translate the rotated world frame to the centroid of the data points and an arbitrary distance in the $y$ and $-y$ directions.

7. Assuming the image center is known, solve (5.11) at the three positions of the rotated world frame determined in step 6.

8. If translating the rotated world frame in either direction results in a larger residual sum-squared-error go to step 9, otherwise continue translating the world frame in the direction which produced the smallest residual sum-squared-error until the error increases or ceases to decrease "significantly".

9. Repeat steps 6–8 for the $z$ and $x$ directions.

10. Use (5.46) to determine the Euler angles of the camera rotation matrix, and translate the world frame to the centroid of the data points, adjusting $t_x$ and $t_y$ as necessary.

11. Using the parameters in step 10 as an initial guess, minimize (5.47) as described in Section 5.3.2. If desired the image center can be determined here, otherwise it must be assumed.

12. Using any of the four methods described in Section 5.3.3, determine the stage 2 parameters. If the image center was not calculated in step 11, then it should be calculated here, and steps 1–11 should be repeated until the final residual sum-squared-error increases or ceases to decrease "significantly".

13. Compensate the extrinsic camera parameters for the rotations and translations of the world frame (i.e., *un*-rotate and *un*-translate the camera parameters).

## 5.4 Calibration of a Laser Scanner

A laser is useful in 3-D visual sensing because it provides an active sensing capability. The laser emits a beam of light that a camera can detect as it reflects off of objects within the camera's field of view. An active sensing configuration, such as a laser and camera, can enhance the reliability and flexibility of a vision system since it can generate structured light and "ground truth". In some applications, the laser does not have to be calibrated (see Section 5.2), but calibration is necessary to fully utilize the capabilities of the laser.

The phrase "calibrated laser" is misleading in that the laser itself is not calibrated. It is usually incorporated in an assembly that can direct the laser beam in some well-defined manner. When the laser is calibrated, the entire assembly is actually calibrated. One useful laser assembly is a laser scanner which uses mirrors to deflect the laser beam in a controlled manner.

Calibrating a laser scanner is similar to calibrating a camera in that both devices have intrinsic and extrinsic parameters. For a laser scanner, intrinsic parameters include the distance between the scanner's mirrors and the relationships between the mirror's rotation and the voltage applied to their rotational mechanisms (galvanometers). A laser scanner's extrinsic parameters describe the pose of the laser scanner with respect to some coordinate frame. To better understand the process of laser scanner calibration, it is instructive to examine the internal arrangement of the device and identify the mathematical relationships that describe its operation.

### 5.4.1 The Mathematical Model of a Laser Scanner

The internal arrangement of the laser scanner is depicted in Figure 5.2. The device consists of a laser and two scanning mirrors. Each mirror is connected to a galvanometer that rotates the mirror as a function of the applied voltage. The mirrors are configured such that their rotational axes are mutually orthogonal. The rotational axis of the $\theta_x$ mirror is parallel with the $z$ axis of the laser scanner, and the rotational axis of the $\theta_y$ mirror is parallel to the $x$ axis of the laser scanner.

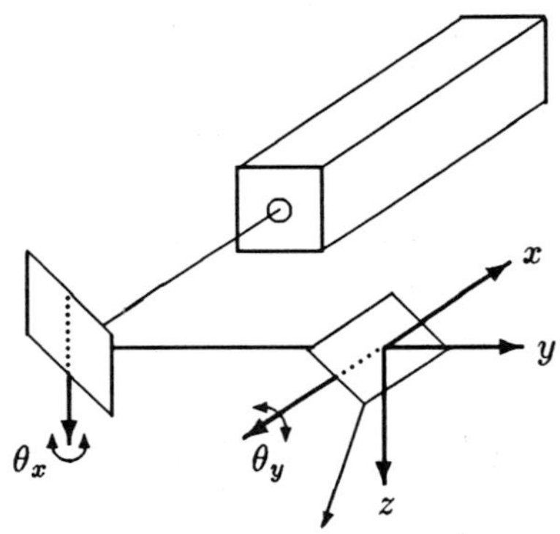


Figure 5.2: Internal Arrangement of a Laser Scanner

Prior research with laser scanners, [19], highlighted the problems associated with creating mathematical relationships between the deflection of the scanning mirrors and the vector of the outgoing laser ray. If the laser and mirrors are placed at arbitrary locations and orientations, these mathematical relationships become intractable. To alleviate these problems, two constraints on the placement of the laser and mirrors must be established. First, the beam emitted by the laser must be parallel to the rotational axis of the $\theta_y$ mirror. Second, the laser beam must intersect each mirror at a point along its rotational axis. These constraints are reasonable and practical when one considers that a laser scanner can be assembled with high precision using current manufacturing technology.

When the laser beam is scanned onto a plane normal to the laser's $z$ axis and at some fixed distance from the laser's origin, it is possible to determine the ($x$, $y$) coordinates of the laser spot as a function of the mirror's angular deflection. In this situation, the $y$ coordinate of the laser spot is strictly a function of $\theta_y$

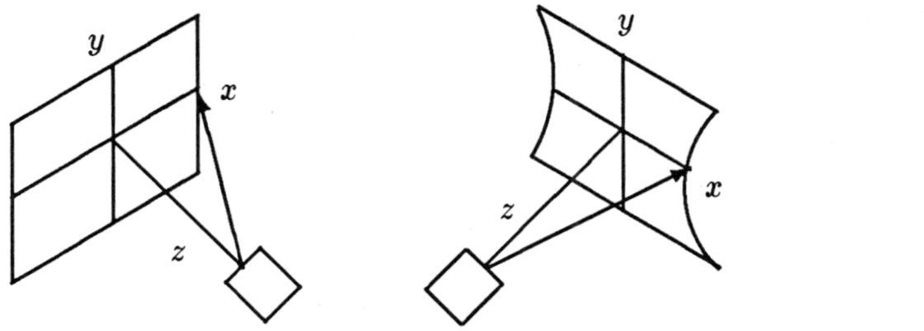


Figure 5.3: Effect of Pincushion Distortion

and the distance to the plane as shown below:

$$y = z \tan \theta_y \tag{5.97}$$

where $z$ is the distance from the laser's origin ($L$) to the plane normal to the laser's $z$ axis.

If the $x$ coordinate of the spot is assumed to be independent of $\theta_y$, then the laser exhibits *pincushion distortion* (as depicted in Figure 5.3). In reality, the $x$ coordinate of the spot is a function of both $\theta_x$ and $\theta_y$. This interdependency is due to the fact that the laser beam strikes the $\theta_y$ mirror after it is deflected by the $\theta_x$ mirror (see Figure 5.2). The expression for the $x$ coordinate of the laser spot is

$$x = (z \sec \theta_y + e) \tan \theta_x \tag{5.98}$$

where $z$ is the distance from $L$ as in (5.97), and $e$ is the distance between the two scanning mirrors. Note that the $z \sec \theta_y$ term increases as $|\theta_y|$ increases. Hence, the displacement of the laser spot "flares out" away from the origin in the $x$ direction as the spot moves away from the origin in the $y$ direction. Equations (5.97) and (5.98) provide the necessary relationships to direct the laser spot to any 3-D point defined with respect to the laser coordinate frame without pincushion distortion.

## 5.4.2 Calibration of the Intrinsic Parameters

The distance between the two mirrors ($e$) can be obtained by direct measurement. The degree to which errors in this measurement will affect the accuracy of the scanner depends on the environment in which it will be used. If the laser scanner is situated at a large $z$ distance from the workspace, then the effects of error in the measurement of $e$ will be reduced. This is the case in the CIRSSE Testbed (see Section 5.6), since the distance between the scanning mirrors is 5 mm and the $z$ distance to the workspace is typically 2000 mm. If the laser scanner is used in situations where $z$ is small, then the distance between the scanning mirrors should be determined analytically. This can be done in conjunction with determining the laser's extrinsic parameters (see Section 5.4.3).

Calibrating the scanning mirrors is critical to proper operation of the laser scanner. Each mirror is rotated with a galvanometer, which transforms a voltage into an angular rotation of the mirror. Zero volts is usually assumed to correspond to a mirror angle of zero degrees (e.g., the laser beam is assumed to coincide with the laser's $z$ axis when both galvanometers have zero input). The $\theta_y$ mirror is calibrated by directing the beam onto a plane at a fixed $z$ distance from the laser with respect to the laser's origin ($L$). With $\theta_x$ fixed at zero, the $\theta_y$ mirror is rotated with a fixed voltage and the amount of $y$ displacement on the plane is recorded. Using this information and (5.97) it is possible to determine $\theta_y$. Assuming the relationship between voltage and mirror rotation is linear, mirror rotation is determined by dividing $\theta_y$ by the voltage applied to the mirror. To confirm the linearity of the galvanometers, the $y$ mirror should be displaced to several different positions and the relationship should be verified to not change within measurement error. The $\theta_x$ mirror is calibrated in the same manner except that $\theta_y$ is fixed at zero (so there will be no pincushion distortion) and (5.98) is used to determine $\theta_x$.

### 5.4.3 Calibration of the Extrinsic Parameters Using an LSE Method

The extrinsic parameters of a laser scanner can be obtained using an LSE method as follows. Direct the laser at a set of 3-D points and record the scanning mirror angles at each point. These points and associated mirror angles can then be used to solve an overdetermined system of linear equations to obtain the laser's extrinsic parameters. The method described in this section is analogous to the LSE approach presented in Section 5.3.1.

The extrinsic parameters of the laser scanner consist of the rotation and translation of the laser coordinate frame with respect to some other fixed coordinate frame. This rotation and translation should ultimately be represented as a $4 \times 4$ homogeneous transformation of the form

$$
{}^{a}_{b}T = \begin{bmatrix} r_1 & r_2 & r_3 & t_x \\ r_4 & r_5 & r_6 & t_y \\ r_7 & r_8 & r_9 & t_z \\ 0 & 0 & 0 & 1 \end{bmatrix} \tag{5.99}
$$

The homogeneous transformation ${}^{a}_{b}T$ is primarily composed of a $3 \times 3$ rotation matrix and a $3 \times 1$ translation vector which define the orientation and position of frame $b$ with respect to frame $a$. To calibrate the laser we need to find ${}^{l}_{w}T$, which is the transformation from the laser coordinate frame to a desired world coordinate frame. What is required is a mathematical relationship that will determine these parameters given a set of points defined in the world coordinate frame and a set of corresponding scanning mirror angles. A point $\vec{P}_w$ defined in the world coordinate frame is transformed to the laser coordinate frame using (5.100) to produce $\vec{P}_l$ ($\vec{P}_l$ and $\vec{P}_w$ are $3 \times 1$ vectors):

$$\vec{P}_l = \begin{bmatrix} x_l \\ y_l \\ z_l \end{bmatrix} = {}^l_wT\vec{P}_w = {}^l_wT \begin{bmatrix} x_w \\ y_w \\ z_w \end{bmatrix} = \begin{bmatrix} r_1x_w + r_2y_w + r_3z_w + t_x \\ r_4x_w + r_5y_w + r_6z_w + t_y \\ r_7x_w + r_8y_w + r_9z_w + t_z \end{bmatrix} \tag{5.100}$$

Also, $x_l$ and $y_l$ can be expressed in terms of the scanning mirrors using (5.97) and (5.98) with (5.100) as shown below:

$$(z_l \sec\theta_y + e)\tan\theta_x = r_1x_w + r_2y_w + r_3z_w + t_x \tag{5.101}$$

$$z_l \tan\theta_y = r_4x_w + r_5y_w + r_6z_w + t_y \tag{5.102}$$

Substituting the expression for $z_l$ in (5.100) into (5.101) and (5.102) we have:

$$((r_7x_w + r_8y_w + r_9z_w + t_z)\sec\theta_y + e)\tan\theta_x = r_1x_w + r_2y_w + r_3z_w + t_x \tag{5.103}$$

$$(r_7x_w + r_8y_w + r_9z_w + t_z)\tan\theta_y = r_4x_w + r_5y_w + r_6z_w + t_y \tag{5.104}$$

Simplifying (5.103) and (5.104) and then dividing through by $t_z$ ($t_z \neq 0$), we have:

$$\frac{\tan\theta_x}{\cos\theta_y}x_w\frac{r_7}{t_z} + \frac{\tan\theta_x}{\cos\theta_y}y_w\frac{r_8}{t_z} + \frac{\tan\theta_x}{\cos\theta_y}z_w\frac{r_9}{t_z} + \frac{\tan\theta_x}{\cos\theta_y} + \tan\theta_x\frac{e}{t_z} = x_w\frac{r_1}{t_z} + y_w\frac{r_2}{t_z} + z_w\frac{r_3}{t_z} + \frac{t_x}{t_z} \tag{5.105}$$

$$x_w\tan\theta_y\frac{r_7}{t_z} + y_w\tan\theta_y\frac{r_8}{t_z} + z_w\tan\theta_y\frac{r_9}{t_z} + \tan\theta_y = x_w\frac{r1}{t_z} + y_w\frac{r_2}{t_z} + z_w\frac{r_3}{t_z} + \frac{t_y}{t_z} \tag{5.106}$$

Equations (5.105) and (5.106) can be expressed in the form $Ax = b$, where $A$ is $2n \times 12$, $b$ is $2n \times 1$, $n$ is the number of data points collected, and $x$ is a $12 \times 1$ vector of unknowns. Note that this system of equations not only determines the laser's extrinsic parameters, but also the distance between the scanning mirrors ($e$). The final form (for $t_z \neq 0$) is presented below:

$$\begin{bmatrix} \tan\theta_{x_1} & -x_{w_1} & -y_{w_1} & -z_{w_1} & 0 & 0 & 0 & \frac{\tan\theta_{x_1}}{\cos\theta_{y_1}}x_{w_1} \\ 0 & 0 & 0 & 0 & -x_{w_1} & -y_{w_1} & -z_{w_1} & x_{w_1}\tan\theta_{y_1} \\ \vdots & \vdots & \vdots & \vdots & \vdots & \vdots & \vdots & \vdots \\ \tan\theta_{x_n} & -x_{w_n} & -y_{w_n} & -z_{w_n} & 0 & 0 & 0 & \frac{\tan\theta_{x_n}}{\cos\theta_{y_n}}x_{w_n} \\ 0 & 0 & 0 & 0 & -x_{w_n} & -y_{w_n} & -z_{w_n} & x_{w_n}\tan\theta_{y_n} \end{bmatrix}$$

$$
\begin{bmatrix}
\frac{\tan\theta_{x_1}}{cos\theta_{y_1}} y_{w_1} & \frac{\tan\theta_{x_1}}{cos\theta_{y_1}} z_{w_1} & -1 & 0 \\
y_{w_1}\tan\theta_{y_1} & z_{w_1}\tan\theta_{y_1} & 0 & -1 \\
\vdots & \vdots & \vdots & \vdots \\
\frac{\tan\theta_{x_n}}{cos\theta_{y_n}} y_{w_n} & \frac{\tan\theta_{x_n}}{cos\theta_{y_n}} z_{w_n} & -1 & 0 \\
y_{w_n}\tan\theta_{y_n} & z_{w_n}\tan\theta_{y_n} & 0 & -1
\end{bmatrix}
\begin{bmatrix} e \\ r_1 \\ r_2 \\ r_3 \\ r_4 \\ r_5 \\ r_6 \\ r_7 \\ r_8 \\ r_9 \\ t_x \\ t_y \end{bmatrix}
\frac{1}{t_z} =
\begin{bmatrix}
\frac{\tan\theta_{y_1}}{\cos\theta_{x_1}} \\
\tan\theta_{y_1} \\
\vdots \\
\frac{\tan\theta_{y_n}}{\cos\theta_{x_n}} \\
\tan\theta_{y_n}
\end{bmatrix}
\tag{5.107}
$$

After solving (5.107) (using SVD), $|t_z|$ can be found from

$$
|t_z| = \frac{1}{\sqrt{\left(\frac{r_7}{t_z}\right)^2 + \left(\frac{r_8}{t_z}\right)^2 + \left(\frac{r_9}{t_z}\right)^2}}
\tag{5.108}
$$

Once $|t_z|$ is determined, it is a straightforward process to obtain $e$, $r_1$, through $r_9$, $t_x$ and $t_y$. With several hundred data points, this method produces reliable results provided that the data points are measured accurately (i.e., measurement error is $\leq$ 1 mm). The major flaw in this calibration method is that it treats the twelve parameters as being independent (which is obviously incorrect) and thereby fails to meet the constraints inherent in the rotation matrix.

### 5.4.4 Direct Geometric Method for Calibrating a Laser Scanner

The LSE approach described in the previous section requires a large number of accurate data points to generate an accurate solution. There are situations where it is impractical to collect a large number of data points. However, it is possible to measure the laser scanner's extrinsic parameters directly since the laser emits a beam of light that can be measured with respect to a reference point. The method presented in this section treats laser scanner calibration as a geometrical problem in which the laser scanner's Euler angles and location with respect to the world coordinate frame are measured directly. The objective of this calibration method is identical to that of the LSE method: to determine the homogeneous transform from the laser coordinate frame to some world coordinate frame.

The laser coordinate frame ($L$) is located at the center of the $\theta_y$ mirror and is oriented as in Figure 5.2. The pose of the world coordinate frame ($W$) is arbitrary. To assist in calibrating the laser, an intermediate coordinate frame $F$

is defined whose origin is located directly below the center of the laser scanner's aperture. The frame $F$ is located by suspending a plumb line from the laser scanner to some fixed plane (the floor is used in the CIRSSE system). This plumb line constitutes the $z$ axis of the $F$ coordinate frame. Note that while $F$'s origin is on the floor, $F$'s $x$ and $y$ axes are not necessarily coplanar with the floor. The location of $F$ is selected in this manner to simplify the measurement of the laser scanner's orientation. The calibration procedure involves measuring twelve parameters, which are depicted in Figures 5.4 and 5.5 and defined below:

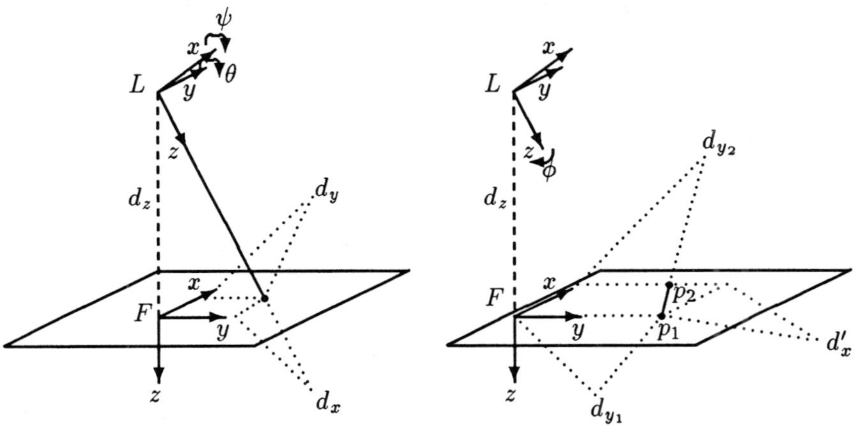


Figure 5.4: Determination of a Laser Scanner's Euler Angles. (Note: $F$ is not necessarily coplanar with the floor).

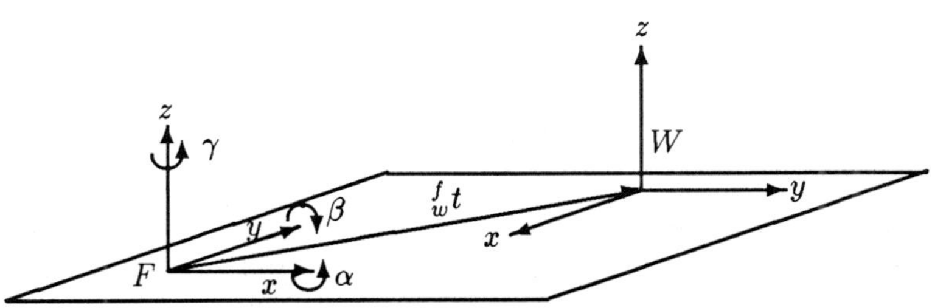


Figure 5.5: Transformation Parameters from Frame $F$ to Frame $W$. (Note: $F$ is not necessarily coplanar with the floor).

$d_z$ The distance from $F$ to $L$ along $F$'s $z$ axis.

$d_x$ The $x$ coordinate of the point where the undeflected laser beam intersects the floor, measured with respect to the origin of $F$.

$d_y$ The $y$ coordinate of the point where the undeflected laser beam intersects the floor, measured with respect to the origin of $F$.

$d_{y_1}$ The $y$ coordinate of the point $p_1$ on the floor, measured with respect to the origin of $F$.

$d_{y_2}$ The $y$ coordinate of the point $p_2$ on the floor, measured with respect to the origin of $F$.

$d'_x$ The length of the projection of the line segment joining $p_1$ and $p_2$ onto the vector formed by projecting the $x$ axis of frame $F$ onto the plane of the floor.

$\rho$ The rotation about $F$'s $x$ axis from $F$'s $xy$ plane to the floor.

$\delta$ The rotation about $F$'s $y$ axis from $F$'s $xy$ plane to the floor.

${}^{f}_{w}t$ The translation vector from $F$ to $W$.

$\alpha$ The pitch angle about $F$'s $x$ axis from frame $F$ to frame $W$

$\beta$ The yaw angle about $F$'s $y$ axis from frame $F$ to frame $W$

$\gamma$ The roll angle about $F$'s $z$ axis from frame $F$ to frame $W$

The above twelve parameters provide all the information required to determine the transformation from the laser coordinate frame to the world coordinate frame: ${}^{l}_{w}T$. The calculations are broken into two steps:

1. Determining ${}^{l}_{f}T$
2. Determining ${}^{f}_{w}T$

Once these transformations are known, it will be possible to determine ${}^{l}_{w}T$ from

$$ {}^{l}_{w}T = {}^{l}_{f}T \cdot {}^{f}_{w}T \tag{5.109} $$

${}^{l}_{f}R$ can be determined by deriving the orientation of frame $F$ with respect to the laser in terms of the Euler angles pitch, yaw and roll ($\psi$,$\theta$,$\phi$) about the laser's $x$, $y$, and $z$ axes, respectively. Figure 5.4 shows how the Euler angles are measured independently of one another. The laser's scanning mirrors are set to zero and the coordinates of the point where the laser beam intersects the floor are measured ($d_x$, $d_y$). It is assumed that any offset of the scanning mirrors from zero is negligible compared to the magnitude of the Euler angles being measured. Since the mirrors are assumed to be in their undeflected state, the beam is coincident with the laser's $z$ axis, and the point ($d_x$, $d_y$) is invariant to rotations about $L$'s $z$ axis. We will therefore assume that the laser coordinate frame is rotated about $z$ by an amount $\phi$ (to be determined) so that the projection of

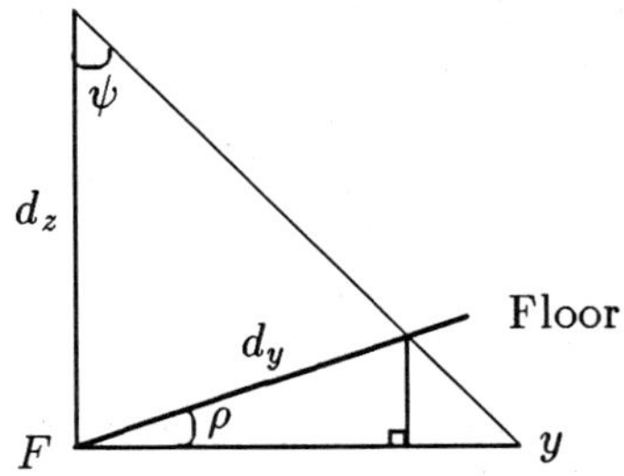


Figure 5.6: Relationship Between the Floor and $F$'s $xy$ Plane

the laser's $y$ axis onto $F$'s $xy$ plane is coincident with $F$'s $y$ axis. Such an orientation implies that $d_x$ is only a function of $\theta$ and $d_y$ is only a function of $\psi$.

When calculating $\psi$ and $\theta$, it cannot be assumed that the floor correctly defines the $xy$ plane of $F$. Indeed, since $d_z$ was measured with a plumb bob, $F$'s $z$ axis is aligned with Earth's gravity vector, but the floor may not be orthogonal to this vector. By using a level it is possible to determine the angles $\rho$ and $\delta$ about $F$'s $x$ and $y$ axes, respectively, between the plane of the floor and the $xy$ plane of $F$. The situation is depicted in Figure 5.6 for the Euler angle $\psi$. The formulas for determining $\psi$ and $\theta$ can be derived directly from the figure and are presented below:

$$\psi = \tan^{-1}\left[\frac{d_y \cos\rho + d_y \sin\rho \tan\rho}{d_z}\right] \tag{5.110}$$

$$\theta = \tan^{-1}\left[\frac{d_x \cos\delta + d_x \sin\delta \tan\delta}{d_z}\right] \tag{5.111}$$

At this point it is necessary to determine $\phi$ so that the laser's $y$ axis can be aligned with $F$'s $y$ axis. The $\theta_x$ mirror is repeatedly rotated by an arbitrary amount while the $\theta_y$ mirror is set to zero, resulting in a line segment traced on the floor. The slope of this line determines $\phi$ and is found by measuring two arbitrary points ($p_1$ and $p_2$) on the line segment (this yields the parameters $d_{y_1}$, $d_{y_2}$, and $d'_x$).

The primary issue with determining $\phi$ in this manner is ensuring the slope of the line segment to be a function only of $\phi$ and not of other variables, such as $\psi$, $\theta$, or $\theta_y$ (which causes pincushion distortion). Because we have assumed $\theta_y = 0$, the slope of the line segment will not be altered by pincushion distortion. The Euler angle $\theta$ affects the laser's $x$ and $z$ components of the beam direction. A change in the $z$ direction will not affect the slope of the line since the line ultimately will lie in $F$'s $xy$ plane. Further, distortion in $x$ will tend to move the line segment by some constant value, leaving the slope unchanged. The Euler angle $\psi$ affects the laser's $y$ and $z$ components of the beam direction. The distortion in $z$ will not affect the measurements for the same reasons stated for $\theta$. The distortion in $y$ due to $\psi$ consists of a constant translation of the line segment along $F$'s $y$ axis. The slope of the line is a function of the relative change in $y$ from points $p_1$ to $p_2$, so the effects of $\psi$ will not change the slope since $p_1$ and $p_2$ will be translated in $y$ by the same amount. Hence, the slope of the line segment is a function only of $\phi$.

Since the measurements used to determine $\phi$ were taken from the floor, they will have to be corrected for the effects of $\rho$ and $\delta$ for the same reasons that the corrections were necessary for determining $\psi$ and $\theta$. The final equation for $\phi$ is presented below:

$$\phi = -tan^{-1}\left[\frac{(d_{y_2}\cos\rho + d_{y_2}\sin\rho\tan\rho) - (d_{y_2}\cos\rho + d_{y_2}\sin\rho\tan\rho)}{d'_x\cos\delta + d'_x\sin\delta\tan\delta}\right] \tag{5.112}$$

At this point the Euler angles $\psi$, $\theta$, $\phi$ have been determined, and it is now possible to create the rotation matrix ${}^l_f R$ from the laser to the F coordinate frame. Each Euler angle changes four parameters in the rotational matrix. The most straightforward approach is to determine separate rotational matrices for each Euler angle and then multiply them together to obtain ${}^l_f R$. The individual matrices for R(X,$\psi$), R(Y,$\theta$) and R(Z,$\phi$) are:

$$R(X,\psi) = \begin{bmatrix} 1 & 0 & 0 \\ 0 & \cos\psi & -\sin\psi \\ 0 & \sin\psi & \cos\psi \end{bmatrix} \tag{5.113}$$

$$R(Y,\theta) = \begin{bmatrix} \cos\theta & 0 & \sin\theta \\ 0 & 1 & 0 \\ -\sin\theta & 0 & \cos\theta \end{bmatrix} \tag{5.114}$$

$$R(Z,\phi) = \begin{bmatrix} \cos\phi & -\sin\phi & 0 \\ \sin\phi & \cos\phi & 0 \\ 0 & 0 & 1 \end{bmatrix} \tag{5.115}$$

$${}^l_f R(\phi,\theta,\psi) = R(X,\psi)R(Y,\theta)R(Z,\phi) \tag{5.116}$$

As is apparent from (5.116), ${}^l_f R$ is obtained by combining the rotation matrices in the order roll, yaw, pitch. This ordering is essential for this calibration procedure. Applying the roll rotation first will align the projections of the laser's $y$ axis to the $F$ coordinate frame's $y$ axis. This condition was assumed when the Euler angles $\psi$ and $\theta$ were determined.

At this point, ${}^l_f R$ has been determined, but to obtain ${}^l_f T$ it is necessary to determine the translational component ${}^l_f t$ from the laser's origin to $F$ with respect to the laser. There is no method of directly measuring this value with any degree of accuracy. However, the translation vector ${}^f_l t$ from $F$ to the laser can be determined since $F$ is located directly below the aperture of the laser, and $d_z$ is known. The resulting value of ${}^f_l t$ is $[0, 0, d_z]^T$.

At this point ${}^f_l T$ can be determined by combining the inverse of ${}^l_f R$ (for a rotation matrix, $R^{-1} = R^T$) with the translation vector ${}^f_l t$. It is then possible to obtain ${}^l_f T$ by taking the inverse of ${}^f_l T$. This relationship is defined as

$$ {}^l_f T = \begin{bmatrix} & {}^l_f R^T & & {}^f_l t \\ 0 & 0 & 0 & 1 \end{bmatrix}^{-1} \tag{5.117} $$

With ${}^l_f T$ defined, the next step is to determine ${}^f_w T$. $W$ should be chosen so that the Euler angles $\alpha$, $\beta$, and $\gamma$ can be easily measured. If $W$ is chosen such that its $z$ axis is plumb (as in the case with $F$), then the Euler angles from $F$ to $W$ can be readily determined. By applying (5.113) through (5.116), it is possible to determine ${}^f_w R$, and this matrix can be combined with ${}^f_w t$ to obtain ${}^f_w T$. At this point, ${}^l_w T$ can be determined by matrix multiplication:

$$ {}^l_w T = {}^l_f T \cdot {}^f_w T \tag{5.118} $$

This concludes the calibration of the laser's extrinsic parameters. It is now possible to transform points defined in the world coordinate frame into points defined in the laser coordinate frame. Additionally, points in the laser frame can be transformed into the world frame using the inverse of ${}^l_w T$.

When presented with two alternatives for calibrating the laser's extrinsic parameters, the question arises as to which method is best. The answer depends on how and where the laser scanner will be used. The transform obtained using the direct geometrical method typically results in errors of less than 0.5%. This level of accuracy is sufficient for many visual sensing tasks. Further, the geometric approach achieves its results in a simple systematic manner. Hence, calibrating the laser using this method requires less effort than the LSE method.

Since the LSE method generates a solution based on a large set of data points, measurement errors among individual points should have less affect on the LSE solution. This is in contrast to the geometric method which uses a small number of measurements to obtain its results. If it is possible to collect a large number of points with high accuracy, the LSE could produce more accurate results than the direct geometric method.

### 5.4.5 Locating a Laser Spot in a Camera Image

To use a calibrated laser scanner in concert with a camera, it is necessary to be able to locate the laser spot in the camera image. This is a simple problem if one can guarantee that the laser spot is the brightest region in the image.

However, such an assumption restricts the utility of a calibrated laser by placing illumination constraints on the image. If methods can be used to locate the laser spot in the presence of "noise" of similar intensity, then a calibrated laser can be used in a wide variety of situations. The method used to locate the laser spot in a noisy image is usually a heuristic approach whereby regions in the image are successively eliminated based on a set of criterion tests.

The first step in locating the laser spot is to perform region growing over some selected area of the image. This results in a list of regions, their area, and their centroids. The laser spot should be one of the regions in this list. To isolate the laser spot it is necessary to eliminate all those regions that are not attributable to the laser. There are four different tests that can be applied to the region list to perform this elimination. Each test returns a list of the regions that passed the test. The laser spot should be the only region that passes all four tests.

The first test is to eliminate all regions that do not fall within a specified intensity range. Since the laser will appear as a small bright spot, it will be one of the brighter regions in the image. However, there is no guarantee that the laser spot will be among the brightest. Indeed, if the laser beam is illuminating a matte (low reflectivity) object, such as a piece of cloth, then the intensity of the laser spot will be lower than if the beam was reflecting off a piece of metal. Additionally, specular reflections of ambient light off of high reflectivity objects can possess the same intensity as the laser spot.

A second test that can be applied to the region list is to eliminate regions that do not fall within a certain range of sizes. The laser spot typically occupies between two pixels and twenty pixels depending on the reflectance of the object the laser beam is striking. The tests for size and intensity can detect the presence of a laser spot in the image in most cases. Problems arise when there are other regions in the image that have the same size and intensity characteristics as the laser, such as specular reflections.

If after application of the intensity and size tests the region list still has more than one candidate region, two more tests can be applied to further reduce the list. One of these tests is to take the centroid coordinates of each region and the known scanning mirror angles, run them through a point estimation algorithm, and eliminate the regions that generate solutions that fall outside of the workspace. This method does assume that some *a priori* knowledge exists about the expected location of the laser spot. The more that is known about the expected location of the laser spot, the greater the chance of identifying the laser spot. Usually, very little *a priori* knowledge is required to locate the laser spot, since only those regions that lie along the laser ray will generate results that are reasonable.

If all the previous tests have failed to return a unique solution for the laser spot, then a final test can be employed. The laser beam can be moved and another image acquired. The new image is passed through the region growing algorithm just as the first image. If the scene is static, the only region that should have moved is the laser spot. Hence, all regions that have the same centroid coordinates across both images should be eliminated. This final test usually yields a unique solution for the laser spot. It is the most computationally expensive test since it requires capturing two images. As such, it may not be the best approach when efficiency is required.

It is important to note that it is not necessary to use all four tests. If a subset of tests yields one region, then the remaining tests do not have to be run. Further, there are situations where it may be impossible to locate the laser spot. If the laser spot is within the bounds of a bright region the camera may not be able to distinguish it. This problem is particularly acute if the camera's aperture is too wide, since bright regions could then saturate the camera's CCD element. The four tests described above provide a set of tools that can work under many conditions.

## 5.5 Three-Dimensional Point Estimation Methods

Once the image correspondence problem is solved and each camera is calibrated with respect to some arbitrary world coordinate frame, it is possible to determine an estimate for the 3-D location of the point being viewed with respect to this arbitrary world coordinate frame.

Two methods of estimating 3-D points using two calibrated cameras are discussed. The first method develops an overdetermined system of equations to calculate an estimate for the location of the point. The second method determines an estimate for the location of the point by calculating the midpoint of the common normal to the rays on which the point is expected to lie. A comparison of these two methods is given in Section 5.5.3.

### 5.5.1 Overdetermined System of Linear Equations Method

Consider the lens-centered coordinate frame with a front image plane shown in Figure 5.7.

Since $\vec{x}$ is a scaled version of $\vec{x}_u$ we may write

$$\vec{x} = k\vec{x}_u \quad \Rightarrow \quad \begin{bmatrix} x \\ y \\ z \end{bmatrix} = k \begin{bmatrix} x_u \\ y_u \\ f \end{bmatrix} \quad \Rightarrow \quad k = \frac{z}{f}, \quad x_u = \frac{f}{z}x, \quad y_u = \frac{f}{z}y \tag{5.119}$$

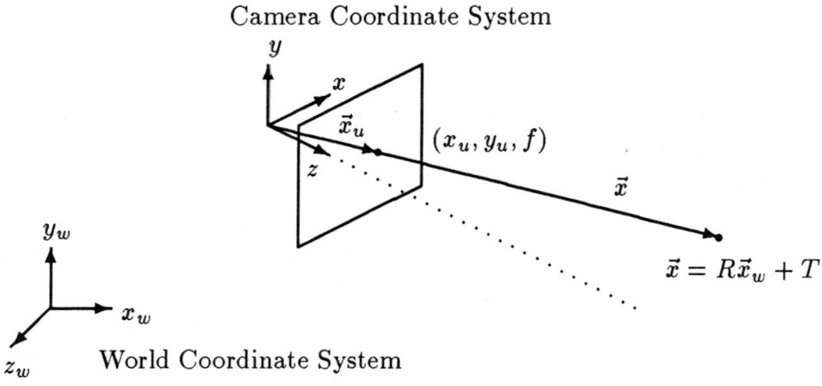


Figure 5.7: Single Image Plane

Using homogeneous coordinates

$$\begin{bmatrix} x_u \\ y_u \\ 1 \end{bmatrix} = \begin{bmatrix} 1 & 0 & 0 & 0 \\ 0 & 1 & 0 & 0 \\ 0 & 0 & \frac{1}{f} & 0 \end{bmatrix} \begin{bmatrix} x \\ y \\ z \\ 1 \end{bmatrix} = \begin{bmatrix} x \\ y \\ \frac{z}{f} \end{bmatrix} \tag{5.120}$$

so,

$$\begin{bmatrix} x_u \\ y_u \\ 1 \end{bmatrix} = P \begin{bmatrix} x \\ y \\ z \\ 1 \end{bmatrix} \tag{5.121}$$

Further,

$$\begin{bmatrix} x \\ y \\ z \end{bmatrix} = R \begin{bmatrix} x_w \\ y_w \\ z_w \end{bmatrix} + T \tag{5.122}$$

or, in homogeneous coordinates,

$$\begin{bmatrix} x \\ y \\ z \\ 1 \end{bmatrix} = \begin{bmatrix} & R & & T \\ & & & \\ 0 & 0 & 0 & 1 \end{bmatrix} \begin{bmatrix} x_w \\ y_w \\ z_w \\ 1 \end{bmatrix} = \begin{bmatrix} r_1 & r_2 & r_3 & t_x \\ r_4 & r_5 & r_6 & t_y \\ r_7 & r_8 & r_9 & t_z \\ 0 & 0 & 0 & 1 \end{bmatrix} \begin{bmatrix} x_w \\ y_w \\ z_w \\ 1 \end{bmatrix} = H \begin{bmatrix} x_w \\ y_w \\ z_w \\ 1 \end{bmatrix} \tag{5.123}$$

Therefore,

$$\begin{bmatrix} x_u \\ y_u \\ 1 \end{bmatrix} = PH \begin{bmatrix} x_w \\ y_w \\ z_w \\ 1 \end{bmatrix} = \begin{bmatrix} r_1 & r_2 & r_3 & t_x \\ r_4 & r_5 & r_6 & t_y \\ \frac{r_7}{f} & \frac{r_8}{f} & \frac{r_9}{f} & \frac{t_z}{f} \end{bmatrix} \begin{bmatrix} x_w \\ y_w \\ z_w \\ 1 \end{bmatrix}$$

$$= \begin{bmatrix} (r_1 x_w + r_2 y_w + r_3 z_w + t_x) \\ (r_4 x_w + r_5 y_w + r_6 z_w + t_y) \\ \left(\frac{r_7 x_w}{f} + \frac{r_8 y_w}{f} + \frac{r_9 z_w}{f} + \frac{t_z}{f}\right) \end{bmatrix} \tag{5.124}$$

Hence,

$$x_u = \frac{f(r_1 x_w + r_2 y_w + r_3 z_w + t_x)}{r_7 x_w + r_8 y_w + r_9 z_w + t_z} \tag{5.125}$$

$$y_u = \frac{f(r_4 x_w + r_5 y_w + r_6 z_w + t_y)}{r_7 x_w + r_8 y_w + r_9 z_w + t_z} \tag{5.126}$$

For camera number n,

$$x_{u_n} = \frac{f_n(r_{1_n} x_w + r_{2_n} y_w + r_{3_n} z_w + t_{x_n})}{r_{7_n} x_w + r_{8_n} y_w + r_{9_n} z_w + t_{z_n}} \tag{5.127}$$

$$y_{u_n} = \frac{f_n(r_{4_n} x_w + r_{5_n} y_w + r_{6_n} z_w + t_{y_n})}{r_{7_n} x_w + r_{8_n} y_w + r_{9_n} z_w + t_{z_n}} \tag{5.128}$$

which leads to

$$x_w(r_{7_n} x_{u_n} - f_n r_{1_n}) + y_w(r_{8_n} x_{u_n} - f_n r_{2_n}) + z_w(r_{9_n} x_{u_n} - f_n r_{3_n}) = (f_n t_{x_n} - t_{z_n} x_{u_n}) \tag{5.129}$$

$$x_w(r_{7_n} y_{u_n} - f_n r_{4_n}) + y_w(r_{8_n} y_{u_n} - f_n r_{5_n}) + z_w(r_{9_n} y_{u_n} - f_n r_{6_n}) = (f_n t_{y_n} - t_{z_n} y_{u_n}) \tag{5.130}$$

So, each camera provides two equations in three unknowns. With $n$ cameras we have $A\vec{x} = \vec{b}$ where $A$ is $2n \times 3$, $\vec{x}$ is $3 \times 1$, and $\vec{b}$ is $2n \times 1$.

So, for $n = 1, 2$ we have

$$\begin{bmatrix} (r_{7_1} x_{u_1} - f_1 r_{1_1}) & (r_{8_1} x_{u_1} - f_1 r_{2_1}) & (r_{9_1} x_{u_1} - f_1 r_{3_1}) \\ (r_{7_1} y_{u_1} - f_1 r_{4_1}) & (r_{8_1} y_{u_1} - f_1 r_{5_1}) & (r_{9_1} y_{u_1} - f_1 r_{6_1}) \\ (r_{7_2} x_{u_2} - f_2 r_{1_2}) & (r_{8_2} x_{u_2} - f_2 r_{2_2}) & (r_{9_2} x_{u_2} - f_2 r_{3_2}) \\ (r_{7_2} y_{u_2} - f_2 r_{4_2}) & (r_{8_2} y_{u_2} - f_2 r_{5_2}) & (r_{9_2} y_{u_2} - f_2 r_{6_2}) \end{bmatrix} \begin{bmatrix} x_w \\ y_w \\ z_w \end{bmatrix} = \begin{bmatrix} (f_1 t_{x_1} - t_{z_1} x_{u_1}) \\ (f_1 t_{y_1} - t_{z_1} y_{u_1}) \\ (f_2 t_{x_2} - t_{z_2} x_{u_2}) \\ (f_2 t_{y_2} - t_{z_2} y_{u_2}) \end{bmatrix} \tag{5.131}$$

The closed-form solution to (5.131) is quite cumbersome and does not easily lend itself to simplification, especially when we are unsure of the camera calibration parameters. This can occur when imperfect calibration data causes the chosen calibration procedure to produce an $R$ matrix that is not a perfect rotation matrix. Even though the calibration procedure developed by the authors does produce orthonormal rotation matrices, (5.131) does not simplify well. However, (5.131) may be solved numerically using SVD [11].

The overdetermined system of linear equations approach for dual cameras can readily be adapted to a camera and a laser scanner. What is needed is a relationship for the laser between the scanning mirror angles and the coordinates of a 3-D point ($\vec{P}_w$) defined in the world frame. Such a relationship was derived in Section 5.4.3 as part of the LSE method for calibrating the laser (see (5.103) and (5.104)).

In the context of laser calibration, the unknown variables were the distance between the scanning mirrors ($e$) and the rotation and translation components of the transformation from the laser coordinate frame to the world coordinate frame. Since the laser is assumed to be calibrated at this point, all these values are known. Additionally, the angles of the scanning mirrors are known. What is not known are the $x$, $y$, and $z$ coordinates of the world point. The terms in (5.103) and (5.104) can be regrouped into a form that is more suitable for point estimation:

$$\left(\frac{\tan\theta_x}{\cos\theta_y}r_7 - r_1\right)x_w + \left(\frac{\tan\theta_x}{\cos\theta_y}r_8 - r_2\right)y_w + \left(\frac{\tan\theta_x}{\cos\theta_y}r_9 - r_3\right)z_w = t_x - t_z\frac{\tan\theta_x}{\cos\theta_y} - e\tan\theta_x \tag{5.132}$$

$$(r_7\tan\theta_y - r_4)x_w + (r_8\tan\theta_y - r_5)y_w + (r_9\tan\theta_y - r_6)z_w = t_y - t_z\tan\theta_y \tag{5.133}$$

Therefore, the laser contributes two equations with the same three unknowns as the camera equations (assuming that the laser and camera are calibrated to the same world coordinate frame). Thus (5.132) and (5.133) can be substituted into (5.131) in place of the equations for the second camera.

### 5.5.2 Midpoint of the Common Normal Method

Corresponding points in the image planes of two calibrated cameras define two 3-D rays. In theory, these rays should intersect, and the point of their intersection defines the 3-D location of the point in the two image planes. In practice, the calibration parameters of the cameras are not exactly known. This causes the 3-D rays to be skew. Given that the 3-D point in question is supposed to lie at the intersection of these rays (which do not intersect), and given that it is not known which (if either) ray lies closest to the point, a reasonable estimate of the location of the point would be where the rays come closest together. This point of closest approach occurs at the midpoint of the common normal to the rays.

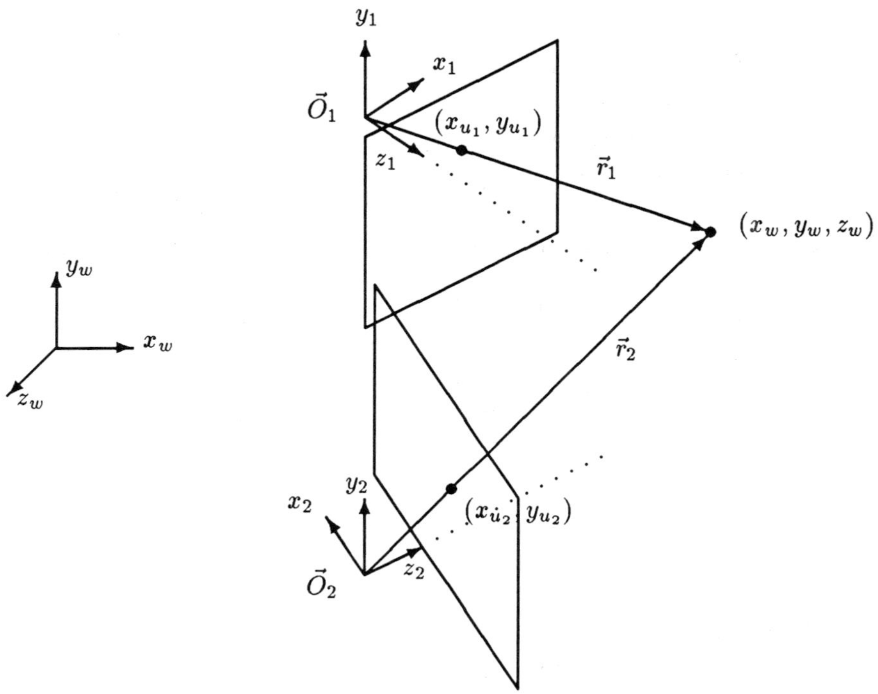


Figure 5.8: Two Front Image Planes

From (5.122) we know that a point expressed in the local coordinate frame of a camera is given by

$$\begin{bmatrix} x \\ y \\ z \end{bmatrix} = R \begin{bmatrix} x_w \\ y_w \\ z_w \end{bmatrix} + T \tag{5.134}$$

Therefore, the origins of the lens-centered coordinate frames for cameras one and two in world coordinates are, (see Figure 5.8),

$$\vec{O}_1 = -R_1^T T_1 = - \begin{bmatrix} (r_{1_1} t_{x_1} + r_{4_1} t_{y_1} + r_{7_1} t_{z_1}) \\ (r_{2_1} t_{x_1} + r_{5_1} t_{y_1} + r_{8_1} t_{z_1}) \\ (r_{3_1} t_{x_1} + r_{6_1} t_{y_1} + r_{9_1} t_{z_1}) \end{bmatrix} \tag{5.135}$$

$$\vec{O}_2 = -R_2^T T_2 = - \begin{bmatrix} (r_{1_2} t_{x_2} + r_{4_2} t_{y_2} + r_{7_2} t_{z_2}) \\ (r_{2_2} t_{x_2} + r_{5_2} t_{y_2} + r_{8_2} t_{z_2}) \\ (r_{3_2} t_{x_2} + r_{6_2} t_{y_2} + r_{9_2} t_{z_2}) \end{bmatrix} \tag{5.136}$$

The directions (in world coordinates) of the 3-D rays $\vec{r}_1$ and $\vec{r}_2$ are

$$\vec{d}_1 = R_1^T \begin{bmatrix} x_{u_1} - t_{x_1} \\ y_{u_1} - t_{y_1} \\ f_1 - t_{z_1} \end{bmatrix} + R_1^T T_1 = \begin{bmatrix} (r_{1_1} x_{u_1} + r_{4_1} y_{u_1} + r_{7_1} f_1) \\ (r_{2_1} x_{u_1} + r_{5_1} y_{u_1} + r_{8_1} f_1) \\ (r_{3_1} x_{u_1} + r_{6_1} y_{u_1} + r_{9_1} f_1) \end{bmatrix} \quad (5.137)$$

$$\vec{d}_2 = R_2^T \begin{bmatrix} x_{u_2} - t_{x_2} \\ y_{u_2} - t_{y_2} \\ f_2 - t_{z_2} \end{bmatrix} + R_2^T T_2 = \begin{bmatrix} (r_{1_2} x_{u_2} + r_{4_2} y_{u_2} + r_{7_2} f_2) \\ (r_{2_2} x_{u_2} + r_{5_2} y_{u_2} + r_{8_2} f_2) \\ (r_{3_2} x_{u_2} + r_{6_2} y_{u_2} + r_{9_2} f_2) \end{bmatrix} \quad (5.138)$$

Therefore, the 3-D rays themselves are expressed in world coordinates as

$$\vec{r}_1 = R_1^T \left( s_1 \begin{bmatrix} x_{u_1} \\ y_{u_1} \\ f_1 \end{bmatrix} - T_1 \right) = s_1 \vec{d}_1 + \vec{O}_1, \qquad s_1 \geq 0 \quad (5.139)$$

$$\vec{r}_2 = R_2^T \left( s_2 \begin{bmatrix} x_{u_2} \\ y_{u_2} \\ f_2 \end{bmatrix} - T_2 \right) = s_2 \vec{d}_2 + \vec{O}_2, \qquad s_2 \geq 0 \quad (5.140)$$

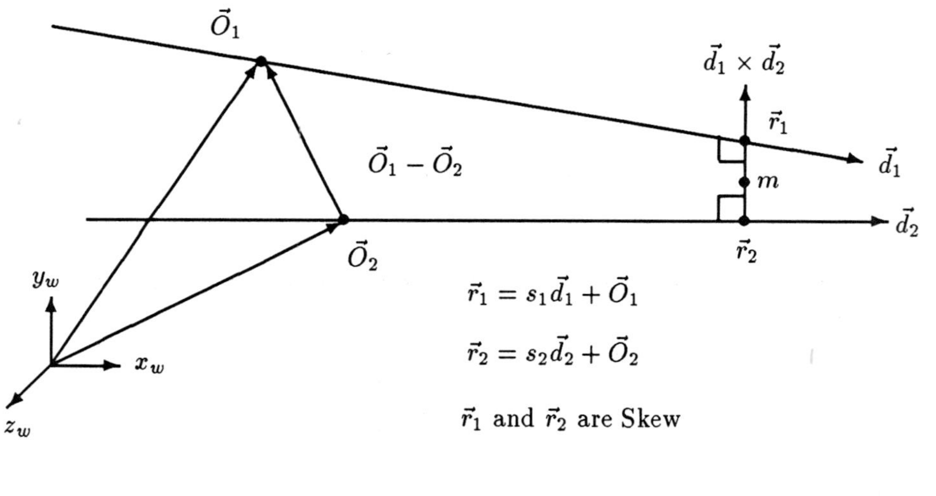


Figure 5.9: Midpoint of the Common Normal

Clearly, $\vec{d}_1 \times \vec{d}_2$ is normal to both $\vec{r}_1$ and $\vec{r}_2$ (see Figure 5.9). The unit direction (in world coordinates) of the common normal to these rays is

$$\hat{n} = \frac{\vec{d}_1 \times \vec{d}_2}{\left|\vec{d}_1 \times \vec{d}_2\right|}, \qquad |\vec{d}_1 \times \vec{d}_2| \neq 0 \quad (5.141)$$

The shortest distance between $\vec{r}_1$ and $\vec{r}_2$, $l$, is given by projecting $\vec{O}_1 - \vec{O}_2$ in the $\hat{n}$ direction. Note, to ensure that $l$ is unique, $\vec{r}_1$ and $\vec{r}_2$ must be skew, or equivalently, $|\vec{d}_1 \times \vec{d}_2| \neq 0$.

$$l = \frac{(\vec{O}_1 - \vec{O}_2) \bullet (\vec{d}_1 \times \vec{d}_2)}{\left|\vec{d}_1 \times \vec{d}_2\right|} = (\vec{O}_1 - \vec{O}_2) \bullet \hat{n}, \quad |\vec{d}_1 \times \vec{d}_2| \neq 0 \tag{5.142}$$

In order to determine the location of the midpoint of the common normal to $\vec{r}_1$ and $\vec{r}_2$ in world coordinates, the values of the scalars $s_1$ and $s_2$ must first be determined. This is accomplished using (5.144).

$$(s_1\vec{d}_1 + \vec{O}_1) - (s_2\vec{d}_2 + \vec{O}_2) = l\hat{n} \tag{5.143}$$

or,

$$\begin{bmatrix} d_{x_1} & -d_{x_2} \\ d_{y_1} & -d_{y_2} \\ d_{z_1} & -d_{z_2} \end{bmatrix} \begin{bmatrix} s_1 \\ s_2 \end{bmatrix} = \begin{bmatrix} ln_x - (O_{x_1} - O_{x_2}) \\ ln_y - (O_{y_1} - O_{y_2}) \\ ln_z - (O_{z_1} - O_{z_2}) \end{bmatrix} \tag{5.144}$$

It can be shown that

$$s_1 = \frac{1}{e^2}\left[g(cd_{y_2} - bd_{z_2}) + h(ad_{z_2} - cd_{x_2}) + i(bd_{x_2} - ad_{y_2})\right] \tag{5.145}$$

$$s_2 = \frac{1}{e^2}\left[g(cd_{y_1} - bd_{z_1}) + h(ad_{z_1} - cd_{x_1}) + i(bd_{x_1} - ad_{y_1})\right] \tag{5.146}$$

where

$$a = (O_{x_1} - O_{x_2}) \tag{5.147}$$

$$b = (O_{y_1} - O_{y_2}) \tag{5.148}$$

$$c = (O_{z_1} - O_{z_2}) \tag{5.149}$$

$$e = \left|\vec{d}_1 \times \vec{d}_2\right| \tag{5.150}$$

$$\begin{bmatrix} g \\ h \\ i \end{bmatrix} = \vec{d}_1 \times \vec{d}_2 = \begin{bmatrix} (d_{y_1}d_{z_2} - d_{y_2}d_{z_1}) \\ (d_{x_2}d_{z_1} - d_{x_1}d_{z_2}) \\ (d_{x_1}d_{y_2} - d_{x_2}d_{y_1}) \end{bmatrix} \tag{5.151}$$

Once $s_1$ and $s_2$ are determined from (5.145) and (5.146), the midpoint of the common normal to $\vec{r}_1$ and $\vec{r}_2$ is given in world coordinates by

$$\vec{m} = \vec{r}_1 \mid_{s_1} - \tfrac{1}{2}l\hat{n} = \vec{r}_2 \mid_{s_2} + \tfrac{1}{2}l\hat{n} \tag{5.152}$$

Since

$$\vec{r}_k = s_k\vec{d}_k + \vec{O}_k \quad k = 1, 2, \quad \hat{n} = \frac{1}{e}\begin{bmatrix} g \\ h \\ i \end{bmatrix}, \quad \text{and,} \quad l = \frac{1}{e}(ag + bh + ci) \tag{5.153}$$

the coordinates of the midpoint of the common normal $(m_x, m_y, m_z)$ are given by

$$\begin{aligned} m_x &= \frac{1}{e^2}\left\{[g(cd_{y_2} - bd_{z_2}) + h(ad_{z_2} - cd_{x_2}) + i(bd_{x_2} - ad_{y_2})]\, d_{x_1} \right. \\ &\quad \left. - \frac{(ag + bh + ci)g}{2}\right\} + O_{x_1} \qquad (5.154) \\ &= \frac{1}{e^2}\left\{[g(cd_{y_1} - bd_{z_1}) + h(ad_{z_1} - cd_{x_1}) + i(bd_{x_1} - ad_{y_1})]\, d_{x_2} \right. \\ &\quad \left. + \frac{(ag + bh + ci)g}{2}\right\} + O_{x_2} \qquad (5.155) \\ m_y &= \frac{1}{e^2}\left\{[g(cd_{y_2} - bd_{z_2}) + h(ad_{z_2} - cd_{x_2}) + i(bd_{x_2} - ad_{y_2})]\, d_{y_1} \right. \\ &\quad \left. - \frac{(ag + bh + ci)h}{2}\right\} + O_{y_1} \qquad (5.156) \\ &= \frac{1}{e^2}\left\{[g(cd_{y_1} - bd_{z_1}) + h(ad_{z_1} - cd_{x_1}) + i(bd_{x_1} - ad_{y_1})]\, d_{y_2} \right. \\ &\quad \left. + \frac{(ag + bh + ci)h}{2}\right\} + O_{y_2} \qquad (5.157) \\ m_z &= \frac{1}{e^2}\left\{[g(cd_{y_2} - bd_{z_2}) + h(ad_{z_2} - cd_{x_2}) + i(bd_{x_2} - ad_{y_2})]\, d_{z_1} \right. \\ &\quad \left. - \frac{(ag + bh + ci)i}{2}\right\} + O_{z_1} \qquad (5.158) \\ &= \frac{1}{e^2}\left\{[g(cd_{y_1} - bd_{z_1}) + h(ad_{z_1} - cd_{x_1}) + i(bd_{x_1} - ad_{y_1})]\, d_{z_2} \right. \\ &\quad \left. + \frac{(ag + bh + ci)i}{2}\right\} + O_{z_2} \qquad (5.159) \end{aligned}$$

The midpoint to the common normal method can also be used with a calibrated camera and laser scanner. To do this, it is necessary to derive a parametric equation for the 3-D ray of the laser beam. This equation is expressed as

$$\vec{r_l} = s_l \vec{d_l} + \vec{O_l} \qquad (5.160)$$

where $\vec{O_l}$ is the origin of the laser beam in terms of the world coordinate frame, $\vec{d_l}$ is the direction vector of the laser ray and $s_l$ is a parameter.

Equation (5.160) is of the same form as (5.139) and (5.140). To use the midpoint to the common normal method with a calibrated laser, it is necessary to derive expressions for $\vec{O_l}$, $\vec{d_l}$ and $s_l$. Once these values are determined, the mathematical relationships for the midpoint of the common normal for two cameras will also work for a calibrated camera and laser.

$\vec{O}_l$ can be determined based on the values of the intrinsic and extrinsic parameters of the laser scanner. Specifically, the required parameters are the distance between the scanning mirrors, and the transformation ${}^{l}_{w}T$ from the laser coordinate frame to the world coordinate frame. The laser calibration assumes the laser's origin to be at the center of the $\theta_y$ mirror. The coordinates of the laser's origin with respect to the world coordinate frame can be derived by extracting the translational component of the inverse of the ${}^{l}_{w}T$. This translation vector is defined as ${}^{w}_{l}t$.

The origin of the laser has one additional component. Recalling the arrangement of the laser scanner described in Section 5.4.1, the $\theta_x$ mirror deflects the laser beam along the rotational axis of the $\theta_y$ mirror. Since the origin of the laser scanner is defined to be on the $\theta_y$ mirror, the origin of the laser ray is translated along the $x$ axis of the laser scanner (this is the rotational axis of the $\theta_y$ mirror) by the rotation of the $\theta_x$ mirror. The translation of the laser's origin as a function of $\theta_x$ with respect to the laser's coordinate frame can be expressed as a $3 \times 1$ vector:

$$\vec{t}_{\theta_x} = \begin{bmatrix} e \tan\theta_x \\ 0 \\ 0 \end{bmatrix} \tag{5.161}$$

The term $e \tan\theta_x$ from (5.98) defines the $x$ coordinate of the laser beam with respect to the laser's coordinate frame given a set of mirror angles and a specific $z$ coordinate. The vector $\vec{t}_{\theta_x}$ is defined in terms of the laser coordinate frame; hence it must be transformed into a translation with respect to the world coordinate frame. This is accomplished by multiplying $\vec{t}_{\theta_x}$ by the rotation matrix ${}^{w}_{l}R$ contained in the homogeneous transformation ${}^{w}_{l}T$. This yields a new translation vector defined in the world coordinate frame:

$$\vec{t}_{w_{\theta_x}} = {}^{w}_{l}R \cdot \vec{t}_{\theta_x} \tag{5.162}$$

Therefore, the final value for $\vec{O}_l$ can be expressed as

$$\vec{O}_l = {}^{w}_{l}t + \vec{t}_{w_{\theta_x}} \tag{5.163}$$

The next step is to determine $\vec{d}_l$ from (5.98) and (5.97). Since the expression for $\vec{O}_l$ already compensates for the translation of the laser's origin due to rotation of the $\theta_x$ mirror, the $e \tan\theta_x$ term in (5.98) can be removed. Hence, the direction of the laser ray with respect to the laser coordinate frame can be expressed as a $3 \times 1$ vector of the form

$$\vec{d}_l = \begin{bmatrix} z \sec\theta_y \tan\theta_x \\ z \tan\theta_y \\ z \end{bmatrix} \tag{5.164}$$

The direction vector is currently defined with respect to the laser coordinate frame. In order to use it in determining the midpoint of the common normal, it will have to be transformed with respect to the world coordinate frame. This can be done simply by multiplying $\vec{d_l}$ by the rotation matrix ${}^w_l R$. Further, the $z$ term in $\vec{d_l}$ can be factored out and used as the variable parameter $s_l$. Therefore, the parametric equation for the 3-D laser ray can be expressed as

$$\vec{r_l} = {}^w_l Rz\vec{d_l} + \vec{O_l} \tag{5.165}$$

Equation (5.165) can be used in place of the parametric equation for the second camera to determine the midpoint of the common normal for a camera and a laser scanner. The remaining mathematical expressions for the midpoint of the common normal calculation remain valid.

### 5.5.3 Comparison of Point Estimation Methods

The LSE point estimation method of Section 5.5.1 uses an overdetermined system of linear equations very similar to the method suggested by Schalkoff [20]. However, it should be stressed that the LSE estimation method gives an $\hat{x}$ such that $(A\hat{x} - b)^T(A\hat{x} - b)$ is minimum *but does not minimize the error between the estimate and the point.* It is easy to see that minimizing the error between the estimate and the point is not possible since calculating this error requires knowledge of the location of the point. However, if the location of the point was known, it would not have to be estimated.

The point estimation technique of Section 5.5.2 begins with the same information as the LSE approach of Section 5.5.1, but uses a purely geometric technique to determine its estimate. Note that this technique degenerates for parallel rays, but in a 3-D vision system, the rays projected from the cameras *should* never be parallel. If the 3-D rays are parallel, the vision system would have to conclude that the 3-D point in question was located at infinity.

Given perfect camera calibration, the 3-D point estimation methods of Sections 5.5.1 and 5.5.2 yield the same location for the 3-D point. *However, the estimate of the 3-D point determined by an LSE solution to* (5.131) *is not, in general, the estimate of the 3-D point determined by the midpoint of the common normal to the two rays projected from the cameras given by* (5.152).

For example, consider the case where the first camera's lens-centered coordinate frame has its origin at $\vec{O_1} = (1000, 10, -1000)$ and the origin of the second camera's lens-centered coordinate frame is at $\vec{O_2} = (0, 0, 0)$. Also, let the normal to the first camera's image plane be in the world $z$ direction, and the normal to the second camera's image plane be in the world $x$ direction. Finally, let $x_{u_1} = y_{u_1} = 0$ and $x_{u_2} = y_{u_2} = 0$. For this case,

$$R_1 = \begin{bmatrix} 1 & 0 & 0 \\ 0 & 1 & 0 \\ 0 & 0 & 1 \end{bmatrix}, \quad T_1 = \begin{bmatrix} -1000 \\ -10 \\ 1000 \end{bmatrix}, \text{ and } \vec{d_1} = \begin{bmatrix} 0 \\ 0 \\ f_1 \end{bmatrix} \tag{5.166}$$

$$R_2 = \begin{bmatrix} 0 & 0 & -1 \\ 0 & 1 & 0 \\ 1 & 0 & 0 \end{bmatrix}, \quad T_2 = \begin{bmatrix} 0 \\ 0 \\ 0 \end{bmatrix}, \quad \text{and } \vec{d}_2 = \begin{bmatrix} f_2 \\ 0 \\ 0 \end{bmatrix} \tag{5.167}$$

For any focal lengths $f_1, f_2 > 0$, the midpoint of the common normal is $\vec{m} = (1000, 5, 0)$. However, using LSE to solve (5.131) gives a value for $\hat{x}$ that is equal to $\vec{m}$ only when $f_1 = f_2$. When $f_1 \neq f_2$, $\hat{x}$ is moved from $\vec{m}$ towards the ray emanating from the camera with the larger focal length in the direction of the common normal to the rays, $\hat{n}$. For example, when $f_1 = 50$ and $f_2 = 35$, $\hat{n} = (0, 1, 0)$ and $\hat{x} = (1000, 6.7, 0)$.

When the point estimation methods are compared with actual data, they typically give comparable results. If the rotation matrices used are orthonormal, the midpoint method very often gives results slightly superior to the LSE method. However, if the rotation matrices used are not orthonormal, the accuracy of the midpoint method degrades sharply.

### 5.5.4 Point Estimation Using More than Two Calibrated Cameras

Since the 3-D vision system described in Section 5.6 has five cameras, the question of determining an estimate of the location of a 3-D point using more than two calibrated cameras is raised.

One method to obtain an estimate of the location of a 3-D point using $n > 2$ cameras is to project 3-D rays from each of the $n$ cameras and calculate the midpoints of the common normals to all pairwise combinations of the $n$ rays. The estimate of the 3-D point could then be declared to be at the centroid of these midpoints.

Another method to obtain an estimate of the location of a 3-D point using $n > 2$ cameras is to increase the dimension of (5.131) and solve the resulting system using the LSE technique described above for an $\hat{x}$ that is the estimate of the 3-D point. Since each camera contributes two equations to the system of equations, $n$ cameras would result in an overdetermined system of equations of the following form:

$$\begin{bmatrix} (r_{7_1} x_{u_1} - f_1 r_{1_1}) & (r_{8_1} x_{u_1} - f_1 r_{2_1}) & (r_{9_1} x_{u_1} - f_1 r_{3_1}) \\ (r_{7_1} y_{u_1} - f_1 r_{4_1}) & (r_{8_1} y_{u_1} - f_1 r_{5_1}) & (r_{9_1} y_{u_1} - f_1 r_{6_1}) \\ (r_{7_2} x_{u_2} - f_2 r_{1_2}) & (r_{8_2} x_{u_2} - f_2 r_{2_2}) & (r_{9_2} x_{u_2} - f_2 r_{3_2}) \\ (r_{7_2} y_{u_2} - f_2 r_{4_2}) & (r_{8_2} y_{u_2} - f_2 r_{5_2}) & (r_{9_2} y_{u_2} - f_2 r_{6_2}) \\ \vdots & \vdots & \vdots \\ (r_{7_n} x_{u_n} - f_n r_{1_n}) & (r_{8_n} x_{u_n} - f_n r_{2_n}) & (r_{9_n} x_{u_n} - f_n r_{3_n}) \\ (r_{7_n} y_{u_n} - f_n r_{4_n}) & (r_{8_n} y_{u_n} - f_n r_{5_n}) & (r_{9_n} y_{u_n} - f_n r_{6_n}) \end{bmatrix} \begin{bmatrix} x_w \\ y_w \\ z_w \end{bmatrix}$$

$$= \begin{bmatrix} (f_1 t_{x_1} - t_{z_1} x_{u_1}) \\ (f_1 t_{y_1} - t_{z_1} y_{u_1}) \\ (f_2 t_{x_2} - t_{z_2} x_{u_2}) \\ (f_2 t_{y_2} - t_{z_2} y_{u_2}) \\ \vdots \\ (f_n t_{x_n} - t_{z_n} x_{u_n}) \\ (f_n t_{y_n} - t_{z_n} y_{u_n}) \end{bmatrix} \quad (5.168)$$

However, once again the two estimation methods described above are not equivalent. This can be seen by extending the example used in Section 5.5.3 to include a third ray. Keeping all parameters for the first two rays the same as before, a third ray may be added whose origin is at $\vec{O}_3 = (0, 5, -1000)$. The corresponding pixel occurs at $x_{u_3} = y_{u_3} = 0$. The other vital parameters are

$$R_3 = \frac{1}{\sqrt{2}} \begin{bmatrix} 1 & 0 & -1 \\ 0 & \sqrt{2} & 0 \\ 1 & 0 & 1 \end{bmatrix}, \quad T_3 = \frac{1}{\sqrt{2}} \begin{bmatrix} -1000 \\ -5\sqrt{2} \\ 1000 \end{bmatrix}, \text{ and } \vec{d}_3 = \frac{1}{\sqrt{2}} \begin{bmatrix} f_3 \\ 0 \\ f_3 \end{bmatrix} \quad (5.169)$$

With this third ray, and the two rays defined in (5.166) and (5.167), the midpoint between $\vec{r}_1$ and $\vec{r}_2$ is $\vec{m}_{12} = (1000, 5, 0)$, the midpoint between $\vec{r}_1$ and $\vec{r}_3$ is $\vec{m}_{13} = (1000, 7.5, 0)$, and the midpoint between $\vec{r}_2$ and $\vec{r}_3$ is $\vec{m}_{23} = (1000, 2.5, 0)$. Therefore, the centroid of these midpoints is $\vec{m} = (1000, 5, 0)$ for any $f_1, f_2, f_3 > 0$. But when (5.168) is solved using the parameters for these same three rays, the estimate $\hat{x}$ is the same as $\vec{m}$ (as determined above) only when $f_1 = f_2 = f_3 > 0$. For example, if $f_1 = f_3 = 50$ and $f_2 = 35$, then (5.168) yields $\hat{x} = (1000, 6, 0)$.

## 5.6 Description of the CIRSSE 3-D Vision System

The Center for Intelligent Robotic Systems for Space Exploration (CIRSSE) Testbed is designed to facilitate research in the area of space-based robotic systems. The base of the system was custom built for CIRSSE by Aronson Corporation and consists of a 12 foot railed platform with two carts. Each of the carts carries a 6-degree-of-freedom (DOF) PUMA robot arm (specifically one PUMA 560 and one PUMA 600). Each cart is capable of linearly translating its robot arm a maximum of 8 feet, rotating it ±135°, and tilting it ±45°. The combination of the platform with the robot arms results in two 9-DOF manipulators. The CIRSSE Testbed hardware is described in Chapter 1.

In order for the CIRSSE Testbed to interact with its environment, it was decided that a 3-D machine vision system would be required. This 3-D machine vision system serves both as a tool for the testbed system and as a research area in itself.

### 5.6.1 Five Camera Vision System

Mounted on the ceiling above the testbed are two black and white Javelin video cameras (model JE2362). These cameras are fixed and provide a general view of the robot workspace. Mounted at joint six of one of the robot manipulators are two Toshiba micro-miniature remote sense-head black-and-white CCD video cameras (model IK-M30M). This "eyes-on-hand" configuration is useful for providing mobile 3-D vision to the testbed system. A third Toshiba micro-miniature remote sense-head camera (also model IK-M30M) is mounted at joint six of the second robot manipulator. This third "eye-on-hand" camera augments the views provided by the two cameras on the ceiling and the two cameras on the first robot manipulator. The information from the five cameras is processed using a high speed vision system from Datacube, Inc., which is capable of taking input from any two of the five cameras simultaneously. The Datacube vision system is controlled by a Motorola 68030-based MVME 147SA-1 computer which runs VxWorks, a real-time operating system. The multiple views from the five cameras are used to provide information to identify objects which may not be recognizable from a single view.

### 5.6.2 Structured Light Source

Also, fixed to the ceiling over the robotic testbed is a laser scanning system made by General Scanning, Inc. (model EDK 500M). This system is composed of two orthogonal mirrors whose orientations are controlled by galvanometers, and a ferro-electric liquid crystal polarizing filter. The laser used is a Uniphase 5 mW HeNe laser (model 1125P). The scanning of the laser is controlled by a Motorola 68020-based MVME 135 computer which runs VxWorks. The MVME 135 computer is capable of changing the orientation of the laser mirrors in such a way that the laser spot is deflected as desired. The polarizing filter is used to "shutter" the laser beam so that the laser scanner can generate unconnected points. With this system it is possible to scan any arbitrarily complex two-dimensional pattern.

The laser scanner is used to solve the image correspondence problem inherent to 3-D vision. It also provides structured light patterns that are used as an additional means to sense 3-D objects.

### 5.6.3 Calibration of the CIRSSE 3-D Vision System

The two cameras on the ceiling above the testbed are calibrated using one of the camera calibration techniques explained in Section 5.3. In order to use any of these calibration techniques, a "large" number of calibration data points must first be collected. A data point consists of the world coordinates of a point together with its computer pixel coordinates. The pixel coordinates of a point are a function of the perspective transformation of the camera.

To collect the data, a small circular lightsource is placed in the V-shaped gripper of a robot in the CIRSSE Testbed system. After lowering the room lighting, the lightsource becomes the brightest event in the image. Thresholding this image provides a binary image that is used to determine the centroid of the lightsource's projection. This centroid is used as the computer image coordinates of the calibration point. The world coordinates of the calibration point are determined from the position of the robot's gripper, which is known through the forward kinematics of the robot.

This process was repeated for every calibration point. The total number of calibration points needed to achieve an "accurate" camera calibration is not well defined analytically. With the wide-angle (9mm) lenses placed on the ceiling cameras, approximately 300 calibration points were gathered for the strut insertion demonstration described in Section 5.6.4.

A small error is introduced by assuming the centroid of the projection of the lightsource is the projection of the centroid of the lightsource. A spherical lightsource can be used to reduce this error, but not eliminate it.

The gathering of the calibration points can be somewhat automated. The user may define a cube (with one face parallel to the $xy$ plane) by selecting two diametrically opposed vertices of the cube. The calibration points are then automatically generated from this information by spacing them evenly within this cube. This spacing is also based upon the number of divisions in each direction $(x, y, z)$, which are also specified by the user.

In the interest of efficiency, the cube of calibration points may be selected to fall within the field of view of both ceiling cameras. However, this proved to be non-trivial. The overlap of the fields of view provide one major constraint while the workspace of the robot provides another. Also, obtaining calibration points over the entire field of view of each camera is highly desirable. It should be noted that the calibration points used to calibrate one camera do not need to be in the field of view of the second camera, nor need they all be from the same "cube" of calibration points.

However, to determine an estimate of the resulting accuracy of the camera calibrations, a group of calibration points in the field of view of both cameras is needed. These calibration points may be obtained in the same manner as the other calibration points. The accuracy of the camera calibrations is checked by determining the error between the 3-D position of a calibration point as given by the robot, and the 3-D position of the same calibration point calculated using the methods described in Section 5.5.

### Mobile Cameras

The camera calibration techniques described in Section 5.3 are for fixed cameras. The cameras mounted on the robot arm will be calibrated using a technique based on work originally developed by Tsai [7], [8], [21], [22], [23]. This technique extends the calibration Tsai did for fixed cameras to calibration for mobile cameras. The technique for calibrating a mobile camera is divided into three parts:

1. Calibrating the camera using the calibration technique for fixed cameras.
2. Calibrating the transformation from the end-effector of the robot arm to the camera.
3. Calibrating the robot arm. This step is required if the vision system must relate its 3-D coordinate frame located at the end-effector of the robot arm to the base coordinate frame of the robot arm.

### 5.6.4 Demonstration of the CIRSSE 3-D Vision System

In November 1990, a demonstration of the CIRSSE 3-D vision system involving the completion of a simple strut and node structure by a robot was performed. This structure was in the shape of an equilateral triangle with a missing side (i.e., a "V") and was composed of white nodes with a black bolt in their centers, and white struts. Each strut is approximately 2.5 feet long. The structure was placed in the field of view of both ceiling cameras on a planar black background along with other disconnected struts and nodes. The task of the vision system was to determine where to place the final strut to complete the equilateral triangle (see Figure 5.10).

To determine the necessary position and orientation of the final strut, the vision system carried out the following steps:

1. Grab an image with each ceiling camera.
2. Perform a Sobel-edge detection on each image.
3. Threshold the magnitude of the edge-detected images (to form binary images).
4. Perform an erosion on the images resulting from step 3 with a $2 \times 2$ kernel.
5. Perform a Hough transform on the image resulting from step 4.
6. Grab a new image with each ceiling camera.
7. Threshold these images (form binary images).
8. Locate the centroid of every black object entirely surrounded by white in each image from step 7.
9. Reject all centroids found in step 8 not "near" a line found in step 5.
10. Identify the vertex node of the incomplete triangle in each image based upon the number of lines its centroid is "near".

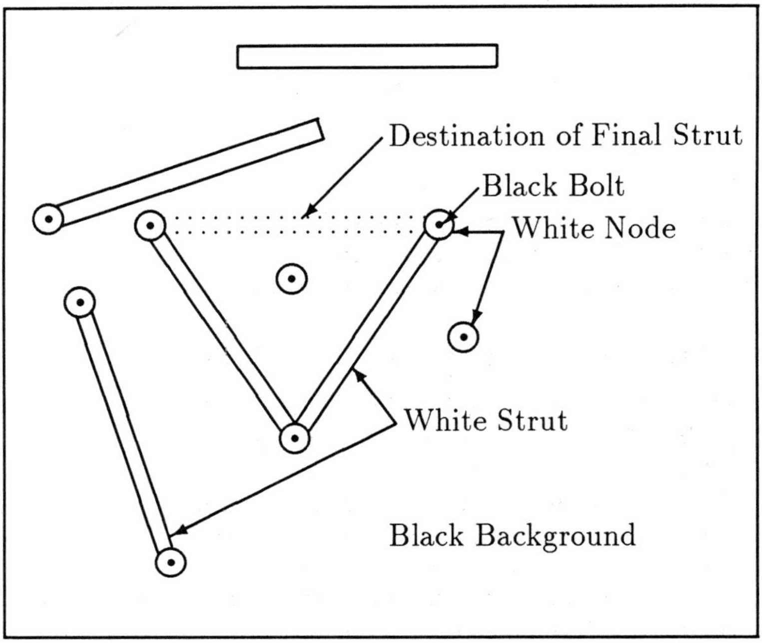


Figure 5.10: Triangle Completion Demonstration

11. Pair vertex nodes from each image and triangulate to determine its 3-D position.

12. Form all pairs of remaining centroids in each image with a centroid from the other image and perform triangulation to obtain their 3-D positions.

13. Select the pairing that results in an equilateral triangle.

Thus, the vision system extracted both line features from the struts, and bolt centroid features from the nodes. The combination of these features was used to determine what struts and nodes made up the incomplete triangle, and what part of the triangle was incomplete.

The robot then picked up the last strut from a known position and placed it 10 cm above the insertion location determined by the vision system. At this point, the operator was given the opportunity to correct the position of the robot, if necessary (due to camera calibration inaccuracies). After making any necessary corrections, the operator then started a compliant force insertion program, which completed the insertion automatically.

## Acknowledgments

This work was supported under NASA grant NAGW-1333.

## 5.7 References

[1] R. A. Jarvis, "A perspective on range finding techniques for computer vision," *IEEE Transactions on Pattern Analysis and Machine Intelligence*, vol. PAMI-5, pp. 122–139, March 1983.

[2] Y. Hung, "A simple real-time method for calibrating a camera mounted on a robot for three dimensional machine vision," in *SPIE Vol. 1005 Optics, Illumination, and Image Sensing for Machine Vision III*, pp. 12–19, 1988.

[3] Y. Yakimovsky and R. Cunningham, "A system for extracting three-dimensional measurements from a stereo pair of TV cameras," *Computer Graphics Image Processing*, vol. 7, pp. 195–210, 1978.

[4] R. Y. Tsai, "A versatile camera calibration technique for high accuracy 3-D machine vision metrology using off-the-shelf TV cameras and lenses," IBM Research Report RC 11413, IBM, October 1985.

[5] R. Y. Tsai, "A versatile camera calibration technique for high-accuracy 3-D machine vision metrology using off-the-shelf TV cameras and lenses," *IEEE Journal of Robotics and Automation*, vol. RA-3, pp. 323–344, August 1987.

[6] R. Y. Tsai, "Synopsis of recent progress on camera calibration for 3-D machine vision," IBM Research Report RC 13954, IBM, August 1988.

[7] R. Y. Tsai and R. K. Lenz, "Overview of a unified calibration trio for robot eye, eye-to-hand, and hand calibration using 3-D machine vision," IBM Research Report RC 14218, IBM, November 1988.

[8] R. Y. Tsai and R. K. Lenz, "Overview of a unified calibration trio for robot eye, eye-to-hand, and hand calibration using 3-D machine vision," in *SPIE Vol. 1003 Sensor Fusion: Spatial Reasoning and Scene Interpretation*, (Cambridge, Massachusetts), November 1988.

[9] R. K. Lenz and R. Y. Tsai, "Techniques for calibration of the scale factor and image center for high accuracy 3-D machine vision metrology," in *Proceedings of the 1987 IEEE International Conference on Robotics and Automation*, (Raleigh, NC), pp. 68–75, 1987.

[10] R. K. Lenz and R. Y. Tsai, "Techniques for calibration of the scale factor and image center for high accuracy 3-D machine vision metrology," *IEEE Transactions on Pattern Analysis and Machine Vision*, vol. PAMI-10, pp. 713–720, September 1988.

[11] W. H. Press, B. R. Flannery, S. A. Taukolsky, and W. T. Vetterling, *Numerical Recipes in C: The Art of Scientific Computing.* Cambridge University Press, 1990.

[12] S. Ganapathy, "Decomposition of transformation matrices for robot vision," in *Proceedings of the 1984 IEEE International Conference on Robotics*, (Atlanta, GA), pp. 130–139, 1984.

[13] J. Funda, R. H. Taylor, and R. P. Paul, "On homogeneous transforms, quaternions, and computational efficiency," *IEEE Transactions on Robotics and Automation*, vol. RA-6, pp. 382–388, June 1990.

[14] K. S. Fu, R. C. Gonzalez, and C. S. G. Lee, *Robotics: Control, Sensing, Vision, and Intelligence.* New York, NY: McGraw-Hill, 1987.

[15] J. J. Craig, *Introduction to Robotics, Mechanics, and Control.* Reading, MA: Addison-Wesley Publishing Company, 1986.

[16] D. Sood, M. C. Repko, and R. B. Kelley, "An implementation of a camera technique to obtain three-dimensional (3-D) vision information for simple assembly tasks," CIRSSE Report 76, Rensselaer Polytechnic Institute, Troy, NY, November 1990.

[17] W. H. Beyer, ed., *CRC Standard Mathematical Tables.* Boca Raton, Florida: CRC Press, Inc., 28th ed., 1988.

[18] R. P. Paul, *Robot Manipulators: Mathematics, Programming, and Control.* Cambridge, MA: MIT Press, 1981.

[19] W. J. Mistretta, "Analysis, enhancement, and application of a three dimensional inspection system in computer integrated manufacturing," Master's thesis, Rensselaer Polytechnic Institute, Electrical, Computer, and Systems Engineering Dept., Troy, NY 12180–3590, December 1989.

[20] R. J. Schalkoff, *Digital Image Processing and Computer Vision.* New York, NY: John Wiley & Sons, Inc., 1989.

[21] R. Y. Tsai and R. K. Lenz, "A new technique for autonomous and efficient 3-D robotics hand/eye calibration," IBM Research Report RC 13212, IBM, September 1987.

[22] R. Y. Tsai and R. K. Lenz, "Real time versatile robotics hand/eye calibration using 3-D machine vision," in *Proceedings of the 1988 IEEE International Conference on Robotics and Automation*, (Philadelphia, PA), pp. 554–561, 1988.

[23] R. Y. Tsai and R. K. Lenz, "A new technique for fully autonomous and efficient 3-D robotics hand/eye calibration," *IEEE Transactions on Robotics and Automation*, vol. RA-5, pp. 345–358, June 1989.

# Chapter 6

# CIRSSE General Purpose Gripper and Controller System

by: Robert B. Kelley
Jodi Tsai
Jeff Bethel
John Peiffer
Electrical, Computer, and Systems Engineering Department
Mechanical Engineering, Aeronautical Engineering, and
Mechanics Department
Center for Intelligent Robotic Systems for Space Exploration
Troy, New York 12180-3590

The Center for Intelligent Robotic Systems for Space Exploration (CIRSSE) has designed a general purpose gripper to be used with a 6-axis-wrist-mounted force-torque sensor as part of an experimental testbed for cooperative robotic manipulation. The gripper system consists of a dedicated controller and a pneumatic powered two-fingered gripper that is servoed in position or grasping force. The gripper has a between-fingertip-light-beam sensor; it is used to detect when a part is introduced between the fingers and provides an automatic grasp function. The major design features of the gripper system are: a 25-to-1 grasping force-to-weight ratio; mechanical compactness of 14 cm base-fingertip length; an overall weight of 0.8 kilograms; a low cost, dedicated slave controller; and a simple host-controller communications protocol. Software functions are shared between the host and the controller processor. This permits the host to select specific servo modes, servo gain-parameter values, and calibration offsets.

Hidden primitive self-test, calibration and servo confidence functions are also implemented.

This chapter presents the details of the mechanical and electronic design. The sensing capabilities and the controller software design are also reported.

## 6.1 Introduction

An important area of research at the Center for Intelligent Robotic Systems for Space Exploration (CIRSSE) involves the assembly of struts for the construction of truss structures. The new CIRSSE two-finger gripper and dedicated controller are the result of an evolution from parallel-jaw grippers originally designed for the bin-picking environment, [1]. These grippers provide the basis for the new design. The CIRSSE gripper is a 6-axis-wrist-mounted force-torque sensor for use as part of an experimental testbed for cooperative robotic manipulation (see Chapter 1). Therefore, the CIRSSE design does not include the integrated mechanical overload mechanism and sensor of the original bin-picking grippers. To accommodate the handling of the cylindrical struts and to provide a centering grasp, the fingertips consist of V-blocks. As a further departure from the past designs, the clamping motion is based on a scissor action rather than a parallelogram action. Parallel clamping surfaces to grasp parallel-faced objects are obtained by a change of fingertips. These design features enable various robotic assembly scenarios to be addressed with a minimum of special fixturing.

## 6.2 Gripper Design Specifications

The major goals for the redesign of the CIRSSE gripper and controller are:

- reduction of the length and weight of the gripper,
- provide real-time interrupt driven control of the gripper,
- permit servo parameter tuning by the host computer,
- simplify the host-gripper communications protocol, and
- incorporate self-check, calibration and confidence routines.

## 6.3 Mechanical Design

The mechanical redesign of the CIRSSE gripper focused on the reduction in length and weight, and on reducing the mechanical complexity of the mechanism. The length of the predecessor gripper was about 23 cm from its base to the fingertip. The CIRSSE gripper is mounted on a force-torque sensor that adds 7 cm to the effective moment arm of the gripper. This resulted in a moment arm of 30 cm for the predecessor gripper, which is considered excessive. Hence, reducing the working length of the CIRSSE gripper to 14 cm results in an effective moment arm of 21 cm, a reduction of 30 percent when mounted on the force-torque sensor. In addition, with the redesign, the weight of the CIRSSE gripper is reduced by 20 percent to 0.8 kilograms. This yields a reduced moment for the force-torque sensor mounted CIRSSE gripper of about 17 kilogram-centimeters, less than 60 percent of the predecessor gripper.

The CIRSSE gripper mechanism has been simplified by reducing the number of moving parts from 7 to 5 and the number of pin joints to ground from 2 to 1. Figure 6.1 provides a sketch of the gripper. The finger travel is approximately 30 degrees (open) and 5 degrees (close) relative to the alignment shown. The 25-to-1 grasping force-to-weight ratio is accomplished by the use of a pneumatic cylinder actuator.

### 6.3.1 Sensors

The **position** sensor is a linear potentiometer mounted on the pneumatic cylinder. It measures the displacement of the piston directly through a spring-loaded probe in contact with the piston adapter bar. Thus, the actual opening of the fingertips must be inferred from the mechanical linkage.

The **grasping force** is measured directly at the fingertips. This eliminates the need to compensate for any frictional effects from the actuator and the linkages. The force sensors are mounted on opposite sides of both fingers. They consist of strain gage pairs in a half-bridge configuration with an integral amplifier.

The crossfire-between-fingertip-infrared-(IR)-light-beam sensor is located near the outer aspect of the finger tips. Figure 6.2 shows the details of the strain gage mounting and the light beam sensor element installation.

## 6.4 Controller Electronic Design

The low level control of the gripper is performed by a dedicated controller. The Motorola M68HC11EVGB is a single board processor using the MC68HC11 8-bit micro-controller that provides a low cost means to implement the low-level-gripper controller, [2]. On-chip features of the MC68HC11 include 8 Kbytes of ROM, 512 bytes of EEPROM and 256 bytes of RAM, [3]. It has a 16-bit timer as well as asynchronous and synchronous serial interfaces, 8 channel 8-bit analog-to-digital converters, and real-time interrupts. The control algorithms along with the interface protocol are handled by the micro-controller. A custom-designed double-sided printed circuit board, the RPI pneumatic hand control electronics board, provides the regulated power for the sensors, the pneumatic

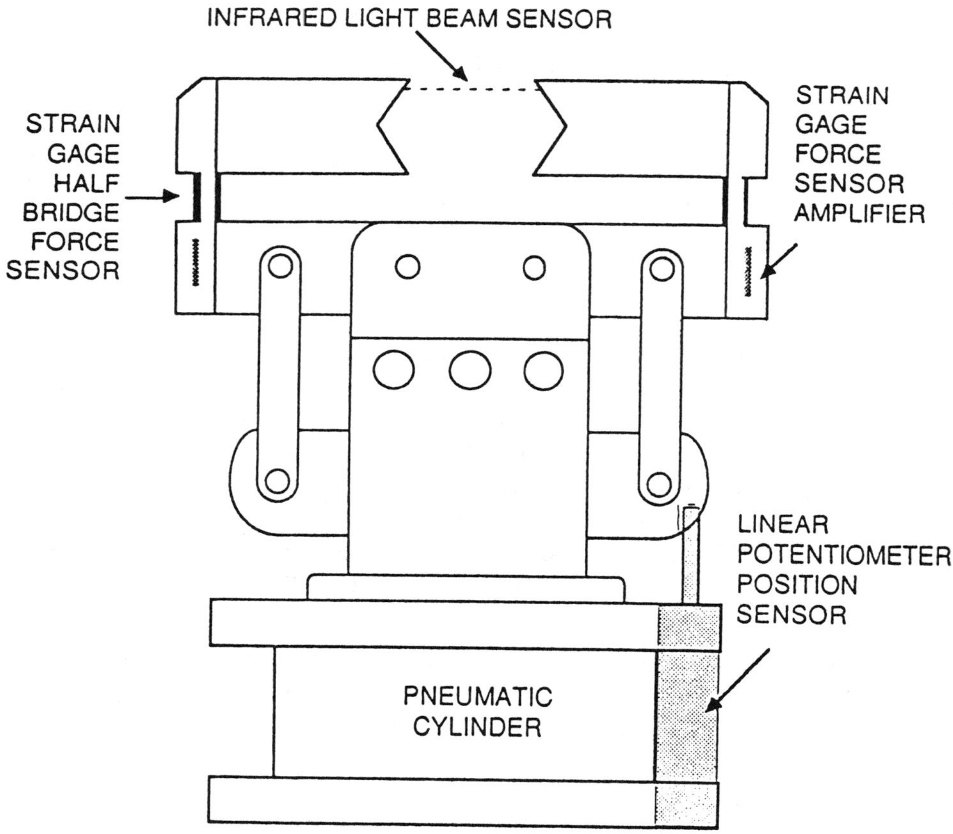


Figure 6.1: The CIRSSE Gripper Shown in Reduced Scale

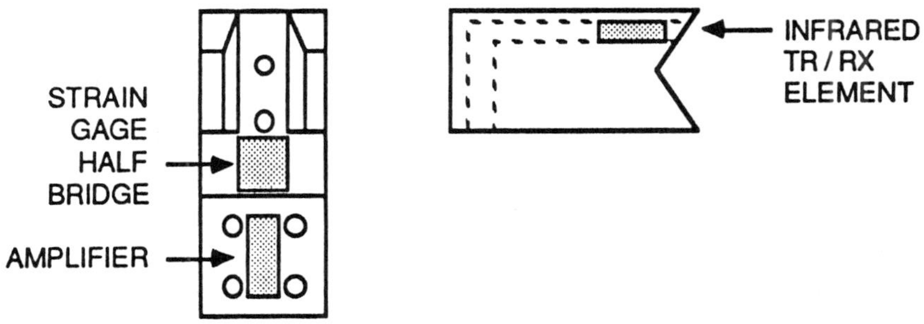


Figure 6.2: Strain Gage and Amplifier on the Finger Mounting Details. Light Beam Sensor Element Internal Fingertip Detail.

valve drive electronics, the drive for the light beam sensor, and the signal conditioning electronics for the position, force, and crossfire sensors. The layout of the custom electronics board is shown in Figure 6.3.

### 6.4.1 Regulated Voltage Supplies

An important function of the control electronics board is to supply the analog electronics for the hand sensors with stable reference voltages. There are three regulated power sources on the board: V1 is the +10 volt regulated supply, V2 the -10 volt supply, and V3 is the +8 volt supply for both strain gage bridges that measure the grasping force. These regulated sources provide the stable voltages needed by the operational amplifiers used to provide necessary signal gains and signal offsets. V1 is also used as $V_{high}$ for the analog-to-digital converter (ADC) on the microprocessor board. An input voltage to the ADC equal to +10 volts results in a digital value of $FF_{16}$, and an input of 0 volts results in $00_{16}$.

### 6.4.2 Pneumatic Valve Drive

The pneumatic valve drive electronics controls a high performance pneumatic servo valve to regulate the air flow that drives the pneumatic cylinder. This

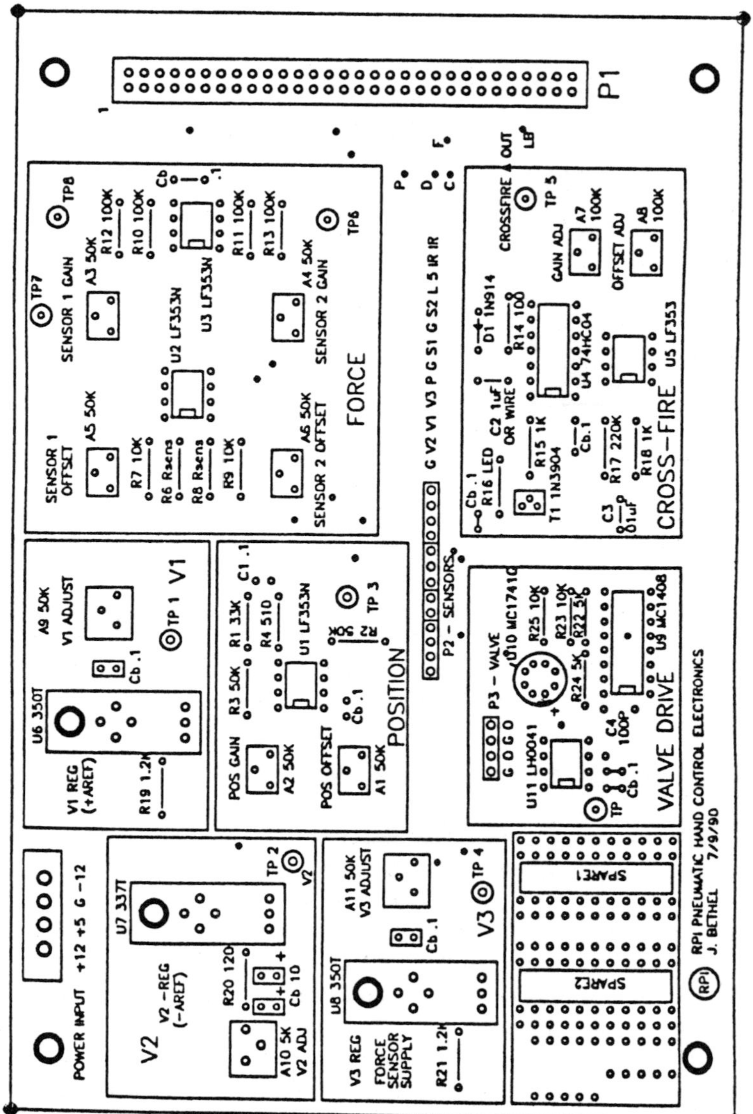


Figure 6.3: RPI Pneumatic Hand Control Electronics Board

valve has been used by the previous pneumatic gripper control systems at Rensselaer Polytechnic Institute and the University of Rhode Island. The valve maintains an output pressure that is linearly proportional to the input voltage. The two output ports of the valve are connected to the opposing ports of the pneumatic cylinder to obtain a direct, proportional force mode actuator. A special digital-to-analog converter (DAC) on the control electronics board changes the micro-controller 8-bit digital signal into an analog control signal ($\pm$10 volt range) that drives the pneumatic servo valve.

### 6.4.3 Light Beam Sensor

The drive for the light beam sensor is capable of operating either pulse or continuous mode. The pulse mode allows the IR transmitter to emit short, more powerful pulses. The maximum power dissipation in pulse mode can be several times the average power limitation (continuous mode). For example, a 10 percent duty cycle pulse (10 percent on, 90 percent off) permits the maximum power to be 10 times the average power limit and provides a 10 dB increase in the radiated output. This mode is useful when a stronger IR signal is needed to accommodate larger separation distances between transmitter and detector or the fall-off of the received signal due to larger angular ranges for the scissor action of the gripper. The present gripper design does not require the light beam sensor to be operated in pulse mode. Hence the light beam sensor is operated in continuous mode not exceeding the device limits.

The ambient IR level detected by the sensor varies as the gripper is reoriented during manipulation. To account for this, two detector readings are taken in sequence: IR transmitter on, IR transmitter off. The difference between the two readings measures the net detected transmitted IR value. A threshold is applied to this value to decide if there is an object between the fingertips.

The signal conditioning electronics for the IR detector are adjusted so that the output is not saturated when the IR detector is aimed at the brightest ambient light source. The gain and offset parameters are adjusted so that the output is close to the saturation level when the transmitter and detector are aligned, as shown in Figure 6.1, and does not swing below 0 volts under zero IR light conditions.

### 6.4.4 Force Sensors

The grasping force sensor consists of two independent strain gage bridges. Each strain gage bridge has an operational amplifier mounted on the finger. This is done to raise the strain gage signal levels by a factor of 150 and, thereby, minimize possible harmful effects on the signals from other nearby high power PWM motor drive systems.

The signal conditioning circuitry amplifies each sensor signal, adds its individual zero force offset value, and sums the two force readings to provide a single total grasping force signal. The sensitivity of the force sensor is about 100 millivolts per kilogram. When the zero force level is set to 2 volts and a guard band of 0.5 volts at both limits is used, the range of forces which can be measured is from -15 kilograms to +75 kilograms.

### 6.4.5 Position Sensor

The linear potentiometer provides a measurement of the displacement of the piston. V1 (+10 volts) and V2 (-10 volts) provide the reference voltages for the potentiometer reading. The gain and offset for the potentiometer is adjusted to allow the full displacement of the piston to be read by the micro-controller ADC. Since the fingertips do not close completely, the fingertip opening must be calibrated by using a known diameter cylinder. This data is used by the software to obtain a zero position reference.

## 6.5 Controller Software Design

The software for the gripper controller is a simple real-time executive which has two responsibilities:

- to servo the gripper to either a desired position or force setpoint, and
- to receive commands for the gripper from the host over an RS-232 serial port.

Since the gripper interfaces to the host through an RS-232 serial port, any host with an RS-232 serial port may be used to control the gripper. The controller software architecture is shown in Figure 6.4.

### 6.5.1 Power-on Hardware Initialization

Upon power-on, the controller software runs tests to ensure that the MC68HC11 evaluation board is operating properly. If a hardware error is detected by the software, an error message is sent through the RS-232 serial port and the MC68HC11 is put into a quiescent state. If no hardware error is detected, the hardware is initialized and the controller is immediately put into position servo mode. To prevent the gripper from moving, the position setpoint is the gripper position that is read immediately after initialization.

### 6.5.2 Base Level Execution

The base level refers to the non-interrupt driven software in the controller. This level of execution performs all the command line processing through the RS-232 serial interface. The software reads a command line from the RS-232 serial port.

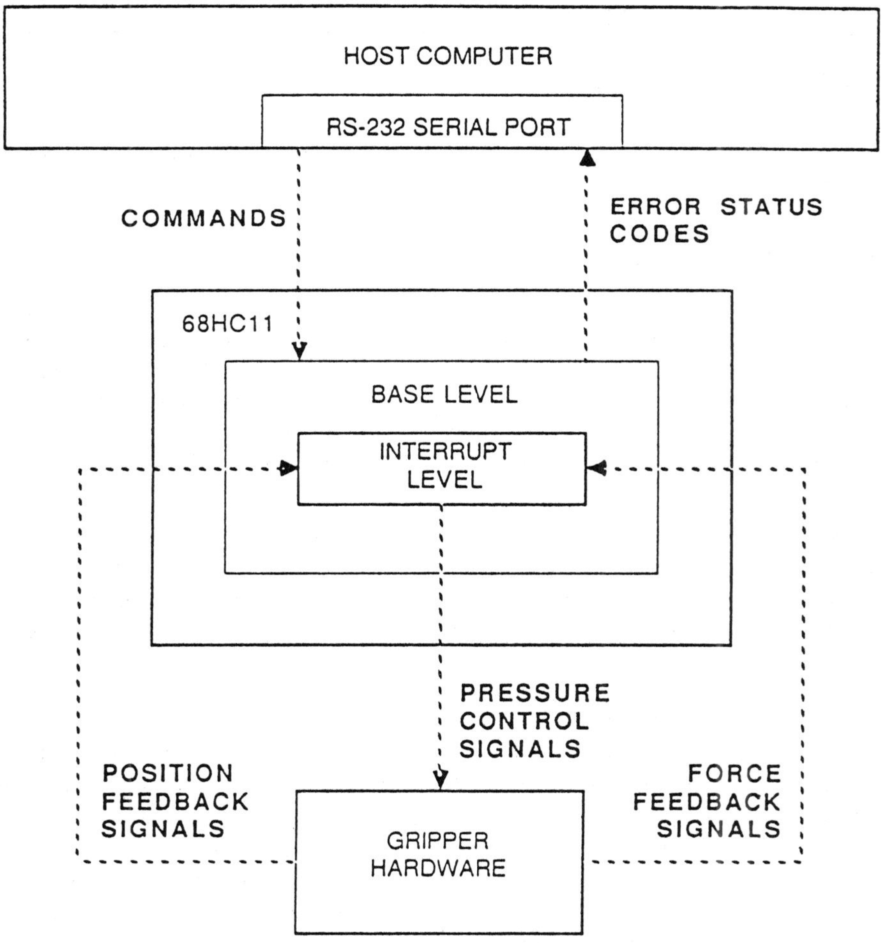


Figure 6.4: CIRSSE Gripper Controller Software Architecture

| Function | Command to MC68HC11 | MC68HC11 Response |
|---|---|---|
| grab servo mode | g | errorCode |
| position servo mode | p | errorCode |
| force servo mode | f | errorCode |
| read register | r *register* | errorCode *registerValue* |
| write register | w *register registerValue* | errorCode |

Table 6.1: Commands: RS-232 Serial Port

When a command line is received, it is parsed and the appropriate processing occurs.

### 6.5.3 Commands

The commands for the gripper controller are listed in Table 6.1. The argument *register* in the **write register** and **read register** commands refers to one of the parameter registers listed in Table 6.2. The *registerValue* parameter is a floating point number. Note that both the position and force setpoints are converted to 8-bit values that are sent to the micro-controller DAC ports. In this case, the range is restricted by the software so the setpoints are in the range 0-255.

After processing each command, a line is returned to the host through the RS-232 serial port. The *errorCode* field contains the error status code. The meaning of the *errorCode* field is listed in Table 6.3. In response to the **read register** command, the *registerValue* field contains a floating point number.

### 6.5.4 Interrupt Level

The interrupt level refers to the interrupt driven software in the controller, which is the interrupt handler code. This level of execution performs the servoing of the fingertips. A periodic interrupt occurs every 8.19 milliseconds. The pneumatic cylinder pressure is adjusted by writing an 8-bit value to a parallel port on the micro-controller. This port is connected to the DAC on the hand control electronics board. The pressure signal is adjusted to maintain the current position or force setpoint. The control law is implemented in a general proportional-integral-derivative controller. All servo controller parameters are writable and may be dynamically adjusted while the gripper controller is servoing. This is done by using the **write register** commands.

### 6.5.5 Controller States

The states of the gripper controller are shown in Figure 6.5. Immediately after initialization, the gripper controller automatically servoes the fingers to maintain the current position.

| Register | Control Parameter | Writable | Definition |
|---|---|---|---|
| 0 | position setpoint | Y | $\mathrm{p}_{setpnt}$ |
| 1 | position proportional gain | Y | $\mathrm{p}_{pgain}$ |
| 2 | position derivative gain | Y | $\mathrm{p}_{dgain}$ |
| 3 | position integral gain | Y | $\mathrm{p}_{igain}$ |
| 4 | position integral minimum | Y | $\mathrm{p}_{imin}$ |
| 5 | position integral maximum | P | $\mathrm{p}_{imax}$ |
| 6 | force setpoint | Y | $f_{setpnt}$ |
| 7 | force proportional gain | Y | $f_{pgain}$ |
| 8 | force derivative gain | Y | $f_{dgain}$ |
| 9 | force integral gain | Y | $f_{igain}$ |
| 10 | force integral minimum | Y | $f_{imin}$ |
| 11 | force integral maximum | Y | $f_{imax}$ |
| 12 | light beam grab threshold | Y | $b_{threshold}$ |
| 13 | light beam grab delay | Y | $b_{delay}$ |
| 14 | light beam on/off | Y | $b_{led}$ |
| 15 | zero pressure point | Y | $z_{pressure}$ |
| 16 | position | N | $p$ |
| 17 | position error | N | $p_{err} = p - p_{setpnt}$ |
| 18 | previous position error | N | $p_{errprev}$ |
| 19 | position proportional term | N | $p_{pterm} = p_{pgain} p_{err}$ |
| 20 | position derivative term | N | $p_{dterm} = (1/T)[p_{dgain}(p_{err} - perrprev)]$ |
| 21 | position integral term | N | $p_{iterm} = p_{igain} p_i$ |
| 22 | position integral | N | $p_i$ |
| 23 | next position integral | N | $p_{inext} = p_i + (T/2)(p_{errprevg} + p_{err})$ |
| 24 | position control | N | $p_{ctl} = p_{pterm} + p_{dterm} + p_{iterm}$ |
| 25 | position control output | N | $p_{out} = min(max(p_{ctl} + z_{pressure}, 0), 255)$ |
| 26 | force | N | $f$ |
| 27 | force error | N | $f_{err} = f - f_{setpnt}$ |
| 28 | previous force error | N | $f_{errprev}$ |
| 29 | force proportional term | N | $f_{pterm} = f_{pgain} f_{err}$ |
| 30 | force derivative term | N | $f_{dterm} = (1/T)[f_{dgain}(f_{err} - f_{errprev})]$ |
| 31 | force integral term | N | $f_{iterm} = f_{igain} f_i$ |
| 32 | force integral | N | $f_i$ |
| 33 | next force integral | N | $f_{inext} = f_i + (T/2)(f_{errprev} + f_{err})$ |
| 34 | force control | N | $f_{ctl} = f_{pterm} + f_{dterm} + f_{iterm}$ |
| 35 | force control output | N | $f_{out} = min(max(f_{ctl}, 0), 255)$ |
| 36 | light beam sensor | N | $b_{sensor}$ |
| 37 | expiration time | N | $t_{expire}$ |
| 38 | real time | N | $t$ |
| 39 | grab state | N | $s_{grab}$ |
| 40 | grab mask | N | $s_{mask}$ |

Table 6.2: Parameter Registers

| errorCode | Error |
|---|---|
| 0 | no error |
| 1 | invalid argument |
| 2 | invalid command |

Table 6.3: Error Codes

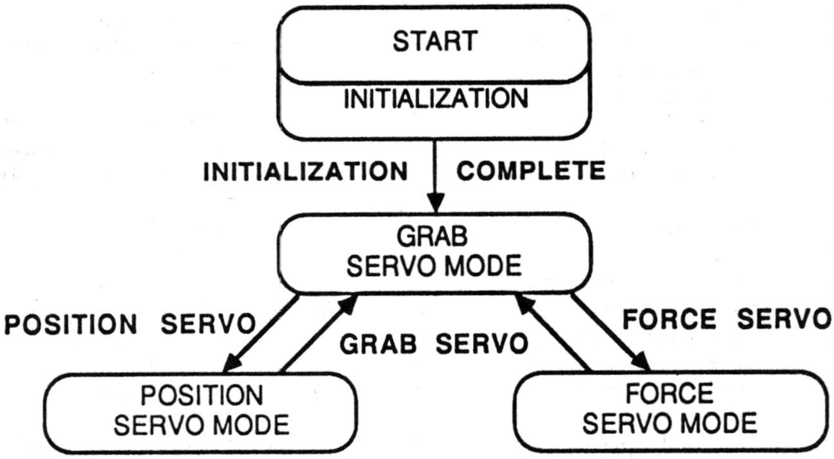


Figure 6.5: Top-Level Controller State Diagram

The gripper controller begins in a grab servo mode upon initialization. When in the grab servo state, the gripper closes whenever an object blocks the light beam between the fingertips. The closing action can be delayed for a specified amount of time, determined by the contents of register 13 (light beam grab delay). The state transition diagram for the grab servo mode is given in Figure 6.6. Position servoing is performed in both the Position_Wait and Position_Delay states while force servoing is performed in the Force state. The Position servo and Force servo modes of operation are activated through commands sent via the RS-232 serial port, (see Table 6.1). The position and force servo modes contain no additional states.

### 6.5.6 Design Issues

The register model is used to set all of the parameters for the CIRSSE gripper controller. This model is used for the following reasons:

- the register model reduces the amount of code necessary to interpret the commands,
- the command syntax does not change with the parameters,
- the register model makes it easy to add or delete parameters in the software to accommodate different control algorithms, and
- the register model makes it easy to identify the critical data and code in the software.

The critical data is the *register area* written and read by the user via the **write register** and **read register** commands.

The use of floating point control parameters introduces critical code elements. This arises because the MC68HC11 has no atomic instruction that can read or write 3 bytes, the length of a floating point number. When a critical control parameter is set, the interrupts are masked while the floating point number is being transferred to the *register area* used by the interrupt handlers. This prevents the interrupt handler from using a partially updated floating point parameter. The amount of critical code is limited to two places in the software. Thus, masking can be performed without any danger of missing any interrupts and without causing any discontinuities in the servoing.

## 6.6 Summary

The CIRSSE general purpose two-finger gripper and controller is designed to be used with a wrist-mounted force-torque sensor. Accordingly, the mechanical redesign achieved a reduction in the length and weight of the gripper, which

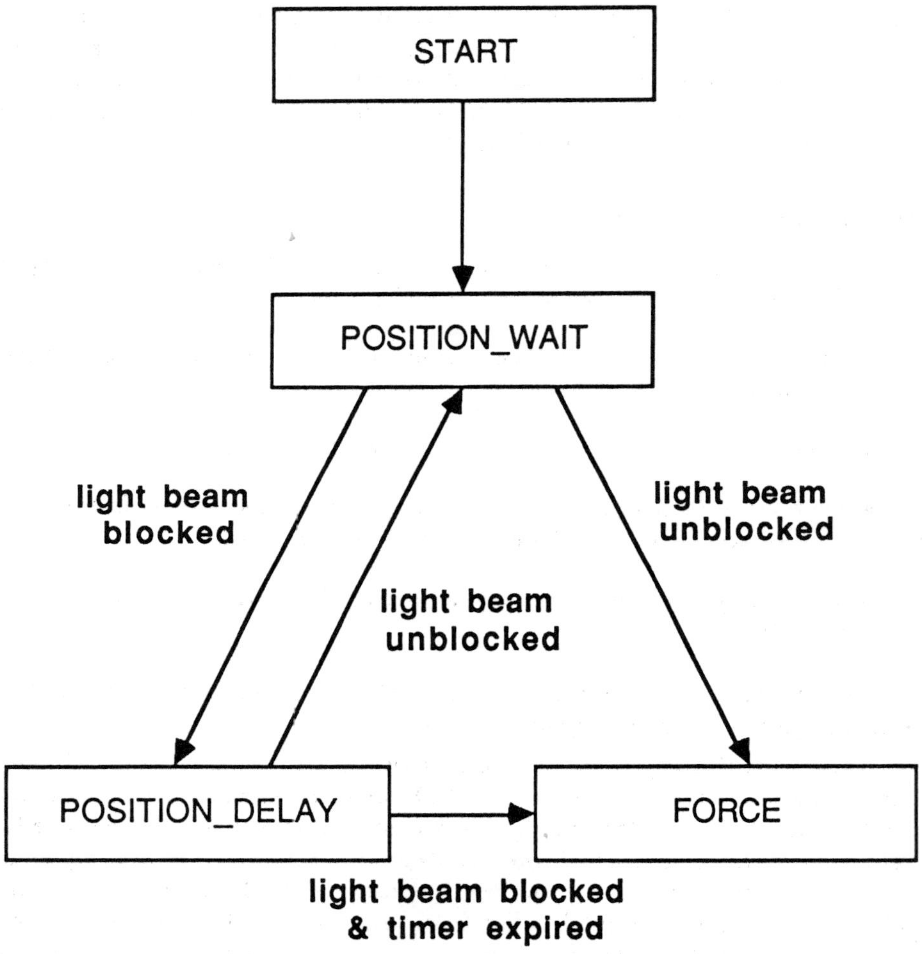


Figure 6.6: Grab Servo Mode State Transition Diagram

resulted in more than a 40 percent reduction of the gripper self-moment over the previous design. The pneumatic powered gripper features a grasping force-to-weight ratio of better than 25-to-1. An MC68HC11EVB evaluation board provides a low cost, effective platform for the controller. The controller software implements a simple real-time executive. On power-on, it performs a hardware self-check and starts the controller in position servo mode to prevent the gripper from moving and then enters the grab servo mode. Either position or grasping force servo modes are software selectable and the control law parameters may be tuned dynamically while servoing by changing parameter register values. The simple host-controller communications protocol based on parameter registers minimizes the command interpretation effort and maximizes the versatility of the controller.

This chapter updates a preliminary version of this material that was presented in [4]. More detailed information is available in [5] and [6].

## Acknowledgments

John Peiffer's work was supervised by Stephen Derby, Professor in the Mechanical Engineering, Aeronautical Engineering, and Mechanics Department at Rensselaer. This research is supported in part by Grant NAGW-1333 from NASA to the Center for Intelligent Robotic Systems for Space Exploration.

## 6.7 References

[1] R. T. J. Birk and R. Kelley, "General purpose hands for bin-picking robots," *IEEE Trans. on Systems, Man and Cybernetics*, vol. SMC-12, November/December 1982.

[2] Motorola, Inc., Phoenix, AZ, *M68HC11EVB Evaluation Board User's Manual*, 1989.

[3] Prentice-Hall, Englewood Cliffs, NJ, *M68HC11 Reference Manual*, 1986.

[4] R. Kelley, J. Tsai, J. Bethel, and J. Peiffer, "Gripper for truss structure assembly," in *SPIE Conference: Cooperative Intelligent Robotics in Space*, vol. 1387 of *38-46*, (Cambridge, MA), November 1990.

[5] J. Tsai, "M68HC11 gripper controller software," CIRSSE Report 90, Rensselaer Polytechnic Institute, Troy, NY, May 1991.

[6] J. Bethel, "M68HC11 gripper controller electronics," CIRSSE Report 92, Rensselaer Polytechnic Institute, Troy, NY, May 1991.

# Chapter 7

# Simulation of Space Manipulators

by: Stephen H. Murphy
Electrical, Computer, and Systems Engineering Department
Center for Intelligent Robotic Systems for Space Exploration
Rensselaer Polytechnic Institute
Troy, New York 12180-3590

The simulation of robot manipulators in space must satisfy many requirements. One of these requirements is the need for fast and accurate simulations. This chapter presents results on work towards the simulation of space-based manipulators. An algorithm which is order $n$ in the number of links is presented and it includes effects such as flexible joints and motor gyroscopic forces. The approach is extended beyond single manipulators in space to the simulation of cooperating manipulators. Efficient simulation is maintained as well as producing a model for control development. Two simulation case studies are presented: a simulation of a single manipulator in space complete with the gyroscopic forces of the motors, and a simulation of two cooperating manipulators on a common base.

## 7.1 Introduction

The future implementation of robotic manipulators in space will require extensive modeling, simulation and analysis. This work will occur during the construction of the manipulators, development and verification of the controllers, training of the operators, and during the actual operation of the manipulators. Each one of these areas has different needs for modeling and simulation.

The model for control development and analysis requires the model equations in a closed-form that can be analyzed together with the controller to produce information on the stability and tracking. Both the controller development and the manipulator construction requires information on the forces of interaction between the manipulator and the base or between manipulators during cooperation. Real-time simulations of the Remote Manipulator System of the space shuttle have already proven to be extremely valuable in the training of robot operators. More advanced systems will require equally extensive real-time simulations for training. The operation of advanced robotic systems may also encompass the use of real-time models and simulations for the prediction of motions under large time delays.

For earth-bound manipulators, the dynamic model used for the control analysis has been effectively used for simulation. However the demands on the simulation have not been as high as those for space manipulators. Earth-bound manipulators seldom have more than seven degrees-of-freedom which places smaller requirements on the solution of the dynamic equations of motion. Additionally, the manipulators are quite rigid, with low flexibilities in the links and motor-link couplings, resulting in relatively simpler nonlinear dynamic models. In contrast, space-based manipulators have a much larger number of degrees-of-freedom due to the interactions of the manipulators and their mounting platform. The lightweight structure of the manipulators and the large masses accentuate the flexibilities in the motor couplings and links which must be included in the model. The performance of simulation methods developed for earth-based manipulators degrades rapidly as the number of degrees-of-freedom increase and the demands for more accurate models increase.

Complex assembly and disassembly in space entails the cooperation of two or more manipulators. Almost all efficient robot simulation methods have not been extended to the case where the manipulators form a closed kinematic chain. The system becomes more complicated in space where the robot motions are coupled through the base motion.

The purpose of this chapter is to present a method for the modeling and simulation of robot manipulators which can be used for both the control development and real-time simulation of single or multiple robot manipulators in space or on earth. The real-time simulation requirement is met through the development of a simulation method which is order $n$ in the the number of links. The simulation method is extended to accommodate the complex models of motor/gyroscopic forces and motor-link flexible couplings. In this manner, more accurate models of manipulators are encompassed in the order $n$ simulation method. The simulation model is readily transformed into the closed-form model necessary for control development and analysis. This chapter summarizes one area of work in the simulation and modeling of space manipulators. A more in-depth treatment of this work can be found in [1], [2].

In the literature on space robotics there is a large difference between the models created for control development, path planning, and simulation. Control development models are often based upon the Lagrange-Euler dynamics formulation which produces closed-form equations for development and analysis, [3], [4], [5], [6], [7]. Path planning models rely upon the conservation of momentum to produce a kinetic formulation for the movement of the manipulator and platform, [8], [9], [10], [11], [12], [13], [14]. These models have been successfully used for control and simulation but do not contain all the dynamic interactions as the control dynamic models. More accurate control models which include the effect of spinning motors and flexible motor couplings have been developed for fixed-base manipulators, [15], [16], [17], [18], [19].

Recently, there has been a large number of developments towards simulation models and methods of simulation that grow only linearly in the number of links, [20], [21], [22], [23], [24], [25], [26], [27]. These methods are more efficient for systems with large numbers of degrees-of-freedom and manipulator links. The effects of motor coupling flexibility, link flexibility, and motor gyroscopic forces have also been added to the efficient simulation methods, [28], [29], [30].

Research on the cooperation of multiple robot manipulators has concentrated on earth-based manipulators. There is a large and continually growing body of literature discussing the problems in control and planning that is beyond the scope of this chapter. For the interested reader, references may be found in [31], [32], [33], [34], [35], [36], [37], [38], [39], [40], [41], [42].

This chapter is organized as follows: Section 2 presents the notation used to describe the manipulators, motors, and flexible couplings. In Section 3, the simulation model and control model of a single manipulator on a mobile platform is presented. Results on the impact of motor gyroscopic forces in space are included as a case study on the simulation. Cooperation between two manipulators on a space platform is the topic of Section 4 with a simulation case study of two PUMA manipulators in a cooperative motion. The chapter is concluded in Section 5.

## 7.2 Notation

The notation in this work derives from the Spatial Operator Algebra developed at JPL by Rodriguez, Kreutz, and Jain, [25], [43], [22], [44]. The spatial operator algebra uses the concept of coordinate independent spatial vectors which are assigned to a specific coordinate frame only at the time of actual implementation. Spatial vectors are the vectors of combined angular and linear components of either moment/force, angular/linear velocity, or angular/linear acceleration. The presentation in this chapter will differ in that all operators and vectors will be represented in specific coordinate frames. The purpose is to allow the reader to compare this approach with other methods of modeling and simulation.

Coordinate frames are assigned using the modified Denavit-Hartenberg convention as described in Craig, [45]. Figure 7.1 shows the overall picture of a

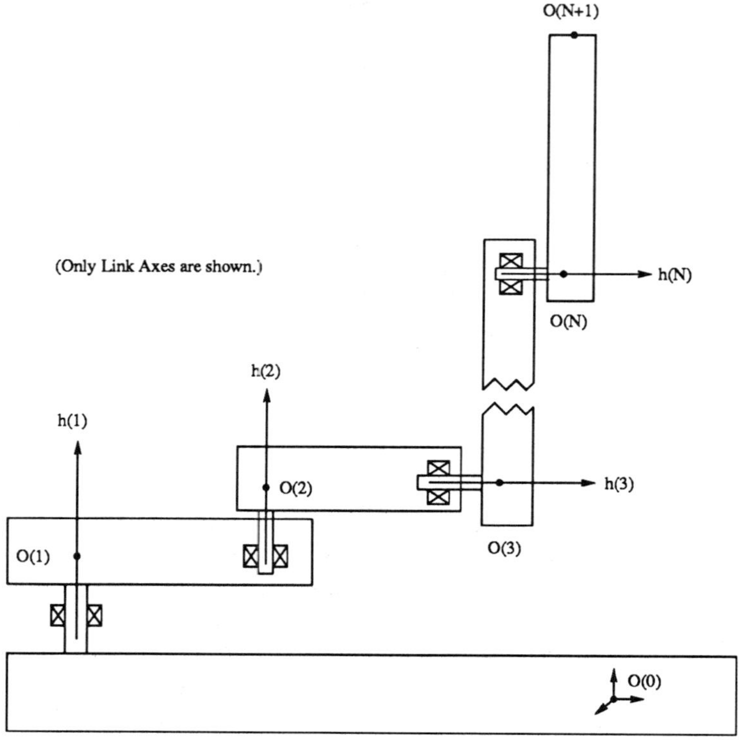


Figure 7.1: Overall Robot Link Structure

single manipulator on a platform. A coordinate frame is established at each point $O(k)$ according to the convention. All the axes of each frame are not shown in the figure.

The vectors $h(k)$ denote the axis of rotation for link $k$ and by the modified Denavit-Hartenberg convention the $z$ axis of the frame $k$ would point along the direction of $h(k)$. The origin of the frame $k$ is denoted by $O(k)$.

The creation of coordinate frames also establishes the $3 \times 3$ rotation matrices that define the orientation of one frame with respect to another. Let ${}^i_jR$ be the rotation matrix that relates the orientation of frame $j$ to frame $i$. The rotation matrices are functions of the joint angles, $\theta(k)$. All vectors are elements of $\Re^3$, $3 \times 1$ column vectors. Spatial quantities are the result of stacked vectors and are elements of $\Re^3 \times \Re^3$. Spatial vectors are not elements of $\Re^6$. With each vector is the associated coordinate frame in which the elements are written.

Given a vector, ${}^jx \in \Re^3$, where the $j$ indicates which frame the vector $x$ is written. Then

$$ {}^ix = {}^i_jR\,{}^jx \tag{7.1} $$

where the rotation matrix ${}^{i}_{j}R$ takes a vector $x$ written in frame $j$ to the same vector $x$ written in frame $i$.

The operator, $*$, is the matrix transpose operator. Given vectors $x, y \in \Re^3$, the dot product operator is defined, $x \cdot y = x^T y$. The cross product operator becomes, $x \times y = \tilde{x}y = z$, where $\tilde{x}$ is the $3 \times 3$ skew-symmetric matrix representing $x\times$ in coordinate frame notation, $z \in \Re^3$ and $z^T y = 0$, $z^T x = 0$.

$l(k, k+1)$ The vector from the origin of the frame at $O(k)$ to the origin of the frame at $O(k+1)$. $l(k, k+1)$ is written in the frame at $O(k)$ and is constant.

$\omega(k)$ The $3 \times 1$ vector of angular velocity of link $k$ at point $O(k)$, written in the frame at $O(k)$.

$\omega_m(k)$ The $3 \times 1$ vector of angular velocity of the motor at $O_m(k)$. written in the frame at $O_m(k)$.

$v(k)$ The $3 \times 1$ vector of linear velocity of link $k$ written in the frame at $O(k)$.

$v_m(k)$ The $3 \times 1$ vector of linear velocity of the motor at $O_m(k)$, written in the frame at $O_m(k)$.

$V(k)$ $6 \times 1$ vector of spatial velocity of link $k$, written in the frame at $O(k)$, $V(k) = \begin{bmatrix} \omega(k) \\ v(k) \end{bmatrix}$

$V_m(k)$ $6 \times 1$ vector of spatial velocity of the motor written in the frame at $O_m(k)$, $V_m(k) = \begin{bmatrix} \omega_m(k) \\ v_m(k) \end{bmatrix}$

$\alpha(k)$ $6 \times 1$ vector of spatial acceleration of $O(k)$, written in the frame at $O(k)$, $\alpha(k) = \begin{bmatrix} \dot{\omega}(k) \\ \dot{v}(k) \end{bmatrix}$.

$\alpha_m(k)$ $6 \times 1$ vector of spatial acceleration of $O_m(k)$, written in the frame at $O_m(k)$, $\alpha_m(k) = \begin{bmatrix} \dot{\omega}_m(k) \\ \dot{v}_m(k) \end{bmatrix}$.

$F(k)$ $3 \times 1$ vector of forces acting upon link $k$ at point $O(k)$, written in the frame at $O(k)$.

$N(k)$ $3 \times 1$ vector of torques or moments acting upon link $k$ at point $O(k)$, written in the frame at $O(k)$.

$f(k)$ $6 \times 1$ vector of spatial force acting upon link $k$ at point $O(k)$, written in the frame at $O(k)$, $f(k) = \begin{bmatrix} N(k) \\ F(k) \end{bmatrix}$.

$p(k)$ The $3 \times 1$ vector from $O(k)$ to the center of mass of link $k$, written in the frame at $O(k)$ and is constant in that frame.

$p_m(k)$ The $3 \times 1$ vector from $O_m(k)$ to the center of mass of the motor/gear driving link $k$, written in the frame at $O_m(k)$, and is constant in that frame.

$l_m(k-1,k)$ — $3 \times 1$ vector from $O(k-1)$ to $O_m(k)$, written in the frame at $O(k-1)$.

$\phi(k+1,k)$ — A $6\times6$ matrix which translates a spatial velocity at point $O(k)$ written in the frame at $O(k)$ to a spatial velocity at $O(k+1)$ written in the frame at point $O(k+1)$.

$$\phi(k+1,k) = \begin{bmatrix} {}^{k+1}_{k}R & 0 \\ -{}^{k+1}_{k}R\tilde{l}(k,k+1) & {}^{k+1}_{k}R \end{bmatrix}$$

$\phi^*(k+1,k)$ — Transpose of the $6\times6$, $\phi(k+1,k)$ matrix. $\phi^*(k+1,k)$ translates a spatial force acting at $O(k+1)$ written in the frame at $O(k+1)$ to a spatial force acting at $O(k)$ written in the frame at $O(k)$. $\phi^*(k+1,k) = \begin{bmatrix} {}^{k}_{k+1}R & \tilde{l}(k,k+1){}^{k}_{k+1}R \\ 0 & {}^{k}_{k+1}R \end{bmatrix}$.

${}^{m_k}_{k-1}R$ — The rotation matrix which relates the orientation of the frame at $O_m(k)$ to the frame at $O(k-1)$.

$\phi_m(k,k-1)$ — A $6 \times 6$ matrix which translates a spatial velocity at point $O(k-1)$ written in the frame at $O(k-1)$ to a spatial velocity at $O_m(k)$ written in the frame at $O_m(k)$. $\phi(k,k-1) = \begin{bmatrix} {}^{m_k}_{k-1}R & 0 \\ -{}^{m_k}_{k-1}R\,\tilde{l}_m(k-1,k) & {}^{m_k}_{k-1}R \end{bmatrix}$

$\phi^*_m(k,k-1)$ — $\phi^*_m(k,k-1) = \phi^T_m(k,k-1)$.

$m(k)$ — The mass of link $k$.

$m_m(k)$ — The mass of the motor/gear assembly that drives link $k$.

$I(k)$ — The $3\times3$ inertia matrix of link $k$ at $O(k)$, written in the frame at $O(k)$.

$I_m(k)$ — The $3\times3$ inertia matrix of the motor/gear assembly at $O_m(k)$, written in the frame at $O_m(k)$.

$M(k)$ — The $6 \times 6$ spatial mass matrix of link $k$ at $O(k)$ written in the frame at $O(k)$. It incorporates the inertia matrix of link $k$, link mass, and location of center of mass, $p(k)$. $M(k) = \begin{bmatrix} I(k) & m(k)\tilde{p}(k) \\ -m(k)\tilde{p}(k) & m(k)I \end{bmatrix}$, where $I$ is the $3\times3$ identity matrix.

$M_m(k)$ — The $6 \times 6$ spatial mass matrix of the motor and gear at $O_m(k)$ written in the frame at $O_m(k)$.

$$M_m(k) = \begin{bmatrix} I_m(k) & m_m(k)\tilde{p}_m(k) \\ -m_m(k)\tilde{p}_m(k) & m_m(k)I \end{bmatrix}.$$

$a(k)$ — The $6 \times 1$ vector of spatial bias accelerations of link $k$, written in the frame at $O(k)$. For a rotational joint,

$$a(k) = \begin{bmatrix} \left({}^{k}_{k-1}Rw(k-1)\right) \times w(k) \\ {}^{k}_{k-1}R\Big(w(k-1) \times [w(k-1) \times l(k-1,k)]\Big) \end{bmatrix}.$$

$a_m(k)$ The $6 \times 1$ vector of spatial bias accelerations of the motor/gear driving link $k$, written in the frame at $O_m(k)$.

$$a_m(k) = \begin{bmatrix} \left({}^{m_k}_{k-1}Rw(k-1)\right) \times w_m(k) \\ {}^{m_k}_{k-1}R\Big(w(k-1) \times [w(k-1) \times l_m(k-1,k)]\Big) \end{bmatrix}.$$

$b(k)$ The $6 \times 1$ vector of spatial bias forces acting on link $k$ at $O(k)$, written in the frame at $O(k)$. For a rotational joint,

$$b(k) = \begin{bmatrix} w(k) \times I(k)w(k) \\ m(k)w(k) \times [w(k) \times p(k)] \end{bmatrix}.$$

$b_m(k)$ The $6 \times 1$ vector of spatial bias forces acting on the motor/gear at $O_m(k)$, written in the frame at $O_m(k)$.

$$b_m(k) = \begin{bmatrix} w_m(k) \times I_m(k)w_m(k) \\ m_m(k)w_m(k) \times [w_m(k) \times p_m(k)] \end{bmatrix}.$$

$h(k)$ The $3 \times 1$ vector aligned with the axis of rotation for link $k$ and defining the direction of positive rotation. When written in the frame at $O(k)$, $h(k) = [0 \quad 0 \quad 1]^T$ written in the frame at $O(k)$.

$H(k)$ The $6 \times 1$ vector for the spatial axis of motion for link $k$, written in the frame at $O(k)$. For a rotational joint,
$H(k) = [h^T(k) \quad 0 \quad 0 \quad 0]^T$, or equivalently,
$H(k) = [0 \quad 0 \quad 1 \quad 0 \quad 0 \quad 0]^T$.

$H^*(k)$ Transpose of $H(k)$, projects onto spatial axis of motion,
$H^*(k) = H^T(k)$.

$h_m(k)$ The $3 \times 1$ vector of rotation for the motor driving link $k$, written in the frame at $O_m(k)$ and is constant, $h_m(k) = [0 \quad 0 \quad 1]^T$.

$H_m(k)$ $6 \times 1$ vector for spatial axis of motion for the motor driving link $k$, written in the frame at $O_m(k)$.
$H_m(k) = [h_m^T(k) \quad 0 \quad 0 \quad 0]$.

$H_m^*(k)$ The transpose of $H_m(k)$.

## 7.3 Single Manipulator Dynamics

For a single manipulator on a mobile platform, the above notation may be used to efficiently state the relationships describing velocity and acceleration propagation across the links of the manipulator. Moving in a sweep from the base of the manipulator to the tip, the velocities and accelerations may be found. This forms the basis for an order $n$ simulation method.

This section examines the flow of velocities, accelerations, forces, and inertias up and down the links of the manipulator. The result is an order $n$ simulation that can include motor gyroscopic forces or flexible couplings be-

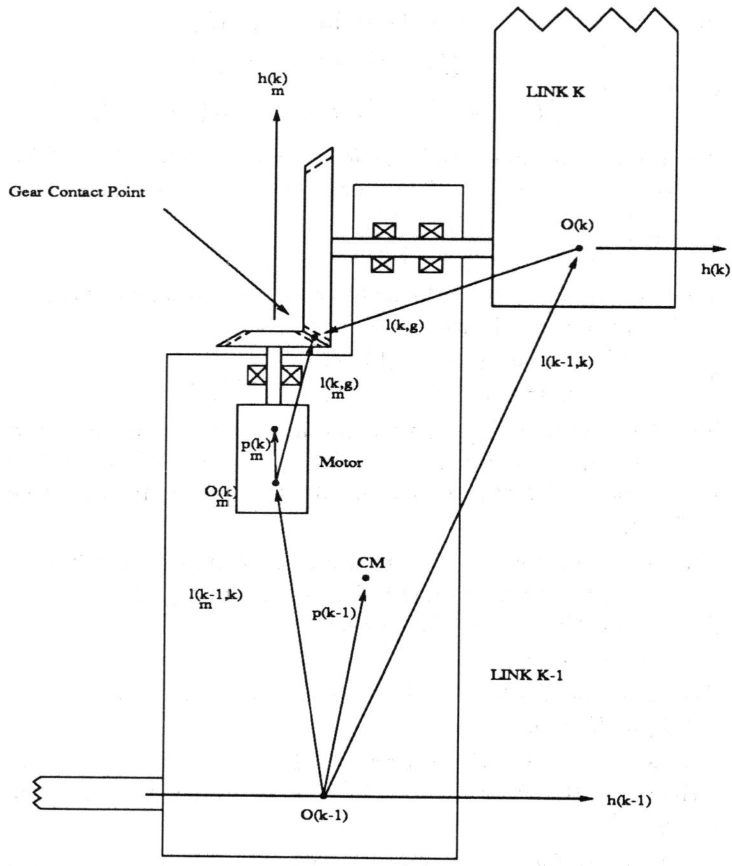


Figure 7.2: Geared Joint Model

tween motor and link. The model of forces and accelerations can also be used to produce the closed-form or Lagrange-Euler model which can be used for control analysis. This will be expanded in the following section on cooperative modeling.

Figures 7.2-7.3 show the basic models of the geared and flexibly jointed manipulators under discussion.

### 7.3.1 Velocity and Acceleration Propagation

The velocity of each link is the sum of the velocity of the previous link and the link's own relative rotational velocity. The velocity of each motor is the velocity of the link upon which it is mounted plus the local rotational velocity

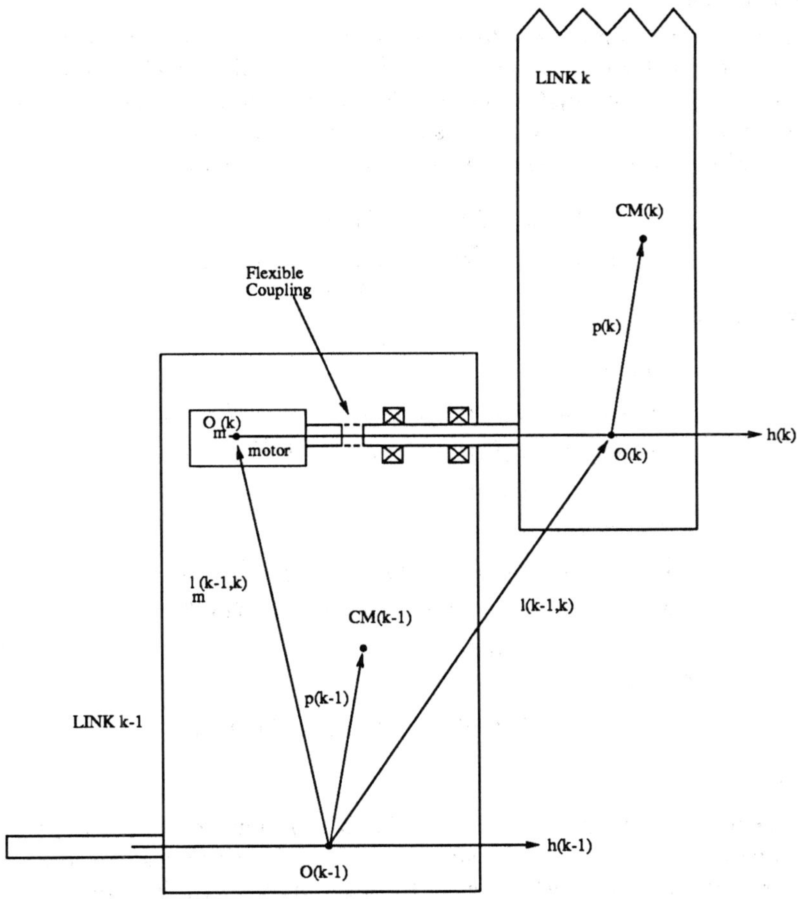


Figure 7.3: Flexible Joint Model

of the motor. For the flexible joint, the motor velocity is independent of the link velocity. For a geared joint, the motor velocity and link velocity are related through the gear ratio, $\Gamma$.

Similar equations hold for the accelerations but with additional bias terms added to account for centrifugal accelerations. The bias accelerations are contained in the $a(k)$, and $a_m(k)$ terms. The base is assumed fixed with respect to an inertial frame but may be considered accelerating to account for gravity. The algorithmic form for the equations is presented below:

$$\begin{aligned} V(0) &= \text{Base Spatial Velocity} \\ \alpha(0) &= \text{Base Spatial Acceleration} \end{aligned}$$

for $k = 1, ..., N$ loop

link:

$$V(k) = \phi(k, k-1)V(k-1) + H(k)\dot{\theta}(k) \tag{7.2}$$

$$\alpha(k) = \phi(k, k-1)\alpha(k-1) + H(k)\ddot{\theta}(k) + a(k) \tag{7.3}$$

motor:

$$V_m(k) = \phi_m(k, k-1)V(k-1) + H(k)\dot{\theta}_m(k) \tag{7.4}$$

$$\alpha_m(k) = \phi_m(k, k-1)\alpha(k-1) + H(k)\ddot{\theta}_m(k) + a_m(k) \tag{7.5}$$

Geared:

$$\ddot{\theta}_m(k) = \Gamma(k)\ddot{\theta}(k) \; ; \quad \dot{\theta}_m(k) = \Gamma(k)\dot{\theta}(k)$$

endloop;

Independent of the geared case, or the flexible joint case, the force acting upon a link can be written as the sum of the d'Alembert forces, the forces from the outer link, and the inertial forces from the motor, [1]. The inertial forces from the motor are a function of the motor acceleration and force biases due to angular velocities.

$$\begin{aligned} f(N+1) &= \text{vector of external torques/forces} \\ f_m(N+1) &= 0 \end{aligned}$$

for $k = N, ..., 1$ loop

link:

$$f(k) = \phi^*(k+1, k)f(k+1) + \phi_m^*(k+1, k)f_m(k+1) + M(k)\alpha(k) + b(k) \tag{7.6}$$

motor:

$$f_m(k) = M_m(k)\alpha_m(k) + b_m(k) \tag{7.7}$$

endloop;

where $f_m(k)$ expresses the d'Alembert forces for the motor. Equation 7.6 shows how the forces acting upon a link are composed of the standard rigid body forces along with additional forces from the spinning motor for both the geared and flexible joint cases.

## Motor and Link Torque

The determination of motor and link torque differs between the flexible joint case and the geared case.

### Flexible Joint Motor Torque

The torque acting on the shaft of link $k$ is the projection of the forces acting on link $k$ along the axis of the shaft,

$$T(k) = H^*(k)f(k). \tag{7.8}$$

This torque is also a function of the state-dependent coupling between the motor and link. For a simple torsional spring model, the torque is a function of the angular displacement between the motor and the link,

$$T(k) = K(k)[\theta_m(k) - \theta(k)] \tag{7.9}$$

where $K(k)$ is the torsional spring constant for the $k^{\text{th}}$ axis.

To account for internal damping in the shaft, the torque may be written,

$$T(k) = K(k)[\theta_m(k) - \theta(k)] + Z(k)[\dot{\theta}_m(k) - \dot{\theta}(k)] \tag{7.10}$$

where $Z(k)$ represents the damping coefficient.

In general, the torque coupling may be expressed as any function of the state of the manipulator,

$$T(k) = f(\theta, \dot{\theta}, \theta_m, \dot{\theta}_m). \tag{7.11}$$

More advanced models, such as backlash, may be included in this function. The function $T(k)$ will be referred to as the coupling function between the motor and link.

Through a force balance on the motor and shaft, the torque on the motor can be expressed as the sum of the projection of the d'Alembert forces along the motor shaft and the torque on the link,

$$T_m(k) = H^*(k)f_m(k) + T(k). \tag{7.12}$$

**Geared Motor Torque**

Assuming that the interaction between the gears is a point contact, and that the bearings exert no torque about the axis of rotation, then the torque on the motor may be written as

$$T_m(k) = \frac{1}{\Gamma(k)} H^*(k) f(k) + H_m^*(k) f_m(k) \tag{7.13}$$

The torque on the motor can be seen as the torque on the link $k$ reduced by the gear ratio plus the torque due to the acceleration of the motor itself. $H_m^*(k)$ projects the forces due to acceleration, $f_m(k)$ onto the axis of rotation of the motor. The geometry of the gear contact is not necessary in the calculation of the motor torque.

### 7.3.2 Simulation

Equations (7.2–7.7), and (7.11–7.13) define the complete inverse dynamics of both flexibly jointed and geared manipulators. The goal for the simulation, or forward dynamics, is to calculate the accelerations of the motor and link when given the motor/link angular positions, velocities, and motor torques. One possibility for the forward dynamics is to define the $2N \times 2N$ mass matrix for the flexibly jointed system, or the $N \times N$ mass matrix for the geared system, and invert the matrix to calculate the link and motor accelerations. As the number of links increases, this inversion becomes increasingly inefficient, [20].

From the propagation of velocities and torques, an algorithm for the calculation of the forward dynamics which grows of order $n$ may be developed. The algorithm is recursive in nature and allows for the use of the general form of the coupling function between motor and link. The algorithm is derived through the observation that for any link $k$, the force on the link may be written as

$$f(k) = P(k)\alpha(k) + \mathcal{B}(k).$$

The complete derivation of the forward dynamics may be found in [1].

The forward dynamics is presented as a three sweep operation:

**Sweep 1:** Sweep up the manipulator.

For each link, calculate the bias forces and accelerations that are due to the velocity of the links. For a flexible joint, calculate the coupling function, $T(k)$.

**Sweep 2:** Sweep down the manipulator.

Calculate the articulated body inertias and bias forces for each link, $P(k)$, and $\mathcal{B}(k)$. These terms are defined recursively from link $N$, and maintain similar structures for both flexibly jointed and geared manipulators. At the base of the manipulator, the acceleration may be found from the equation

$$f(0) = P(0)\alpha(0) + \mathcal{B}(0) \tag{7.14}$$

where $f(0)$ is the controlled torque/force acting upon the base from reaction wheels, thrusters, etc., and $\alpha(0)$ is the resultant angular/linear acceleration.

**Sweep 3:** Sweep up to calculate the accelerations.

From the articulated body inertias, bias forces, and motor torques, the joint acceleration and spatial acceleration of each link is determined starting from the known acceleration of the base and moving outward to the tip.

The three sweeps provide an order $n$ solution to the complete forward dynamics of a geared or flexibly jointed manipulator. For direct-drive manipulators, $P(k)$ is the articulated body inertia of Jain, [46], and Featherstone, [20]. $\mathcal{B}(k)$ is the correction force.

### 7.3.3 A Simulation Case Study

The spinning motors mounted on the links of the robot manipulator act like small gyroscopes during the execution of a robot motion. As gyroscopes, the motors produce torques and forces on the links of the manipulators that are not included in the common dynamic models of manipulators.

There are two main effects of the gyroscopic forces on robot manipulators. The first effect is on the control of the robot link angle. The second effect is on the moments and forces acting upon the base of the manipulator. For earth-based manipulators the second impact is totally negligible as the robot is typically firmly fastened. However, for manipulators which are designed to operate in a space environment, the gyroscopic forces due to the motors will also work to twist the entire robot in space.

In order to demonstrate the simulation method presented above and to examine the effect of motor gyroscopic forces in space, a complete model of a geared PUMA manipulator was created. The PUMA manipulator presented a more difficult case study as this robot has a very non-uniform link structure, as shown in Figure 7.4. However, the algorithm for calculating the $P(k)$ terms for each joint and propagating the articulated body inertias down the links functions in the exact same manner, [1].

### 7.3.4 Simulation Parameters

The dynamic parameters for the PUMA model were taken from model identification work by Armstrong, *et. al.*, [47], and Tarn, *et. al.*, [48]. The mass/inertia

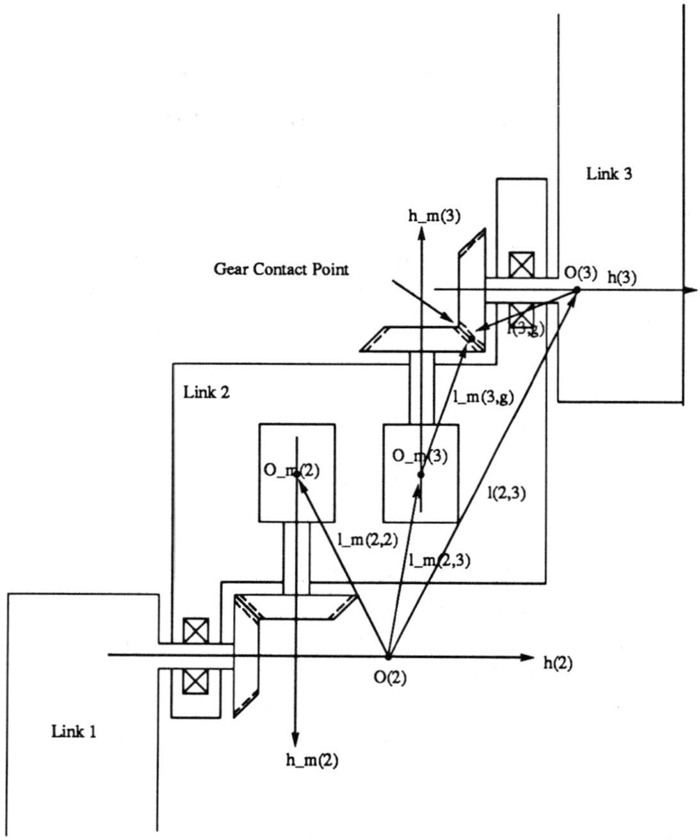


Figure 7.4: Link 2, PUMA 560

| Mass | Center of Mass $x_c$ | $y_c$ | $z_c$ | Inertia $I_{xx}$ | $I_{yy}$ | $I_{zz}$ |
|---|---|---|---|---|---|---|
| kg | m | m | m | kg-m$^2$ | kg-m$^2$ | kg-m$^2$ |
| 15.3 | -0.74 | 0.0 | 0.15 | 0.8163 | 11.34 | 10.72 |

Table 7.1: Mass and Inertia Parameters for Platform

parameters for the platform are given in Table 7.1. The location of the center of mass is in reference to the coordinate frame as shown in Figure 7.5. The PUMA is mounted $-0.80$ m along the $x$ axis, relative to this frame. The motors were assumed to have an $x$ and $y$ axis inertia 110% larger than the $z$ axis inertia.

The starting position for the simulation is shown in Figure 7.5. The motion was chosen to represent a typical pick-and-place operation of the manipulator relative to the platform frame. The trajectory was calculated minimizing the jerk during the motion between endpoints, [49]. The manipulator moves from the start position to the pick location, remains motionless for 2 seconds, moves to the place location, and then moves back to the start. Figures 7.5-7.6 include the two locations. The total time for the motion was 10 seconds.

### 7.3.5 Simulation Experiments

The impact of the motor gyroscopic forces on the position and orientation of the platform was evaluated by modeling a manipulator complete with high gear ratios and motors, and comparing to a model of a manipulator with direct drive motors. The dynamic mass parameters of the two models were identical. No external moments or forces were acting upon the platform. The system was allowed to rotate freely about the center of mass in response to manipulator movements. The complete model was evaluated at the normal gear ratios, and at 5, 10, and 20 times the gear ratios. For the PUMA manipulator, the joint gear ratios range from $62 : 1$ to $107 : 1$. For comparison, the Shuttle Remote Manipulator has gear ratios of $749 : 1$ to $1842 : 1$ but with much larger link masses/inertias, [50].

Two variables were observed to detect differences in platform orientation between the full manipulator model and the direct drive model. The first was the difference in manipulator tip position between the two models over the trajectory. The second is the angle of rotation between the full model platform rotation or attitude matrix, and the gear approximate model platform attitude matrix. Figure 7.7 shows the trace of the tip position for the two models under the motion. Figure 7.8 shows the absolute difference in tip position and Figure 7.9 shows the rotational angle from an axis-angle measurement of the differences in platform orientation during the manipulator motion.

Start                                                After 2 Seconds

Figure 7.5: Initial Position and First Pick Location

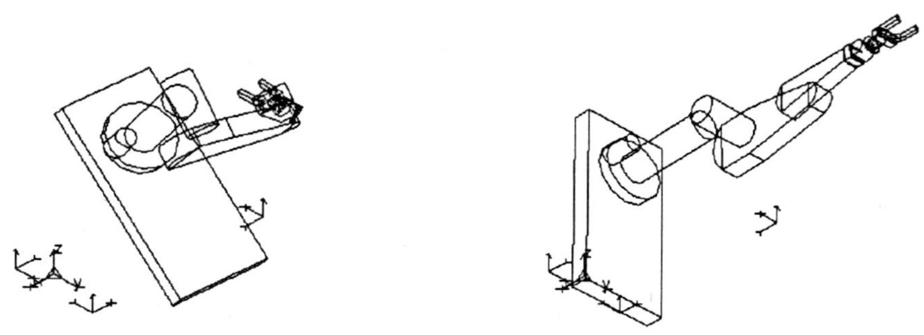

After 7 seconds                                      Final Configuration

Figure 7.6: Place Location and Final Position

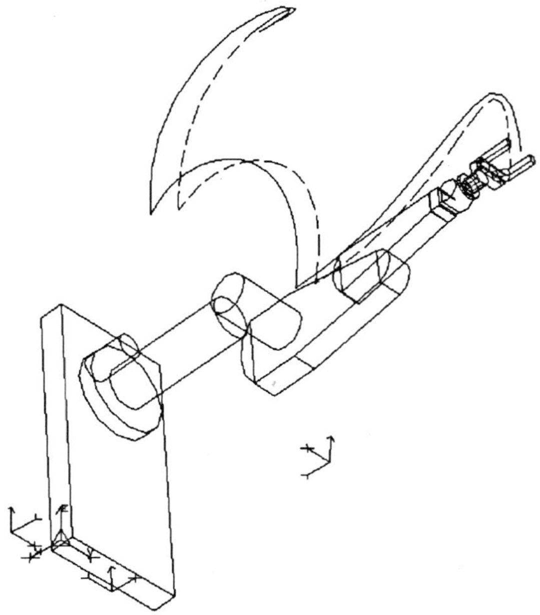

Figure 7.7: Trace of Tip Position During Motions

Because the gyroscopic forces are a function of velocity, the impact of velocity on platform orientation was investigated by slowing down the motion over the desired joint path. Figure 7.10 shows the difference in manipulator tip position between the full gear model and the gear approximation model at 10 times the nominal PUMA gear ratios with two different times for the entire motion.

### 7.3.6 Discussion

A number of observations can be made from the results in Figures 7.8–7.10. Clearly, from all the figures, the gear ratio impacts the positions, and orientations linearly with the increase in gear ratio. Figures 7.8 and 7.9 show the significance of the motor gyroscopic force on the platform orientation. The difference in tip position is a function of the final endpoint of the manipulator, by virtue of the lever arm from platform to tip. The difference in the angle between the full model platform orientation and the gear approximation model platform orientation is independent of the manipulator end position. For the single manipulator on the platform as described in Table 7.1, the difference in orientation is remarkable. Any increase in the platform or manipulator inertia

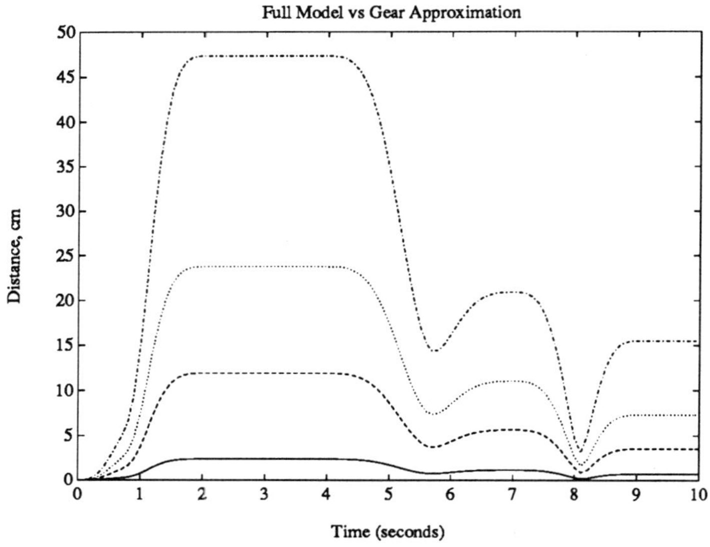


Figure 7.8: Difference in Tip Position During Motions

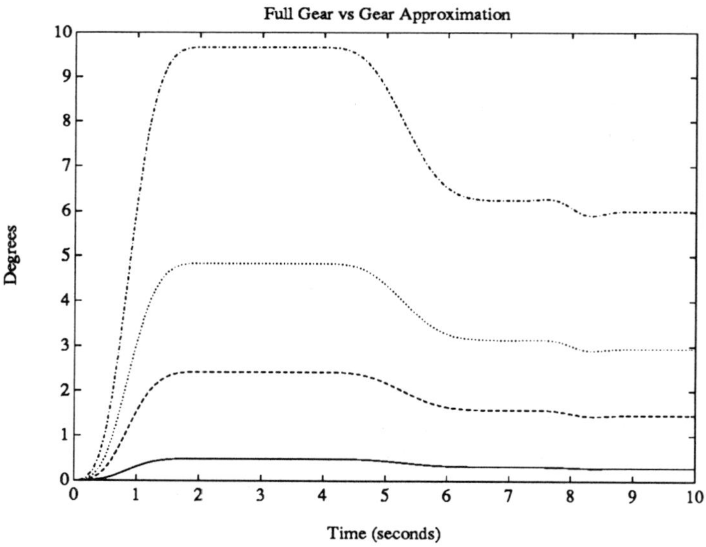


Figure 7.9: Degrees of Rotation Between Orientations

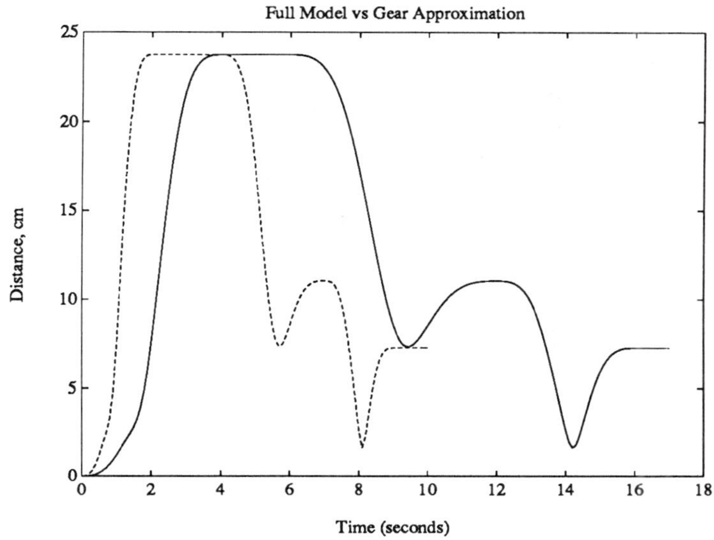


Figure 7.10: Impact of Longer Trajectory Times on Difference in Tip Position, 10x Gear Ratio

will decrease the impact of the motor gyroscopic forces proportionally.

However, one of the possible goals of manipulator motion in space is to change the orientation of the platform. As orientation in space is not conserved during manipulator motion, it is possible to plan a sequence of manipulator motions that can arbitrarily change the platform orientation, [13], [51]. This task becomes increasingly difficult due to the orientation errors created by the gyroscopic forces from the motors. A more massive manipulator and platform would be less affected by these forces but with a greater cost of energy that would be necessary to create the desired motion.

The result of Figure 7.10 may not be immediately expected. This figure shows that slowing down the motion of the manipulator has no impact on the effect of the gyroscopic forces. This is apparent if it is considered that the forces are integrated over the entire motion.

While much of the moment created by the spinning motors may not have any component along a link axis, and thus will not affect link torque, all the gyroscopic moments will be transferred to the platform, where it will affect orientation.

For most manipulators there exists a tradeoff between the gear ratio and the motor size. Larger motors allow smaller gear ratios which will decrease the impact of unmodeled gyroscopic forces. The extreme is a direct drive manipulator with the corresponding problems of load capacity and power consumption.

The results show that the motor gyroscopic forces can be significant depending upon the gear ratios, motor inertias, and the total mass of the manipulator and platform. In the example that was studied, the motor gyroscopic forces created a tip positioning error of 2.5 cm to 47 cm when the effect of the spinning motors was included. This was due to a change in the motion of the platform in space under the influence of the gyroscopic forces from the motors. The change in platform motion was shown to be linearly related to the gear ratios, system mass, and motor inertias. It was independent of the time taken for the manipulator to complete the motion.

## 7.4 Simulation of Cooperating Manipulators

The previous section presented a method for simulating a single manipulator on a space platform. The calculations for the simulation grew linearly in the number of links. An advantage of that method was that it could accommodate more complex models for the motor and link.

The performance of complex tasks in space will require cooperation of manipulators, whether operated as telerobots or as autonomous machines. The simulation of robot cooperation in a space environment consists of two issues: the simulation of robot cooperation, and the simulation of robot motion on a mobile platform. The previous section examined a method of simulation on a mobile platform. This section looks at the simulation of cooperation.

The section is split into several parts. In the first part, the inverse dynamics is evaluated. The results from the inverse dynamics is used to construct the closed-form or Lagrange-Euler model. This model becomes the basis for the simulation of cooperation and is the basis for control development and analysis. Finally, a case study is presented for the simulation of two manipulators on a mobile platform.

### 7.4.1 Recursive Inverse Dynamics

Figure 7.11 shows a model of two arms on a space platform. The notation is exactly the same as the previous section except that in the following derivations, the motor and flexible joint model has been removed. For the clarity of presentation the manipulators are considered to be direct drive robots.

The propagation of velocity, acceleration, forces, and moments up and down the links of the manipulators follows in the same manner as previously. Let link 0 represent the common base, the equations for the propagation of velocities and accelerations from the common base to the tips of the manipulators may be written:

$$\begin{aligned} V_1(0) = V_2(0) = V(0), &\quad \text{Velocity of the Base} \\ \alpha_1(0) = \alpha_2(0) = \alpha(0), &\quad \text{Acceleration of the Base} \end{aligned} \tag{7.15}$$

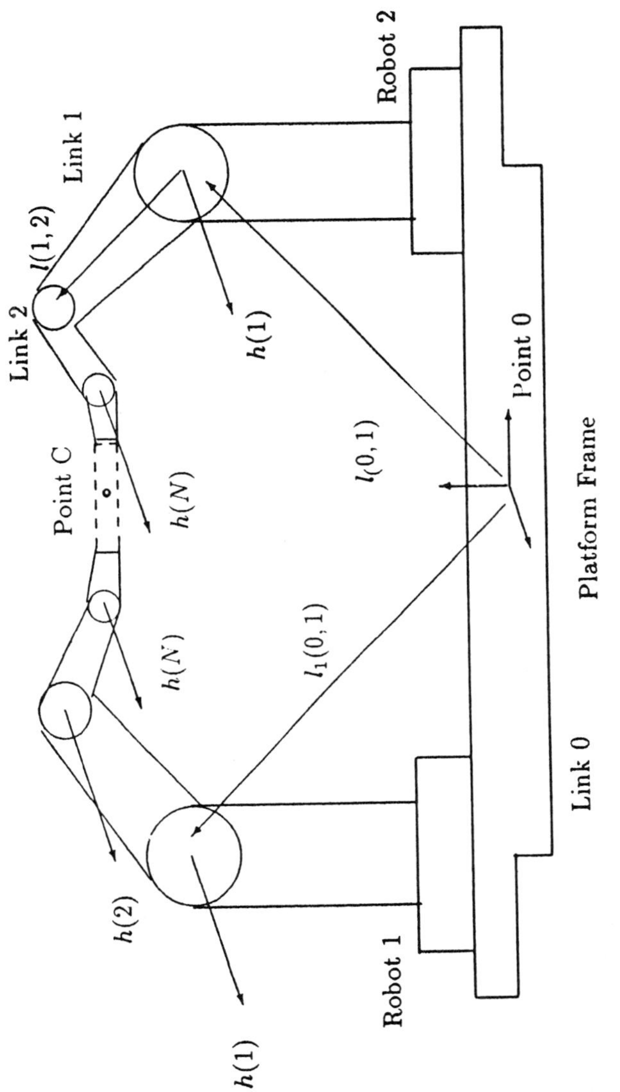


**Figure 7.11: Two Arm Mobile Platform Link Model**

for $k = 0, ..., N$

$$V_1(k) = \phi_1(k,k-1)V_1(k-1) + H_1(k)\dot{\theta}_1(k) \quad (7.16)$$
$$V_2(k) = \phi_2(k,k-1)V_2(k-1) + H_2(k)\dot{\theta}_2(k) \quad (7.17)$$

$$\alpha_1(k) = \phi_1(k,k-1)\alpha_1(k-1) + H_1\ddot{\theta}_1(k) + a_1(k) \quad (7.18)$$
$$\alpha_2(k) = \phi_2(k,k-1)\alpha_2(k-1) + H_2\ddot{\theta}_2(k) + a_2(k) \quad (7.19)$$

At the tips of the manipulators:

$$V_1(N+1) = \phi_1(N+1,N)V_1(N) \quad (7.20)$$
$$\alpha_1(N+1) = \phi_1(N+1,N)\alpha_1(N) + a_1(N+1) \quad (7.21)$$
$$V_2(N+1) = \phi_2(N+1,N)V_2(N) \quad (7.22)$$
$$\alpha_2(N+1) = \phi_2(N+1,N)\alpha_2(N) + a_2(N+1) \quad (7.23)$$

where the subscripts refer to robots 1 and 2, respectively, and $V_1(0) = V_2(0)$ and $\alpha_1(0) = \alpha_2(0)$ because the robots share a common base.

The propagation of torques/forces from the tips of the manipulators to the common base and the calculation of joint torques may be written in terms of the accelerations found above:

$$f_1(N+1) = \text{tip forces exerted on the load from robot 1} \quad (7.24)$$
$$f_2(N+1) = \text{tip forces exerted on the load from robot 2} \quad (7.25)$$

for $k = N, ..., 1$

$$f_1(k) = \phi_1^*(k+1,k)f_1(k+1) + M_1(k)\alpha_1(k) + b_1(k) \quad (7.26)$$
$$f_2(k) = \phi_2^*(k+1,k)f_2(k+1) + M_2(k)\alpha_2(k) + b_2(k) \quad (7.27)$$
$$T_1(k) = H_1^T(k)f_1(k) \quad (7.28)$$
$$T_2(k) = H_2^T(k)f_2(k) \quad (7.29)$$

On the base, $f_1(1)$ and $f_2(1)$ are the forces acting on the base of each robot. The forces on the base may be written as

$$f(0) = \phi_1^*(1,0)f_1(1) + \phi_2^*(1,0)f_2(1) + M(0)\alpha(0) + b(0) \quad (7.30)$$

where $f(0)$ is the 6x1 vector of torques and forces acting on the platform at the platform frame, $O(0)$. This frame may be placed at any point on the platform. The forces $f(0)$ may be created, for example, by reaction jets under the control of a position and orientation control system. If the torque/force $f(0)$ was zero, then the base will move in reaction to the manipulator forces.

### 7.4.2 Closed-Form Inverse Dynamics

The previous subsection gave the form of the dynamics which is useful for model development. Using the same notation, the closed-form, or Lagrange-Euler form of the dynamics may also be found. This form allows for better control analysis and analysis of simulation methods.

The vectors and matrices for robot 1 will be defined using subscripts to indicate the robot. The matrices for the second robot are defined similarly with only the subscript changing:

$$\phi_1 = \begin{bmatrix} \phi_1(1,1) & 0 & \dots & 0 \\ \phi_1(2,1) & \phi_1(2,2) & \dots & \vdots \\ \vdots & \vdots & \ddots & 0 \\ \phi_1(N,1) & \phi_1(N,2) & \dots & \phi_1(N,N) \end{bmatrix} \tag{7.31}$$

$$H_1 = \text{diag}[H_1(1), \dots, H_1(N)]; \quad M_1 = \text{diag}[M_1(1), \dots, M_1(N)]; \tag{7.32}$$

$$B_1 = [0, \dots, 0, \phi_1(N+1, N)]; \quad E_1^T = [\phi_1^*(1,0), 0, \dots, 0] \tag{7.33}$$

$$a_1 = \begin{bmatrix} a_1(1) \\ \vdots \\ a_1(N) \end{bmatrix}; \quad b_1 = \begin{bmatrix} b_1(1) \\ \vdots \\ b_1(N) \end{bmatrix}; \quad \alpha_1 = \begin{bmatrix} \alpha_1(1) \\ \vdots \\ \alpha_1(N) \end{bmatrix} \tag{7.34}$$

$$f_1 = \begin{bmatrix} f_1(1) \\ \vdots \\ f_1(N) \end{bmatrix}; \quad V_1 = \begin{bmatrix} V_1(1) \\ \vdots \\ V_1(N) \end{bmatrix}. \tag{7.35}$$

Here we made use of the operator properties of $\phi_1(i,j)$:

$$\phi_1(i,i) = I; \quad \phi_1^{-1}(i,j) = \phi_1(j,i); \quad \phi_1(i,k)\phi_1(k,j) = \phi_1(i,j) \tag{7.36}$$

Stacking up all the vectors with the above definitions allows the equations for the force propagation (7.26–7.27) to be written:

$$f_1 = \phi_1^* \left(M_1\alpha_1 + b_1 + B_1^T f_1(N+1)\right) \tag{7.37}$$

$$f_2 = \phi_2^* \left(M_2\alpha_2 + b_2 + B_2^T f_2(N+1)\right) \tag{7.38}$$

Now continue the stacking operation for both the robots and the base. Define

$$\phi = \begin{bmatrix} I_{6x6} & 0 & 0 \\ \phi_1 E_1 & \phi_1 & 0 \\ \phi_2 E_2 & 0 & \phi_2 \end{bmatrix} \tag{7.39}$$

where

$$\phi_1 E_1 = [\phi_1^*(1,0), \phi_1^*(2,0), \dots, \phi_1^*(N,0)]^T \tag{7.40}$$

$$\phi_2 E_2 = [\phi_2^*(1,0), \phi_2^*(2,0), \dots, \phi_2^*(N,0)]^T \tag{7.41}$$

Also define

$$M = \begin{bmatrix} M(0) & 0 & 0 \\ 0 & M_1 & 0 \\ 0 & 0 & M_2 \end{bmatrix}; \quad \alpha = \begin{bmatrix} \alpha(0) \\ \alpha_1 \\ \alpha_2 \end{bmatrix}; \quad b = \begin{bmatrix} b(0) \\ b_1 \\ b_2 \end{bmatrix} \tag{7.42}$$

$$B = \begin{bmatrix} 0 & B_1 & 0 \\ 0 & 0 & B_2 \end{bmatrix}; \quad f = \begin{bmatrix} f(0) \\ f_1 \\ f_2 \end{bmatrix}; \quad f(N+1) = \begin{bmatrix} f_1(N+1) \\ f_2(N+1) \end{bmatrix}; \tag{7.43}$$

Then the force propagation equations, (7.37), (7.38), and (7.30), may be written:

$$f = \phi^* \left( M\alpha + b + B^T f(N+1) \right) \tag{7.44}$$

In order to write the velocity and acceleration propagation in the same manner, we make the following definitions:

$$V = \begin{bmatrix} V(0) \\ V_1 \\ V_2 \end{bmatrix}; \quad a = \begin{bmatrix} 0 \\ a_1 \\ a_2 \end{bmatrix}; \quad W = \begin{bmatrix} V(0) \\ \dot{\theta}_1(1) \\ \vdots \\ \dot{\theta}_1(N) \\ \dot{\theta}_2(1) \\ \vdots \\ \dot{\theta}_2(N) \end{bmatrix}; \quad T = \begin{bmatrix} f(0) \\ T_1(1) \\ \vdots \\ T_1(N) \\ T_2(1) \\ \vdots \\ T_2(N) \end{bmatrix} \tag{7.45}$$

$$H = \begin{bmatrix} I_{6x6} & 0 & 0 \\ 0 & H_1 & 0 \\ 0 & 0 & H_2 \end{bmatrix} \tag{7.46}$$

where $V(0)$ is the combined angular and linear velocity of the base, and $I_{6x6}$ is a 6x6 identity matrix. $W$ is the velocity vector for the platform and both manipulators, $T$ is the force/torque vector for forces and torques acting on the platform and around each joint. Note that $a(0) = 0$, as there are no bias accelerations of the base.

With these definitions, the equations (7.16–7.19) may be written:

$$V = \phi H W \tag{7.47}$$

$$\alpha = \phi H \dot{W} + \phi a \tag{7.48}$$

with the boundary conditions from (7.15). Define,

$$\alpha(N+1) = \begin{bmatrix} \alpha_1(N+1) \\ \alpha_2(N+1) \end{bmatrix} \quad ; \quad a(N+1) = \begin{bmatrix} a_1(N+1) \\ a_2(N+1) \end{bmatrix} \tag{7.49}$$

The angular and linear accelerations at the tips of the manipulators may be determined:

$$\alpha(N+1) = B\phi H\dot{W} + B\phi a + a(N+1). \tag{7.50}$$

Combining equations (7.44) and (7.48) gives the equation for the forces on each link of the manipulators and the forces on the base:

$$f = \phi^* M\phi H\dot{W} + \phi^* M\phi a + \phi^* b + \phi^* B^T f(N+1) \tag{7.51}$$

The torques on the links are found by projecting the link torque/forces onto the axis of rotation. Using the definition of $H$:

$$\begin{aligned} T &= H^T f & (7.52)\\ &= H^*\phi^* M\phi H\dot{W} + H^*\phi^* M\phi a + H^T\phi^* b + H^T\phi^* B^T f(N+1) & (7.53)\\ &= \mathcal{M}\dot{W} + C + J^T f(N+1) & (7.54) \end{aligned}$$

where the mass matrix, $\mathcal{M} = H^*\phi^* M\phi H$, the Coriolis and centrifugal forces, $C = H^*\phi M\phi a + H^*\phi^* b$, and the combined tip Jacobians for both robots and the base, $J^T = H^*\phi^* B^T$.

Equation (7.54) gives the closed-form solution to the inverse dynamics of an open chain of two robot manipulators on a mobile platform. The form of the equations is identical to a single manipulator system and could be derived equally as well by the Lagrange–Euler method. However, the advantage of the above approach is the insight it gives into the structure of $\mathcal{M}$.

Expanding $\mathcal{M}$, $C$, and $J^T$ and defining the submatrices gives:

$$\mathcal{M} = \begin{bmatrix} \mathcal{M}_B & \mathcal{M}_{1B}^T & \mathcal{M}_{2B}^T \\ \mathcal{M}_{1B} & \mathcal{M}_1 & 0 \\ \mathcal{M}_{2B} & 0 & \mathcal{M}_2 \end{bmatrix} \tag{7.55}$$

where

$$\begin{aligned} M_B &= M(0) + E_1^T\phi_1^* M_1\phi_1 E_1 + E_2^T\phi_2^* M_2\phi_2 E_2 \\ M_{1B} &= H_1^*\phi_1^* M_1\phi_1 E_1 \\ M_1 &= H_1^*\phi_1^* M_1\phi_1 H_1 \\ M_{2B} &= H_2^*\phi_2^* M_2\phi_2 E_2 \\ M_2 &= H_2^*\phi_2^* M_2\phi_2 H_2 \end{aligned}$$

$$C = \begin{bmatrix} b(0) + E_1^T\phi_1^*(M_1\phi_1 a_1 + b_1) + E_2^T\phi_2^*(M_2\phi_2 a_2 + b_2) \\ H_1^*\phi_1^*(M_1\phi_1 a_1 + b_1) \\ H_2^*\phi_2^*(M_2\phi_2 a_2 + b_2) \end{bmatrix} \equiv \begin{bmatrix} C_B \\ C_1 \\ C_2 \end{bmatrix} \tag{7.56}$$

$$J^T = \begin{bmatrix} E_1^T\phi_1^* B_1^T & E_2^T\phi_2^* B_2^T \\ H_1^*\phi_1^* B_1^T & 0 \\ 0 & H_2^*\phi_2^* B_2^T \end{bmatrix} \equiv \begin{bmatrix} J_{1B}^T & J_{2B}^T \\ J_1^T & 0 \\ 0 & J_2^T \end{bmatrix}. \tag{7.57}$$

The mass matrix for the combined robot and platform in equation (7.55) can be expressed as the mass matrix of each robot, as if it were on a fixed platform, $\mathcal{M}_1$, $\mathcal{M}_2$, plus additional elements due to the effect of platform motion on the robots and robot motion on the platform. The $C$ vector can also be seen as the Coriolis terms for each individual arm where now the motion of the base is included. The $J_1^T$, and $J_2^T$ terms are exactly the Jacobians for the manipulators on a fixed-base platform. Thus, in the calculation of the inverse dynamics, any previously developed solution for the mass matrix and Jacobian matrix of fixed-base manipulators could be used in place of the $\mathcal{M}_1$, $\mathcal{M}_2$ matrices and the $J_1^T$, and $J_2^T$ matrices.

### 7.4.3 Interaction with a Common Load

When the two manipulators carry a load, the forces exerted from the tips of the manipulators act to accelerate and/or compress the load. The motion of a point $C$ on the load may be written, [41]:

$$M(C)\alpha(C) + b(C) = A^T f(N+1) \tag{7.58}$$

where $M(C)$ is the 6x6 mass matrix of the load, $b(C)$ is the 6x1 vector of bias forces due to the angular velocity of the load, and $A^T$ is the grasp matrix:

$$A^T = [\phi_1^*(N+1, C), \phi_2^*(N+1, C)] \tag{7.59}$$

which relates torques and forces at the manipulator tips to net torque/force at the point $C$.

The acceleration at point $C$ can be related to the accelerations at the tips of the manipulator:

$$\begin{aligned} \alpha(N+1) &= J\dot{W} + \dot{J}W && (7.60) \\ &= A\alpha(C) + a(C) && (7.61) \end{aligned}$$

where $a(C)$ are the bias accelerations induced by the angular velocity and differences in coordinate frames.

### 7.4.4 Summary

In summary, equations (7.54), (7.58), (7.60), and (7.61) give the complete closed-form solution to the inverse dynamics of two cooperating robot manipulators on a common, mobile platform, or:

$$\begin{aligned} T &= \mathcal{M}\dot{W} + C + J^T f(N+1) && (7.62) \\ \alpha(N+1) &= J\dot{W} + \dot{J}W = A\alpha(C) + a(C) && (7.63) \\ M(C)\alpha(C) + b(C) &= A^T f(N+1). && (7.64) \end{aligned}$$

### 7.4.5 Motion on a Mobile Base

The above equations present the solution to the inverse dynamics of the mobile closed-chain system. The forward dynamics problem consists of calculating the system acceleration, $\dot{W}$, when given the current state, $W$, $\theta_1$, $\theta_2$, and the input torques and forces on the joints and base, $T$.

Because we have placed the equations of motion for the mobile base in the same form as the equations for a fixed-base system, the same approach as in [41] may be followed. The approach consists of finding the forces exerted by each manipulator on the load, $f(N+1)$. With this force, one can then calculate the motion of the system from equation (7.62).

Using equations (7.62) and (7.63),

$$J\dot{W} = J\mathcal{M}^{-1}T' - J\mathcal{M}^{-1}J^T f(N+1) \tag{7.65}$$

where $T' = T - C$. $\mathcal{M}$ is the combined mass matrix of the system and is always positive definite.

From equations (7.63) and (7.64),

$$\begin{aligned} J\dot{W} &= A\alpha(C) + a(C) - \dot{J}W && (7.66) \\ &= AM^{-1}(C)A^T f(N+1) + a' && (7.67) \end{aligned}$$

where $a' = -AM^{-1}(C)b(C) + a(C) - \dot{J}W$.

Combining (7.65), (7.67)

$$\Omega f(N+1) = J\mathcal{M}^{-1}T' - a' \tag{7.68}$$

where

$$\Omega = J\mathcal{M}^{-1}J^T + AM^{-1}(C)A^T \, . \tag{7.69}$$

If it is assumed that the manipulators are not at a singular configuration then $\Omega$ may be inverted and the forces exerted at the tips of the manipulators can be uniquely determined. A stronger result for the invertibility of $\Omega$ for a fixed-base system was developed by Milman *et al.*, [52]. This result states that the probability that the system is at a configuration where $\Omega$ is uninvertible is zero. For the mobile platform case it is assumed that the manipulators are not at a singular configuration. The forces exerted at the tips may be calculated,

$$f(N+1) = \Omega^{-1}\left(J\mathcal{M}^{-1}T' - a'\right) . \tag{7.70}$$

From the known tip forces, the motion of the system may be calculated.

However, the above equations for the calculation of the tip forces and resultant motion are not advantageous for the efficient calculation of manipulator dynamics in simulation. The forward dynamics can be rewritten using the concept of free dynamics. Define the free dynamics of the system as the dynamics with no tip forces, [41]:

$$\mathcal{M}\dot{W}_f + C = T \tag{7.71}$$

$$\alpha_f(N+1) = J\dot{W}_f + \dot{J}W \tag{7.72}$$
$$= J\dot{W}_f + B^T\phi^* a + a(N+1) \tag{7.73}$$

where the subscript $f$ will be used to denote the free dynamics and equation (7.50) was used to derive the expression for $\dot{J}W$.

From (7.71),

$$\mathcal{M}\dot{W}_f = T - C = T' . \tag{7.74}$$

Substituting into (7.68) and using the definitions of $\alpha_f(N+1)$, and $a'$ gives:

$$\Omega f(N+1) = \alpha_f(N+1) + AM^{-1}(C)b(C) - a(C) \tag{7.75}$$

which gives the tip force in terms of the free accelerations of the tips. Thus, for simulation purposes, the forward dynamics of the system are first calculated as if there was no load, or the manipulators are free from constraints, providing the free motion $\alpha_f(N+1)$ and $\dot{W}_f$. Knowledge of the free motion may then be used to calculate the actual tip forces when the manipulator tips are constrained to the common load. From the actual tip forces and free motion, the true system motion may be easily expressed. Let

$$\Delta\dot{W} = \dot{W} - \dot{W}_f \tag{7.76}$$

then (7.62) and (7.71) yield

$$\Delta\dot{W} = -\mathcal{M}^{-1}J^T f(N+1) \tag{7.77}$$

The result is a solution for the forward dynamics which parallels [41] where now the system includes the motion of the base.

1. Find the free accelerations,

$$\mathcal{M}\dot{W}_f + C = T \tag{7.78}$$

$$\alpha_f(N+1) = J\dot{W}_f + \dot{J}W \tag{7.79}$$

2. Find the tip forces, $f(N+1)$

$$\Omega = \left(J\mathcal{M}^{-1}J^T + AM^{-1}(C)A^T\right) \tag{7.80}$$

$$\Omega f(N+1) = \alpha_f(N+1) + AM^{-1}b(C) - a(C) \tag{7.81}$$

3. Find the correction acceleration, $\Delta\dot{W}$,

$$\mathcal{M}\Delta\dot{W} = -J^T f(N+1) \tag{7.82}$$

4. Find the true accelerations of robots and platform,

$$\dot{W} = \dot{W}_f + \Delta\dot{W} \tag{7.83}$$

With this approach, any solution to the forward dynamics of an open topological tree of rigid links may be used to solve equations (7.78), (7.82). For two 6-DOF manipulators on a freely mobile platform in space, the mass matrix, $\mathcal{M}$, is 18x18. Instead of inverting this mass matrix at every simulation time step, it becomes effective to examine more efficient solutions to the forward dynamics equations. A linear recursive method for the solution of the forward dynamics of an open kinematic chain was presented in Section 2 of this chapter. This method also has the capabilities of including the effects of joint flexibilities and motor gyroscopic forces.

### 7.4.6 Case Study

Using the above equations, we simulated the cooperative motion of two PUMA 560 manipulators on a free-space platform with a load of 3 kg. The solution to equations (7.78–7.83) produces the vector $\dot{W}$. This is an 18x1 vector of angular and linear accelerations of the platform, and joint accelerations for robot 1 and robot 2. The joint angles for both robots were found by numerically integrating the joint accelerations. The linear position of the platform was calculated by numerically integrating the linear acceleration of the platform. The orientation of the platform was determined by integrating the angular acceleration vector and using the resultant angular velocity to calculate the rate of change of the rotation matrix. No joint flexibilities or motor gyroscopic forces were included.

Figure 7.12 shows the initial positions of of the manipulators, platform, and load. The simulation of the dynamics was written in MATLAB, [53], and a 4th-5th order Runge-Kutta routine was used to integrate the equations of motion.

### 7.4.7 Control

As an initial investigation of the methods for cooperative control, we chose to implement an object level controller for the positioning of the frame $C$. This is just one of many possible methods for the control of object position and internal force in a cooperative multi-manipulator system, [54], [55], [56]. The position and orientation of the frame $C$ on the load was specified relative to the platform and not the inertial frame. The controller consisted of a PD tracking control on the Cartesian position of the load and a PD quaternion-based control on the orientation of the load, [57]. No dynamic load balancing or squeeze force controllers were simulated at this time.

The platform was left uncontrolled in order to demonstrate the effects of the robot motion on the position and orientation of the platform during motion.

The motion of the system is shown in Figures 7.12–7.14. The desired motion was to move the load downward 45 cm in 1 second, wait 0.5 seconds and then rotate the load 45 degrees about the load's Z axis in 1 second. After another wait of 0.5 seconds, the motion was reversed.

Figures 7.15-7.16 show the Cartesian position errors relative to the platform frame and the difference in tip forces along the load Y axis. The system exhibits large overshoot due to the choice of gains and the lack of manipulator dynamics in the control algorithm. References to controllers for space manipulators may be found in Spofford and Akin, [12], and Papadopoulos and Dubowsky, [58]. The difference in the tip forces along the Y axis is exactly the compressing or pulling force exerted by the manipulators on the load when the load is not moving. While the load is moving, the difference in tip forces includes the d'Alembert forces. The knowledge of the tip forces is critical in using a simulation for the evaluation of multiple arm controllers. While there are other methods for the simulation of closed kinematic chains, not all of them have the capability of producing the forces at the tips of the manipulators as the simulation progresses.

## 7.5 Conclusion

This chapter has examined the simulation of space-based manipulators. An order $N$ method for the simulation of manipulator dynamics has been developed which includes all gyroscopic forces and motor-link acceleration interactions and can also accommodate flexible joint couplings. This simulation method provides an efficient technique for the simulation of complex robot structures and better modeling techniques for alternate manipulator designs. A case study simulation showed the effect of motor gyroscopic forces on the positioning of a manipulator in space.

The modeling was extended to multiple cooperating manipulators on a space platform. The dynamic equations for such a system have an identical form as for the fixed platform case. The approach is extendable to more than two manipulators by increasing the dimensions of the appropriate vectors and matrices. The solution to the cooperative motion can incorporate the modeling solution which includes the motor forces and flexible couplings. The approach has shown the structure of the system where portions of the existing work on fixed-base simulations may be included in the calculations. The simulation equations were shown to produce the closed-form equations necessary for control analysis and development. A complete simulation with two 6-DOF manipulators on a free-floating platform was presented along with a multiple arm controller to position the common load.

Additional work towards the efficient simulation of robotic manipulators in space and on earth includes the effects of link flexibility, [59] and the requirements for efficiently simulating the robot in contact or constrained by the environment, [60]. These methods, along with the simulation method presented in this chapter show the promise of providing a unified method for the simulation of manipulators and their tasks in the space environment.

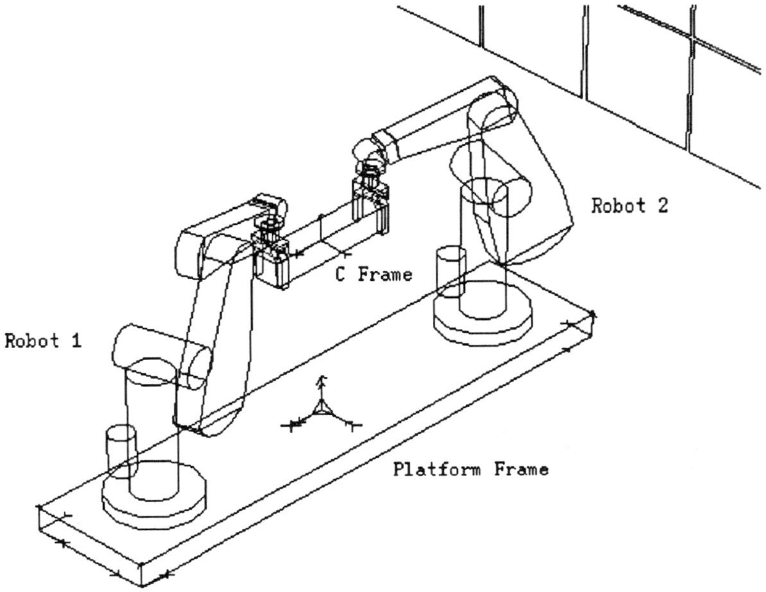


Figure 7.12: Initial Configuration

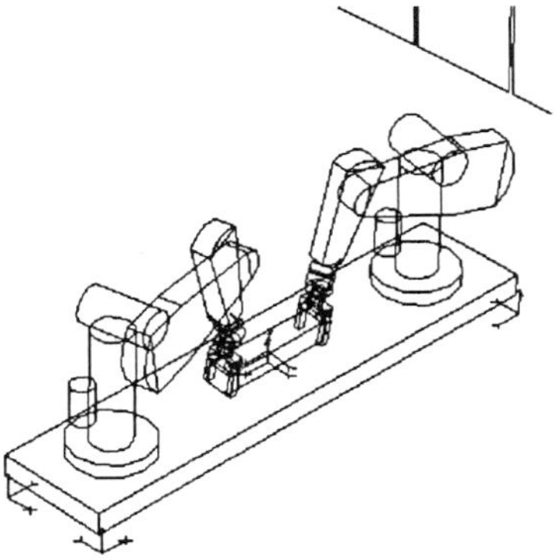

After 1.5 Seconds

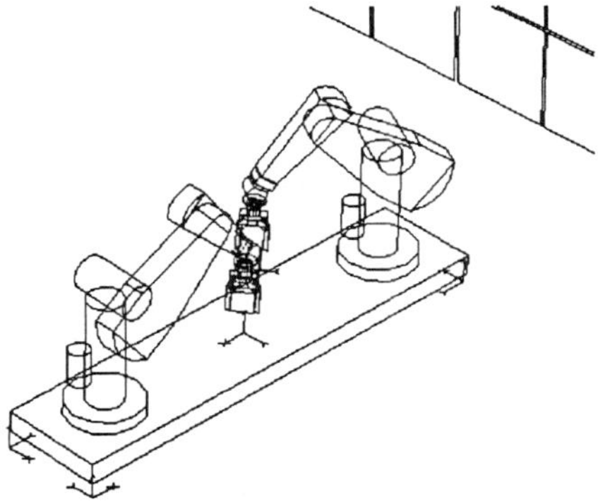

After 3.0 Seconds

Figure 7.13: First Two Motion Sequences

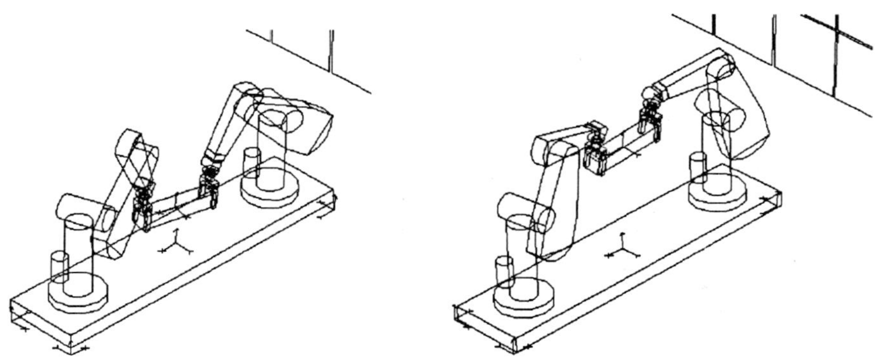

After 4.5 Seconds

After 6 Seconds

Figure 7.14: Last Two Motion Sequences

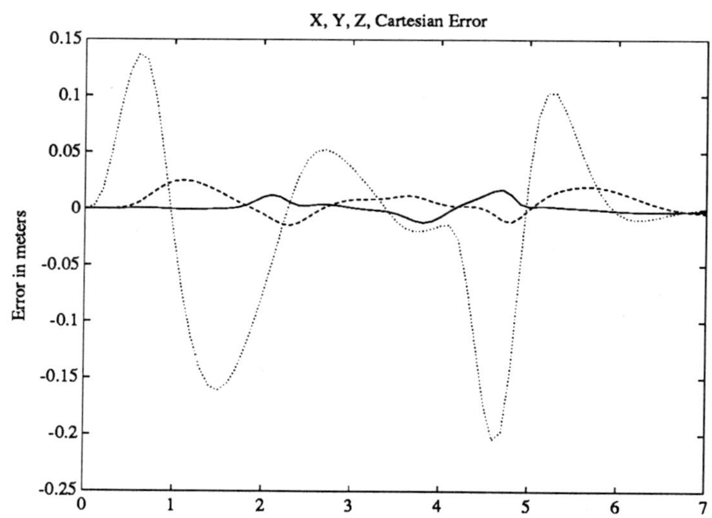


Figure 7.15: Cartesian Error of Load

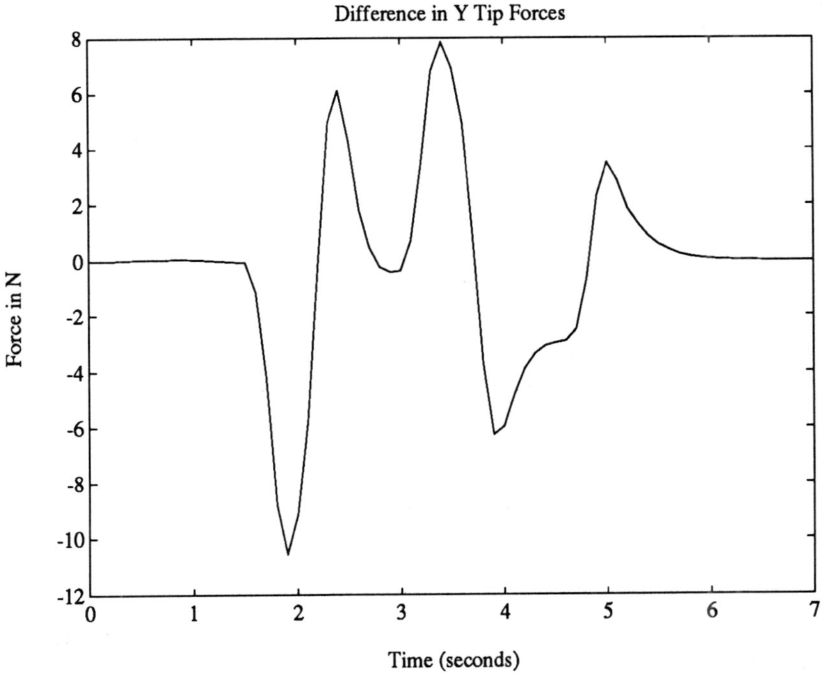


Figure 7.16: Differences in Y Tip Forces

## 7.6 References


[1] S. H. Murphy, *Modeling and Simulation of Multiple Cooperating Manipulators on a Mobile Platform*. PhD thesis, Rensselaer Polytechnic Institute, Troy NY, 1990.

[2] S. H. Murphy, J. T. Wen, and G. N. Saridis, "Simulation of cooperating robot manipulators on a mobile platform," *IEEE Transactions on Robotics and Automation*, vol. 7, pp. 468–478, August 1991.

[3] J. Joshi and A. A. Desrochers, "Modeling and control of a mobile robot subject to disturbances," in *Proceedings of the 1986 IEEE International Conference on Robotics and Automation*, pp. 1508–1513, 1986.

[4] A. B. Tanner, "Study of robotic manipulators subject to base disturbances," Master's thesis, Massachusetts Institute of Technology, January 1987.

[5] H. L. Alexander and R. H. Cannon, "An extended operational-space control algorithm for satellite manipulators," *Journal of the Astronautical Sciences*, vol. 38, pp. 473–486, October-December 1990.

[6] P. K. C. Wang, "Control strategy for a dual-arm maneuverable space robot," in *Proceedings of the Workshop on Space Telerobotics* (G. Rodriguez, ed.), pp. 257–266, JPL Publication 87–13, 1987. Volume 2.

[7] S. Dubowsky, E. E. Vance, and M. A. Torres, "The control of space manipulators subject to spacecraft attitude control saturation limits," in *Proceedings of the NASA Conference on Space Telerobotics*, JPL, Jan 31 – Feb 2 1989.

[8] R. W. Longman, R. E. Lindberg, and M. F. Zedd, "Satellite-mounted robot manipulators – new kinematics and reaction moment compensation," *The International Journal of Robotics Research*, vol. 6, no. 3, pp. 87–103, 1987.

[9] R. E. Lindberg, R. W. Longman, and M. F. Zedd, "Kinematic and dynamic properties of an elbow manipulator mounted on a satellite," *Journal of the Astronautical Sciences*, vol. 38, pp. 397–421, October-December 1990.

[10] Y. Umetani and K. Yoshida, "Experimental study on two dimensional free–flying robot satellite model," in *Proceedings of the NASA Conference on Space Telerobotics*, JPL, Jan 31 – Feb 2 1989.

[11] Y. Umetani and K. Yoshida, "Continuous path control of space manipulators mounted on OMV," *ACTA Astronautica*, vol. 15, no. 12, pp. 981–986, 1987.

[12] J. R. Spofford and D. L. Akin, "Redundancy control of a free-flying telerobot," *AIAA Journal of Guidance and Control*, vol. 13, no. 3, pp. 515–523, 1990.

[13] Y. Nakamura and R. Mukherjee, "Nonholonomic path planning of space robots," in *Proceedings of the 1989 IEEE International Conference on Robotics and Automation*, pp. 1050–1054, 1989.

[14] Z. Vafa and S. Dubowsky, "On the dynamics of space manipulators using the virtual manipulator, with applications to path planning," *Journal of the Astronautical Sciences*, vol. 38, pp. 441–472, October-December 1990.

[15] H. Springer, P. Lugner, and K. Desoyer, "Equations of motion of manipulators including dynamic effects of active elements," in *Proceedings of the 1st IFAC Symposium on Robot Control (SYROCO '85)*, pp. 425–430, 1985.

[16] J. Chen, "The effects of gear reduction on robot dynamics," in *Proceedings of the NASA Conference on Space Telerobotics*, JPL, Jan 31 – Feb 2 1989.

[17] M. W. Spong, "Modeling and control of elastic joint robots," *Journal of Dynamic Systems, Measurement, and Control*, vol. 109, pp. 310–319, December 1987.

[18] A. De Luca, "Dynamic control of robots with joint elasticity," in *Proceedings of the 1988 IEEE International Conference on Robotics and Automation*, pp. 152–158, 1988.

[19] G. Cesareo and R. Marino, "On the controllability properties of elastic robots," in *6th International Conference on Analysis and Optimization of Systems*, June 1984. INRIA, Nice.

[20] R. Featherstone, "The calculation of robot dynamics using articulated-body inertias," *The International Journal of Robotics Research*, vol. 2, pp. 13–29, 1983.

[21] G. Rodriguez and K. Kreutz, "An operator formulated manipulator dynamics," Tech. Rep. 347–88–236, Jet Propulsion Laboratory Engineering Memorandum, April 1988.

[22] G. Rodriquez, "Recursive dynamics for topological trees of rigid bodies by Kalman filtering and Bryson-Frazier smoothing," in *VPI/SU Symposium on Dynamics and Control of Structures*, 1987. Blacksburg, VA.

[23] H. Brandl, R. Johanni, and M. Otter, "A very efficient algorithm for the simulation of robots and similar multibody systems without inversion of the mass matrix," in *Proceedings of the IFAC/IFIP/IMACS International Symposium on Theorie of Robots*, December 1986. Vienna, Austria.

[24] K. W. Lilly, *Efficient Dynamic Simulation of Multiple Chain Robotic Systems.* PhD thesis, The Ohio State University, 1989.

[25] G. Rodriguez, A. Jain, and K. Kreutz, "Spatial operator algebra framework for multibody system dynamics," in *3rd Annual Conference on Aerospace Computational Control*, 1981.

[26] M. G. Nasser, "Recursive Newton-Euler formulation of manipulator dynamics," in *Proceedings of the NASA Conference on Space Telerobotics*, JPL, Jan 31 – Feb 2 1989.

[27] D. E. Rosenthal, "An order $n$ formulation for robotics systems," *Journal of the Astronautical Sciences*, vol. 38, pp. 511–529, October-December 1990.

[28] S. H. Murphy, J. T. Wen, and G. N. Saridis, "Recursive calculation of geared robot manipulator dynamics," in *Proceedings of the 1990 IEEE International Conference on Robotics and Automation*, 1990.

[29] S. H. Murphy, J. T. Wen, and G. N. Saridis, "Simulation and analysis of flexibly jointed manipulators," in *29th IEEE Conference on Decision and Control*, 1990.

[30] A. Jain and G. Rodriguez, "Recursive dynamics for geared robot manipulators," in *29th IEEE Conference on Decision and Control*, December 1990.

[31] J. N. Anderson, "Decoupled simulation of two-arm robot systems," in *Proceedings of the 1986 American Control Conference*, pp. 127–129, June 1986.

[32] J. N. Anderson and M. E. Pittelkau, "Reflex-action position control and two-arm load-sharing force control for robotic manipulators," Tech. Rep. MCTR-1-88-1, Center for Manufacturing Research, Tenn. Tech. Univ, January 1988.

[33] T. J. Tarn, A. K. Bejczy, and X. Yun, "Dynamic coordination of two robot arms," in *Proceedings of the 25th Conference on Decision and Control*, pp. 1267–1270, December 1986.

[34] T. J. Tarn, A. K. Bejczy, and X. Yun, "Design of dynamic control of two cooperating robot arms: Closed chain formulation," in *Proceedings of the 1987 IEEE International Conference on Robotics and Automation*, pp. 7–13, 1987.

[35] S. Hayati, "Hybrid position/force control of multi-arm cooperating robots," in *Proceedings of the 1986 IEEE International Conference on Robotics and Automation*, pp. 82–89, April 1986.

[36] Y. F. Zheng and J. Y. S. Luh, "Control of two coordinated robots in motion," in *Proceedings of the 24th Conference on Decision and Control*, pp. 1761–1765, 1985.

[37] Y. F. Zheng and J. Y. S. Luh, "Joint torques for control of two coordinated moving robots," in *Proceedings of the 1986 IEEE International Conference on Robotics and Automation*, pp. 1375–1380, April 1986.

[38] Y. F. Zheng and J. Y. S. Luh, "Optimal load distribution for two industrial robots handling a single object," in *Proceedings of the 1988 IEEE International Conference on Robotics and Automation*, pp. 344–349, 1988.

[39] M. W. Walker, D. Kim, and J. Dionise, "Adaptive coordinated motion of two manipulator arms," in *Proceedings of the 1989 IEEE International Conference on Robotics and Automation*, pp. 1084–1090, 1989.

[40] S. Y. Oh and D. E. Orin, "Dynamic computer simulation of multiple closed–chain robotic mechanisms," in *Proceedings of the 1986 IEEE International Conference on Robotics and Automation*, pp. 15–20, 1986.

[41] G. Rodriguez and K. Kreutz, "An operator approach formulated dynamics for multiple arms grasping a commonly held object," Tech. Rep. 347–88–237, Jet Propulsion Laboratory Engineering Memorandum, April 1988.

[42] S. H. Murphy and J. T. Wen, "Force decomposition in robot force control," in *Proceedings of the 1991 American Control Conference*, pp. 2533–2538, June 1991.

[43] G. Rodriguez, A. Jain, and K. Kreutz, "A spatial operator algebra framework for manipulator modeling and control," in *Proceedings of the 1989 IEEE International Conference on Robotics and Automation*, May 1989.

[44] A. Jain, "Unified formulation of dynamics for serial rigid multibody systems," *Journal of Guidance, Control and Dynamics*, vol. 14, pp. 531–542, May-June 1991.

[45] J. J. Craig, *Introduction to Robotics, Mechanics & Control.* Reading, Mass.: Addison–Wesley, 1986.

[46] A. Jain, "Unified formulation of dynamics for serial rigid multibody systems," Tech. Rep. 347–88–236, Jet Propulsion Laboratory Engineering Memorandum, April 1989.

[47] B. Armstrong, O. Khatib, and J. Burdick, "The explicit dynamic model and inertial parameters of the PUMA 560 arm," in *Proceedings of the 1986 IEEE International Conference on Robotics and Automation*, pp. 510–518, 1986.

[48] T. J. Tarn, A. K. Bejczy, H. Shuotiao, and X. Yun, "Inertia parameters of PUMA 560 robot arm," Robotics Laboratory Report SSM-RL-85-01, Department of Systems Science and Mathematics, Washington University, September 1985.

[49] K. Kyriakopoulos and G. Saridis, "Minimum jerk path generation," in *Proceedings of the 1988 IEEE International Conference on Robotics and Automation*, April 1988.

[50] C. Trudel, "SRMS joint servo and gearbox math model," spar-tm.1213, SPAR Aerospace Products Ltd., Toronto, Ontario., Canada, 1977.

[51] Y. Nakamura and R. Mukherjee, "Nonholonomic path planning of space robots via bi-directional approach," in *Proceedings of the 1990 IEEE International Conference on Robotics and Automation*, pp. 1764–1769, 1990.

[52] M. Milman, K. Kreutz, and A. Lokshin, "Coordination and load balancing for multi–arm systems," in *Int. Symp. on Robotics*, November 1988. Albuquerque, NM.

[53] C. Moler, J. Little, and S. Bangert, *Pro-Matlab User's Guide.* The MathWorks, Inc., 1987.

[54] J. T. Wen and K. Kreutz, "Stability analysis of PD control for multiple robot manipulators holding a common object," Tech. Rep. 347–88–229, Jet Propulsion Laboratory Engineering Memorandum, February 1988.

[55] K. Kreutz and A. Lokshin, "Load balancing and closed chain multiple arm control," in *Proceedings of the 1988 American Control Conference*, pp. 2148–2155, 1988.

[56] D. E. Orin and S. Y. Oh, "Control of force distribution in robotic mechanisms containing closed kinematic chains," *Journal of Dynamic Systems, Measurement and Control*, vol. 102, pp. 134–141, June 1981.

[57] J. T. Wen and K. Kreutz, "Attitude control of an object commonly held by multiple robot arms: A Lyapunov approach," in *Proceedings of the 1988 American Control Conference*, pp. 1790–1794, 1988.

[58] E. Papadopoulos and S. Dubowsky, "On the nature of control algorithms for space manipulators," in *Proceedings of the 1990 IEEE International Conference on Robotics and Automation*, May 1990.

[59] A. Jain and G. Rodriguez, "Recursive dynamics for flexible multibody systems using spatial operators," Tech. Rep. 90-26, Jet Propulsion Laboratory, December 1990.

[60] K. W. Lilly and D. E. Orin, "Efficient dynamic simulation of a single closed-chain manipulator," in *Proceedings of the 1991 IEEE International Conference on Robotics and Automation*, pp. 210–215, 1991.

# Chapter 8

# Automated Assembly in the JPL Telerobot Testbed

by: J. (Bob) Balaram
Henry W. Stone
Jet Propulsion Laboratory
California Institute of Technology
Pasadena, CA 91109

This chapter describes automated assembly operations in a telerobot testbed developed at JPL for performing task experiments involving assembly and repair of spacecraft mockups in an unstructured environment. The fully operational multi-arm system can operate over a range of modes, from purely hand-controller operated teleoperation to highly automated operations. The autonomous modes range from human-operator supervised execution of autonomous sequences, to shared (mixed) modes in which command execution is provided simultaneously by the operator and the automated portions of the system. While this range of capability enables flexible operations in totally to partially unstructured environments, the emphasis in this chapter will be on the more automated modes of operation.

The system has two 6-degree-of-freedom manipulator arms, which in the teleoperated and shared modes, are operated either independently by two force reflecting hand controllers or in a coordinated fashion for closed-chain dual arm operation using a single hand-controller. Automated operation is currently restricted to simultaneous, but independent, motions of the two manipulator arms. A third arm is utilized to position four end-effector mounted cameras to view the scene and is activated independently of the manipulation arms. An additional three fixed cameras provide an overall view of the operations.

The chapter starts with the background of the JPL Testbed and then describes in a summary fashion, the overall system and various subsystems. A description of a typical servicing assembly operation sequence is then provided. This example sequence will serve to focus the discussion that follows on the specific functional components needed to automate this and other sequences.

## 8.1 Background of the NASA/JPL Telerobot Testbed

The Telerobot Testbed project was initiated by NASA/JPL in 1985 to develop a facility through which various *generic* robot capabilities could be developed, integrated and tested on a realistic set of task experiments in a ground-based laboratory facility [1], [2], (shown in Figure 8.1). A phased series of demonstrations highlighting both the technical capability as well as the usefulness of robotics technology for servicing operations was planned and executed over the course of a 5 year period. In 1990, the facility was closed and the technology results and experience were focussed on more application oriented research projects.

Chosen for study was a realistic task set including typical space related Orbital Replacement Unit (ORU) change-outs; satellite capture; surface cleaning; grasping, bolting operations, etc. The development would aim to build a fully integrated **telerobotic** system, i.e., a synergestic combination of **tele**operation (hand-controller operations) and a **robotic** (autonomous) system. This was perceived as providing an evolutionary path for systems growth with autonomous functions, augmenting or even replacing teleoperated segments of task execution. In addition, the strengths in one area would be able to overcome weaknesses in the other, thereby enabling a more capable system.

The telerobot would not rely on a fully *a priori* known environment, but instead perform tasks in a partially known environment. This meant that the locations of the arms as well as the objects, if known, were only approximate. As such, sensor-based operations and various calibration issues would have to be considered. Further, task elements were not positioned to aid the operator. This was done to understand the difficulties associated with realistic situations that will be encountered when operating the arms from either a space habitation module or from the earth. However, the robotic arms available to the testbed in 1985 (PUMA 560's) complicated such operations because of the limited dexterity of these arms over large regions of the workspace.

It was also desired to understand the engineering issues involved in the integration of a large system whose components were diverse and in different stages of maturity. System demonstrations were to be performed as fast as possible to enhance their visual appeal and lead to a better presentation of the usefulness of the technology to the NASA space-servicing user community.

Figure 8.1: View of the JPL Telerobot Testbed

## 8.2 System Description

### 8.2.1 System Architecture

Architecturally, the telerobot testbed is organized in a traditional hierarchical fashion with task planning levels at the top and the more detailed process, sensor, and control operations at the bottom (Figure 8.2). In addition to the hierarchical aspect, the system is partitioned into a local site where the operator resides and a remote site where the robots are located. Problems associated with time-delayed and bandwidth constrained operations of a telerobot, typical of any ground-based remote operation of a space robot, could be investigated with such an architecture. Also, the issue of partitioning of the operations between a computationally powerful local site and a relatively limited remote site were addressed in the system design. The system is designed so that an operator and much of the associated computing environment can be located remotely relative to the robots [3].

The operator may interact with any of the three hierarchical levels: *Task*, *Process*, and *Servo* (Figure 8.2). In general terms, the higher levels are more suited for fully autonomous operations in more structured environments, while the lower levels are more suitable for teleoperated and shared interaction with more unstructured environments. At the higher levels of operation, various tools are provided so that automated task performance is possible even when objects are partially occluded or their positions are not known *a priori*. At the highest *Task* level, the operator provides the goal condition for the objects in the environment and the system develops all possible sequences of tasks that achieve the desired result. If an unexpected situation arises, such as when an unmodeled object is seen by the operator, the operator interactively modifies the plan and includes a step for building a model for the object using capabilities provided by a *Process* level interface. After the model is built, the system resumes the normal operation from the top level. If certain complex operations cannot be carried out autonomously, such as moving an electrical connector including its attached cable, then the operator can intervene in the lowest *Servo* level and execute the operation either in teleoperation, supervised autonomy, or a mixture of the two [4]. The flexibility provided by the testbed ensures that virtually any operation can be executed by the system provided that the arms and the end-effectors are adequate for the job.

The system implementation utilizes a mix of *ad hoc* and algorithm techniques, in its approach to providing the functionality required for assembly operations. In general, the criteria for selection and inclusion of a particular approach was based on whether the technique was effective in the majority of situations encountered with reasonable execution times. If an algorithmic approach provided this capability then it was included, but if need be, an *ad hoc* "engineered" approach was not precluded.

In keeping with the need for reasonable execution times, planning type op-

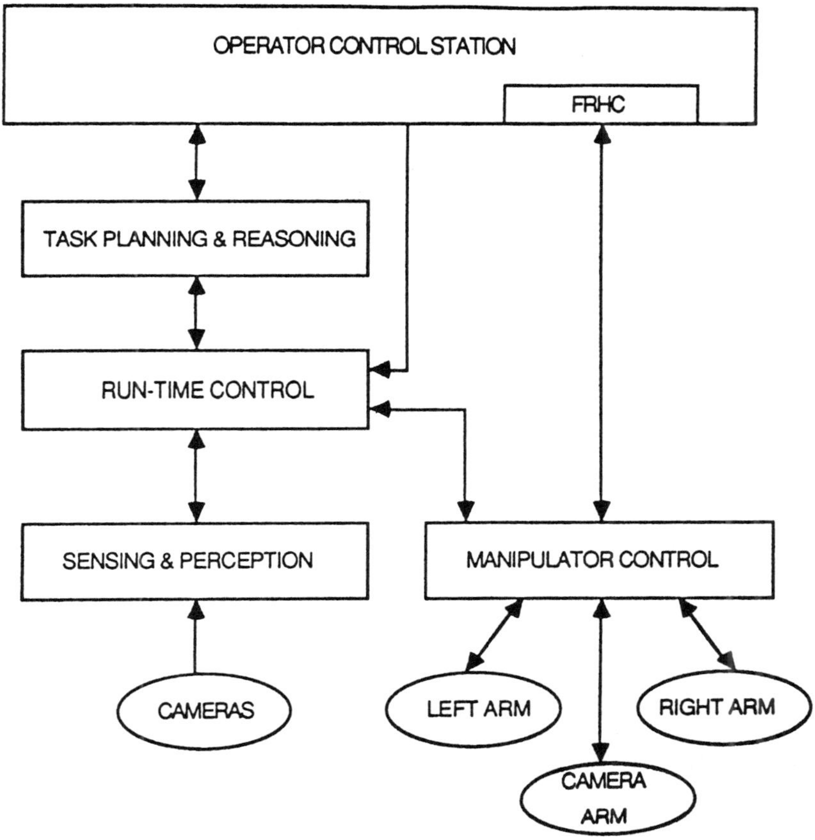


Figure 8.2: Architecture of the Telerobot Testbed

erations in the system have an "optimistic" flavor with limited analysis on the consequences of chained operations. For example, the system will select a grasp point for an object without concerning itself about possible problems during the ungrasp that is to be performed many steps in the future. While nothing inherently in the system precludes this type of analysis, the execution times for performing the detailed analysis and possible backtracking are not warranted in most cases. However, some tasks are then not possible within this framework and the system relies on the operators experience in approving and interactively modifying plan sequences to avert these situations. The degree of optimistic planning also increases with the level in the hierarchy, with the *Task* level incorporating more loosely constrained planning engines. The *Process* level utilizes more physical information to constrain the planning of individual task elements.

Task "intelligence", if it may be termed as such, is derived from world model based simulations, search and reasoning. The world model provides a uniform way for the algorithms and procedures to interact with all objects and parameters in the system and is described in Section 8.4. An object-oriented implementation hides whether the desired functionality and information is provided by table-lookup of engineered data or via algorithms. The world model is updated from sensor data derived from explicit sensing tasks as well as from data accumulated during assembly motions. Many techniques to reduce the overall effects of geometric uncertainty are embedded in the world model. These are designed to make the system more self-calibrating and reduce the inaccuracies in positioning the arms with respect to the objects.

Assembly task sequencing at the *Process* level is via traversal of a finite state graph with task states corresponding to distinct geometrical and logical predicates. The act of traversal invokes corresponding procedures associated with the graph link joining the nodes. These procedures are generic and object-oriented and involve many world model interactions, e.g., an "approach object" procedure works with any object, and utilizes a world model query *to the object itself* to provide the approach positions. The overall method is illustrated for a simple nominal (i.e., error-free) operation in Section 8.7. However, the graph mechanism is generic and suitable for any complex task including error recovery via run-time modifications to task graph structure. Path planning for the procedures is performed via a fast potential field method described in Section 8.6. The planned motions are checked for unwanted collisions using techniques discussed in Section 8.5. Execution of the motions is done via a variety of motion primitives described in Section 8.3.

### 8.2.2 Assembly Example

In order to provide a context for some of the subsequent technical discussion, a typical automated assembly sequence in the testbed is described. It will not, however, be possible to cover the detailed operation of the system for the entirety of the sequence. Instead, we adopt an approach where generic discussions of the capabilities relevant to automated assembly are intermixed with discussions specific to the example sequence.

Consider an ORU on a storage rack in front of the manipulator arms within reach of the left manipulator arm. The module is to be grasped, removed, carried, and then inserted into a platform that is between the robot arms. As the module has an insertion guide pin on its right side, the insertion operation with the left arm would involve an off-centered compliant operation, a difficult task. To accommodate this difficulty it is desired to handoff the module to the right arm which then performs the insertion. Initialization of the module location in the workspace requires the use of the vision editing facilities described in Section 8.4.4. The grasp portion of the sequence is discussed in some detail in Section 8.7 to illustrate the decision making process internal to the system.

### 8.2.3 System Communication

The system is implemented as a collection of subsystems each of which communicate to the others via a Network Interface Package (NIP), [5]. The NIP implements a uniform communication software protocol using a Decnet/Ethernet medium and is resident on each subsystem. More dedicated communication for specialized applications is provided by serial lines and high speed parallel lines. To accommodate a transition from the originally used VAX/VMS based architecture to a Unix based system, TCP/IP socket based communication over Ethernet is also accommodated. A brief description of the various subsystems comprising the system is now provided.

### 8.2.4 Manipulator Control and Mechanization

The Manipulator Control and Mechanization (MCM) subsystem is responsible for servo-level control and monitoring of the devices which relate to manipulation [6]. These devices include three PUMA 560 manipulators, two DC servo controlled parallel-jaw grippers, two 6-axis-wrist force/torque sensors and two six-degree-of-freedom Force Reflecting Hand Controllers (FRHC's) [7]. Two of the arms can be coordinated to move a mutually held object while simultaneously controlling the internal forces and torques in the object [8], [9]. Single arm operations include the real-time execution of servo primitives such as guarded and compliant motions. The subsystem is implemented using the Robot Control C Library (RCCL) [10], [11], modified for dual arm operations [12]. MCM's architecture consists of three levels: *Motion Generation*, *Motion Control*, and *Servo Control*. The *Motion Generation* level provides a robot programming environment for asynchronous operations and uses the ring equation concept for specifying geometric relations as well as general primitives and data structures for calculations related to robotics. The *Motion Control* level is synchronously executed at 100 to 200 Hz, depending on single or dual arm control, and provides Cartesian and joint motion trajectories for the requested motions. User specified sensor-based real-time functions can be run at this level and enable the implementation of force control, visual servoing [13], and other sensor-based control functions. The teleoperation system is developed by essentially treating the hand controllers as real-time sensors. The *Servo Control* level consists of a custom built general motion controller that runs synchronously at 1000 Hz [14]. It takes joint position setpoints and controls the joint motors utilizing a user specified PID control law.

### 8.2.5 Sensing and Perception

The **S**ensing and **P**erception (S&P) subsystem is responsible for controlling, monitoring and processing the video images obtained from the seven CCD video cameras. Four of these cameras are mounted on one of the PUMA manipulators, two with a wide stereo baseline suitable for machine vision and two with a baseline suited for human stereoscopic viewing. The other three cameras are mounted at fixed locations about the workspace. In addition to commercial computers and cameras the computing facility includes a custom built real-time, pipelined image processing system called IMFEX. The S&P subsystem uses information from the cameras to implement stereo vision algorithms [15] that provide position and orientation data of objects using feature extraction, stereo model matching, and object tracking [16]. This enables real-time object localization for stationary objects and tracking for moving objects such as a rotating satellite. Models of previously unknown objects may also be built on-line by the operator. When vision based servoing is required for operations such as grappling a rotating satellite, the object's motion data is provided directly to the real-time robot control system, i.e., MCM. In all other cases, information is routed to RTC (described in the next section) which utilizes it to find parameters for a particular primitive command to be sent to MCM for execution.

### 8.2.6 Run-Time Control

The **R**un-**T**ime **C**ontrol (RTC) subsystem is at the process-level and forms the second level within the testbed hierarchy. Its primary function is to plan and coordinate the execution of tasks using the motion capabilities provided by the MCM subsystem, sensing functions provided by the S&P subsystem, along with its own planning and simulation capabilities [17], [18]. Task primitives can either be executed directly by an operator or by commands from the TPR (described in Section 8.2.7). They perform operations like "grasp Object-X", "insert Object-Y", etc., where the parameter Object-X is a symbolic name for a particular object in the environment. This requires the RTC subsystem to convert the symbolic information into the detailed numerical information required for execution by the lower level subsystems; monitor progress of the subtasks; initiate simple error recovery operations if required; and report back to the operator and/or TPR. While planning at this process level is done on-line it is not done in real-time at servo rates but on an event-driven time scale.

In order to provide the necessary planning capabilities the RTC subsystem contains and maintains the system's detailed "world model". This model contains all of the pertinent physical information and relationships of the objects that comprise the system and the environment. The RTC subsystem also has an extensive kinematic simulation and analysis capability that is used to determine whether or not requested motions are indeed feasible, together with a collision detector capable of detecting collisions between general polyhedra. Finally, the RTC subsystem contains a preliminary implementation of a massively parallel global path planner that can plan complex paths quickly.

### 8.2.7 Task Planning and Reasoning

The highest level subsystem is the **T**ask **P**lanning and **R**easoning (TPR) subsystem. It is an Artificial Intelligence planner at the top of the hierarchy for the supervised autonomy mode of operation of the testbed. The subsystem is a much simplified (and correspondingly faster) version of a very sophisticated Task planner developed earlier at JPL [19], [20]. A typical command executed by the TPR may be a request to "exchange ORU module X with its replacement module located in storage slot 3". Given an appropriate database, in the form of a semantic network, this system develops and graphically presents to the operator all possible sequences for executing high-level operations such as swapping out of modules, demating objects, etc. The operator selects one of the sequences and can then either execute the plan in one step or execute each step and observe the results before proceeding. The planner is interactive and allows on-line modification of automatically generated plans based on operator inputs. The operator has the option of specifying whether *a priori* data in the world model is to be utilized for a given task or whether the data should be refined using the vision-based editing utilities described later in Section 8.4.4.

The internal operations of this task planner are not technically very significant with regards to assembly operations except to the extent of ordering the various assembly tasks via a straightforward implementation of the semantic network. As such, we do not discuss the internal workings of the task planner during the rest of this chapter.

### 8.2.8 Operator Control Station

The **O**perator **C**ontrol **S**tation subsystem (OCS) is the main interface between the operator and the testbed [21], [22]. It is designed to accommodate a primary as well as a secondary operator. The primary operator station is equipped with visual, auditory, computer and kinesthetic media for effective communication between the operator and the testbed system. The available operator interfaces include two computer monitors, a voice recognition and synthesis system, two graphic overlay devices, a video switcher, and associated keyboards and mice. A stereo display system consisting of two NTSC monitors mounted at a 45 degree angle with respect to a half silvered mirror, together with a pair of polarized glasses, provides stereo depth perception for the operator. The inputs for this system are provided by a pair of color cameras mounted on the Camera arm, which is also controlled by the operator to view the task. Seven other monitors are available for the operator to observe the scene through any arbitrary combination of close-up and wide view cameras. The two FRHC's used by the MCM subsystem and a three-degree-of-freedom joystick used by the S&P subsystem are physically located at the control station. A voice recognition system, VERBEX 5000, with a 120 word vocabulary and simple grammar is used to issue voice commands. This is particularly useful when both of the operator's hands are engaged on the hand controllers. A voice synthesizer, DECtalk DTC01, is utilized to provide acknowledgments for operator generated commands as well as announcing important messages. The control station also provides a means for the operator to interrogate and send messages to other subsystems. The secondary operator control station provides for general lower-level health monitoring of the system.

### 8.2.9 Monitoring and Diagnosis

A specialized AI-based subsystem is provided to monitor command execution profiles and perform diagnosis on task failures. As this subsystem has been utilized to a very limited extent in the testbed, no further discussion is provided here. The interested reader is referred to other works that describe the diagnosis capability in more detail [23].

## 8.3 Force and Motion Control Capability

The MCM subsystem implements Guarded, Contact, and Compliant Motions as well as Gripper Action primitives. A Set Parameters primitive updates the MCM subsystem's database of various global control parameters relevant to all primitive executions. A Get Status primitive allows the querying of MCM for arm position/force information. A Teleoperation primitive provides FRHC directed arm motions by passing control of the manipulator actuators to the operator. We provide a brief overview of the Guarded, Contact, and Compliant Motion primitives. For reasons of space, descriptions of the other primitives are not provided.

### 8.3.1 Guarded Motion Primitive

The Guarded Motion primitive provides the most elemental motion capability to the system. By using this primitive, a single continuous motion can be executed that contains segments in which the end-effector is moving in a straight line in either Cartesian or configuration (joint) space. The primitive performs such motions by concatenating a sequence of elemental trajectories together in real-time. Each elemental trajectory is specified by a goal location (i.e., the position and orientation of the end-effector coordinate frame relative to the robot's reference frame), the type of interpolation to be used, the time duration of the motion, and the acceleration/deceleration time to be used at the start and end of the trajectory segment.

The intermediate path goal points are usually referred to as "path via points", thereby signifying that they are points which the arm will use to interpolate its travel on the way to the ultimate goal location. The kinematics of the PUMA 560 manipulators also require the specification of the arm pose (one of 8 possible configurations) in addition to the desired Cartesian location of the end-effector. The selection of a particular pose is typically made by the RTC subsystem to satisfy various task related constraints like avoiding collisions. Further, at singular wrist configurations, the PUMA 560 requires the disambiguation of joints 4 and 6 when they become aligned. This function is also left to the higher level subsystem, which provides the specification for the joint 4 angle at the via point.

The type of interpolation used during a segment determines whether the motion between the starting and ending location will be computed via a linear interpolation in Cartesian space or configuration space. Linear interpolations in Cartesian space result in physically linear motions of the end-effector as long as the motion is kinematically feasible, i.e., it does not cross singularity boundaries. In case the higher level system mistakenly provides an infeasible Cartesian motion, the motion proceeds until the singularity is approached at which point commanded joint velocities build up causing a halt to the motion.

The segment time parameter implicitly defines the speed at which a particular trajectory segment should be executed. The use of a time parameter, as opposed to an explicit specification of speed, was motivated by the desire to make the input parameters to the primitive independent of Cartesian or joint space motions and also independent of the type of representation used to describe the orientation of the end-effector. The specification of an acceleration parameter controls the transitions between consecutive path segments. It affects how closely the realized trajectory comes to passing through the specified via points during the real-time interpolation process.

An important feature of the Guarded Motion primitive is that it is designed to monitor the wrist force/torque sensor values (gravity but not inertial force compensated) against a specified threshold to detect unforeseen collisions. A coordinate frame is specified at which the force/torque measurements are converted for the purpose of thresholding. In the case of a large load, the "Sensing" frame should be located near the load's center of mass. This increases the likelihood that sensed forces and torques are actually being induced by an external object as opposed to the reflected inertial forces and torques of the load.

### 8.3.2 Contact Motion Primitive

The function of the Contact Motion primitive is to control the manipulator to reach out in a particular direction and make contact with an object. The motion is divided into two Cartesian interpolated motions called the "approach" and the "recoil".

In the approach, the end-effector is commanded to move towards a desired goal location. Normally, the goal location is specified to lie beyond the object of interest as will become apparent. During the approach, the primitive monitors the magnitude of the force/torque vector measured with respect to a specified sensing frame. Contact with an object is detected when the magnitudes of the force vector and/or torque vector exceed specified thresholds.

Once contact has been made, the direction of motion of the end-effector is immediately reversed to provide a "recoil" motion. The purpose of the recoil motion is to bring the end-effector back to a location where the forces and torques imposed by the end-effector upon the object, drop from the initial contact force down to a specified value. The recoil phase completes at the moment when the magnitudes of the forces and torques sensed become *less* than specified thresholds. As a safety feature, the magnitudes of the forces and torques are also compared to a pair of upper-bound force/torque thresholds during the recoil motion, and if exceeded, cause the motion to terminate. Such a situation may arise when the initial contact caused the end-effector to "get caught on the object", thus inducing *increasing* forces during recoil.

### 8.3.3 Compliant Motion Primitive

The Compliant Motion primitive is the most sophisticated motion primitive available to the system. It implements a general force control framework called *Generalized Compliance* [24]. It is used extensively by the RTC subsystem in performing tasks such as pin insertion and removal, leveling grippers on grapple lugs, bolt seating and turning, pushing, door opening, and crank turning. All of these tasks require the robot to come in contact with rigid and compliant environments. Many tasks such as opening a door or screwing a bolt also require the control of robot position as well as forces and torques. In the testbed system, the compliant motion capability is geared towards applications in which there exists a reasonable amount of information about the task and the required end-effector trajectory (e.g., crank turning and not random contour following).

A simple but robust force control technique referred to as "position accommodation" was selected to implement a force control capability. In this technique forces and torques are sensed and transformed to a desired control frame. The trajectory of the arm is then modified by scaling the force/torque data to represent a positional increment. The scale factor also serves as a selection matrix. If the scale factor is zero, no force control is performed in that Cartesian direction. Since the whole trajectory generation and inverse kinematics runs at 200 Hz for single arm operations, and 100 Hz for dual arms, this technique has resulted in a very effective force controller.

Three factors influenced the selection of this technique. First, since RCCL forms the low-level programming environment for the MCM subsystem, the compliant control algorithms had to be structured to make use of the real-time trajectory generator inherent to the RCCL environment. Second, since the PUMA 560 manipulators are not capable of measuring joint torques, position-based force control was required (i.e., position accommodation). Third, previous experience with position-based force control strategies indicated that control performance could be enhanced by relegating as much of the control effort as is possible to the positionally controlled degrees-of-freedom. In other words, the deviations from the nominal trajectory produced by forces acting upon the manipulator should be as small as possible. In compliant tasks, the nominal trajectory is an approximation to the trajectory that will be realized. The better the approximation, the better the performance.

In contrast to the standard approach to compliant control, instead of a strict decoupling of the force and position degrees-of-freedom, the Compliant Motion primitive continually tries to control the position/orientation of the end-effector to move along a specified nominal trajectory. At the same time, those components of the sensed forces and torques which are coincident with the directions in which force control is to be applied are extracted and used along with specified control gains to generate perturbations to the nominal trajectory. Virtual spring parameters may also be specified to control the overall deviation from the nominal trajectory along each positionally controlled axis. Using this method, control of the end-effector inherently defaults to that of position control if forces and torques are not present. The advantage of this approach is that it greatly simplifies the manner in which one specifies the overall motion: first, select a nominal trajectory which is as close as possible to the trajectory that will be required to perform the given task; second, select the directions in which to comply to the sensed forces and torques; and third, set the force gains.

In the case of turning a crank, the expected trajectory is a circular arc and can be described as a pure rotation about an axis of a coordinate frame aligned with the crank's degree-of-freedom. Although the exact location of the crank axis is not usually known precisely, its approximate location relative to the manipulator is all that is required to specify the nominal (i.e., approximate) trajectory. The compliant directions are specified by providing the location of a coordinate frame relative to the end-effector, called the comply frame, about which forces and/or torques will be explicitly controlled. Six Boolean flags are used to indicate along which of the six axes of the comply frame forces and torques are to be controlled. For each direction selected, the desired force or torque is also specified. At present, the force and torque profiles are restricted to be constant.

The execution of the Compliant Motion primitive is made up of three phases: Nominal Motion, Compliant Motion, and Ending Motion. Nominal Motion is the motion which satisfies the specified position trajectory. Compliant Motion modifies the nominal motion to control contact forces. At the end of the Nominal Motion, another Ending Motion is executed which has a null position trajectory and the same force control parameters as the Nominal Motion. This terminates on either a time-out or on ending conditions specified in terms of functions of force, torque, position, and orientation variables. When the Compliant Motion terminates, a flag indicating the cause of motion termination is returned to the calling subsystem. A Monitoring component is active during all of the motions and checks that the execution is proceeding safely and terminates the motion prematurely if an anomalous condition is detected.

## 8.4 World Model System

The robot's world model is primarily resident in the RTC subsystem. This model contains a complete geometric and physical description of the objects known to be within the robot's environment, including those elements that make up the robot itself. The information pertaining to each object in the database is organized into a list of data fields. The current implementation uses approximately 50 distinct data fields to completely describe the attributes of a single object. The fact that the database contains approximately 200 objects indicates the extent to which the robot's environment has been modeled.

### 8.4.1 Database Trees

The world model system uses a spanning tree of the general object-to-object spatial relation graph to form the *Database Tree*. Arcs between the nodes define the parent-child relationships between objects. The tree topology is based on the inherent physical connectivity of objects, with parent object motion resulting in implicit motion of all successor objects. A measurement tree traces a sequence of measurements from the telerobot's absolute frame of reference to the location of every major element in the world model. This tree is provided as *a priori* information.

### 8.4.2 World Model Database Fields

The world model supports queries and updates on object location, spatial occupancy, assembly information, object mass properties, payload mass (used to bias the force/torque sensors), camera models, and hand-eye calibration data.

The system also supports the automatic coordination and update of other subsidiary local databases in the MCM and S&P subsystems. Some of the data fields associated with each object that are relevant to assembly operations are discussed below in greater detail.

The *Tool Frame* gives the location of the tool frame associated with a given robot arm. In the absence of a physical tool, this frame is coincident with the end-effector frame object. On affixing a tool or on grasping an object, the *Tool Frame* is automatically moved to the point corresponding to a tool tip. This frame is used for planning of tool tip mating and compliant motions.

An *Object Coordinate Frame* is associated with every object and is typically located at the object's center of volume. Its *Absolute Transform* is a homogeneous transformation that specifies its location in the global telerobot coordinate frame. This transformation is computed recursively from the knowledge of the *Absolute Transform* of the *Parent Object*, which is defined to be the immediate ancestor of the object in the database spanning tree. The computation uses the *Parent Relative Transform* (which provides the transformation from the parent's coordinate frame to the location of action of a hinge-type parameterized functional transform), the *Hinge Transform* itself, and an *Offset Transform* that provides the transformation from the place of application of the hinge to the *Object Coordinate Frame*. The parameterized hinge relationship giving the *Hinge Transform* is computed from the *Hinge Function Name* (e.g., rotation about a Z-axis) and a scalar *Hinge Parameter Value*.

The *Database Tree* permits only one parent for each object, and in those cases where multiple parents may be necessary (e.g., an object is multiply grasped by more than one arm during a transfer operation from one arm to another), *ad hoc* mechanisms are used to compute the *Absolute Transform*.

### 8.4.3 Object Geometry and Properties

Objects have a polygonal representation consisting of four elements: *Vertex Lists*, lists of 3-vectors describing the object vertices in the *Object Coordinate Frame*; a *Volumetric Triangulation* that describes a set of convex volumetric triangulation elements; a *Surface Triangulation*, which triangulates the exterior of the object into convex facets; and finally an *Edge Triangulation* that gives a triangulation of the edges of the object in the form of straight lines. The various triangulations are described in the form of a list of sublists with each sublist containing indices into the *Vertex List*. Triangulation elements are not restricted to simplices but can be any convex element. The use of the *Volumetric Triangulation* permits rapid calculation of object-to-object distances as well as directly providing volumetric information. The surface triangulation information is used for face, line, and vertex penetration and contact determination. The various triangulations are independently stored, even though in principle, some may be derived from others. All of the polygonal objects are associated with a default bounding CSG primitive [25] such as a rectangular box, cylinder or a sphere.

The world model system stores the *Mass Property* of each object, i.e., the mass and a center of mass vector described in the *Object Coordinate Frame*. This information is used to calculate manipulator load handling capabilities as well as other properties such as the *Grasp Mass Property*. This gives the mass and center of mass of the payload and end-effector with reference to the Force/Torque Sensor frame associated with a given arm. The value is computed from the individual *Mass Property* values of all of the objects grasped by the end-effector and in the direct subtree below the Force/Torque sensor object. The value obtained is used to update the MCM subsystem's database with bias values used for safety force/torque threshold monitoring, and compliant motion control. In those cases where an object is multiply grasped, the object's mass is partitioned between the manipulator devices.

Object mating information is provided by means of sets of male and female mating frames associated with the objects. Each graspable feature has an associated set of *Grasp/Approach Points* that defines, in the *Object Coordinate Frame*, the allowable approach frames and grasp frames. These frames when coincided with the manipulator's *Tool Frame* described earlier, define the relative geometry between the end-effector and the object at approach and grasp, respectively. The information in these routines is provided mostly by table-lookup of data generated by off-line software or by hand.

The primary vision information is resident in the S&P subsystem's database. This information concerns such details as object reflectivity, vision feature models, object symmetry properties, etc. It is used by the machine vision system for the purpose of object verification position refinement. Currently, perception models are available for the robot grippers, grapple lugs, certain ORU's, certain standardized interface connector bars, cranks, and a satellite mockup treated as a composite hexagonal shaped body.

A *Camera Position* field gives a location of the camera arm/platform to view the object. Experimentally obtained table-lookup methods are mostly used to determine this information. An *Object Covariance* field is used to store the estimated error covariance of the object location. Currently, the data in the covariance matrix are updated solely from the results of machine vision position refinement, and are used to make a simple Yes/No decision on whether to accept or reject the results.

### 8.4.4 World Model Database Update

The world model is updated at run-time on the reception of sensor data from either the MCM or S&P subsystems. In some cases, the update process represents a simple copying of sensor values into the corresponding world model fields (e.g., arm joint angles). In other cases, sensor data are extensively processed before the update is performed (e.g., updating an object location based on a combination of touch and vision data).

The operator also has the capability to change world model information, using certain *Vision Based Model Editing* features. This provides a capability to obtain the location of a known rigid object via a process called *Object Designation* [22] available at the operator control station. *Object Designation* is a procedure whereby a wire frame image of an object is overlaid on one or more raw video images being sent to the operator control station. Using a mouse and graphical interface the operator then selects vertices of the wire frame and physically indicates where in the image they are actually located (i.e., overlay the wireframe vertices with the image vertices). Since the monitors, overlay generators, and cameras are all calibrated and the camera model parameters are known *a priori*, the system can then compute using triangulation and a least-squares algorithm, where in the space of the robots, the object is located. The wire-frame image is updated and the procedure is repeated by the operator until the wire-frame coincides with the actual object image. If high-accuracy data are required and the images are relatively clutter free then the S&P subsystem can be invoked to further refine the object location.

In order to correct distorted object models of rigid objects, a capability to edit the shape of the object is also provided. This capability utilizes *Object Designation* on parts of the object to obtain the correct relative locations between the parts. The database tree corresponding to the object and its parts is then modified to form the correctly shaped object model.

If the object has not been previously modeled, then an object model can be built on-line using a 3-D stereo cursor overlaid on the stereo monitors [26]. This is accomplished by building faces from several vertices by overlaying the 3-D cursor on vertices of the object as seen via the stereo display system. When objects have curved surfaces, approximate models are built.

The operator can also interactively specify new semantic and spatial relationships between objects using a menu interface. This capability is especially useful after teleoperated tasks, as it provides a means to inform the autonomous system of any new object relationships (e.g., a module has been inserted into a slot using teleoperation).

### 8.4.5 World Model and Calibration

There are various sources of geometric error that require some form of calibration in order for their effects not to adversely affect assembly operations. These errors include those in arm kinematic calibration, uncertainty in the location of the various robot arms with respect to a common reference frame in the task workspace, and orientation errors in the Measurement Tree (see Section 8.4.1). Errors in camera model parameters also result in uncertainty of object locations. Since the camera platform is itself movable, various kinematic uncertainties of the Camera arm propagate into the camera model. Small orientation errors in the camera models can lead to large errors when estimating the object location.

In the testbed, for example, the stereo cameras were mounted on a PUMA arm and despite better than 3.0 mm performance in depth estimation at calibrated locations, they performed with 30 to 50 mm error for camera platform locations away from the nominal locations.

Further, there are errors in the object models themselves, typically introduced during manufacturing or as a result of wear and tear. Any resulting orientation errors can magnify positional errors at other parts of the model because of the long "lever-arm" effects. An example encountered in the testbed was that of an Orbital Replacement Unit (ORU) with two handles that was required to be handed-off from one arm to another. A small twist on one of the handle mounts resulted in the position of the other handle being off by many centimeters from the modeled value, making the handoff process more difficult.

The problem is significant when the location of objects in global coordinates has to be derived from the information provided by the vision system. Unless the camera is in a precise location corresponding to a calibrated camera position, the data from the vision system are essentially useless in determining global locations. We do not discuss the various sources of uncertainty here, but instead briefly indicate the ways in which the system maintains its calibration despite these uncertainties. The interested reader is referred to [27] for a further discussion on these issues as well as the related issue of maintaining world model consistency.

Given a globally and mutually calibrated arm/camera system, the vision-based object world model editing facilities described earlier suffice to allow manipulation of objects even when they are displaced from their nominal locations. The residual error, if any, can usually be handled by means of compliance-based manipulation operations. However, as described earlier, achievement of these globally calibrated situations is all but impossible except for a few highly engineered and calibrated regions of the task space. Two approaches are possible to improve the overall performance.

The first technique attempts to unify the independent calibrations performed on the manipulation and vision systems and correct for the various other uncertainties and errors introduced into each of these systems. A direct map between vision and manipulation coordinates for different regions of the task/vision space is obtained off-line by a process of *Hand-Eye Training.* This involves building a disparity map between the predicted hand location, as would be seen from the camera platform, and the actual hand location as obtained by the sensor (using *Designation*, say). The disparity map encodes the residual error arising from all of the sources described earlier, and it can be used to compensate for errors prior to commanding the motion of the arm to a location derived from the vision system. In the testbed, such a disparity map was built for each of the manipulator arms for about six camera view locations and the task space volumes corresponding to the view volumes at each of these camera locations. A table-lookup scheme embedded in the world model database utilizes this hand-eye table to make automatic compensations for vision sensor derived motions in the corresponding task space regions. However, these hand-eye tables do not encompass all of the space and also do not handle the propagation of uncertainties.

An alternative approach termed *Local Calibration* is adopted instead to provide a more generic capability over the whole workspace. In this approach, using the same camera platform location the location of an object feature and a suitable feature on the end-effector are obtained using *Object Designation* (and machine vision, if required). Using this information, the relative location of the end-effector with respect to the object is obtained, presumably now free of any calibration biases. In effect, errors in the arm's absolute location are ignored and reliance is placed on the arm's relative motion performance (usually very accurate) to achieve the final relative positioning required. The utilization of this method is described in Section 8.7.3 for a grasp operation.

### 8.4.6 Database Coordination

As information in the system is distributed, a number of "validity" flags are used to tell the local world model if the information is locally available or if an access to a remote database is required. An example is the camera model whose validity flag is set to False in the RTC subsystem's world model every time the Camera arm is moved. The flag then triggers a re-query of the S&P subsystem database, based upon the new Camera arm configuration (i.e., camera platform location), in order to obtain recomputed camera model data. Similarly, a number of "has changed" flags are used to tell the RTC subsystem's world model if locally made changes need to be propagated to other databases. A number of database fields are also reserved to store old and/or default parameter values and allows the world model to "restore" itself to an earlier configuration if desired.

### 8.4.7 Database Implementation

The world model implements a straightforward object-oriented programming system that allows any arbitrary procedure to be invoked whenever an object field is accessed/stored or when a message is sent to the object. No class/instance mechanism is directly supported by the system, but equivalent functionality is provided by allowing the user to overlay various methods on top of the existing methods. This overlay process proved to be particularly convenient during the rapid prototyping phase of the development by not locking the code development into any predetermined class and inheritance structure.

## 8.5 Collision Detector

### 8.5.1 Overview

The RTC subsystem includes a capability to determine whether or not objects contained in the world model collide (i.e., intersect). The collision detector, as it is called, operates by analyzing the spatial occupancy of objects contained in the RTC database at a single point in time. Through repeated application, the collision detector can predict whether or not a desired motion of a manipulator will cause a collision. As such, the collision detector supports both the RTC path planning and trajectory generation/simulation capabilities. Independently, the collision detector can be used as a stand alone, on-line analysis tool.

In principle, every object is checked with every other object. However, to keep the combinatorics from getting out of hand, a number of features are implemented to prune the combinatorics.

### 8.5.2 Entities and Object Modeling

The RTC database contains two basic types of objects, which are referred to as "elemental objects" and "bounding objects". Elemental objects, by definition, correspond to physical objects in the environment, whereas bounding objects do not. Bounding objects can be thought of as imaginary objects that spatially enclose one or more elemental objects. As such, they represent an approximation to the spatial extent of a collection of objects. Normally, the set of objects which are enclosed is referred to as an entity, analogous to the concept of an assembly in manufacturing systems. Note, too, that if an entity consists of only one simple object, then that object can serve as its own bounding object. At present bounding objects define a one-deep tree. Hence, bounding objects themselves are not bounded by other bounding boxes. Bounding objects and the concept of entities provide a simple means to reduce the number of collision checks required since objects within an entity, by definition, need not be checked against one other.

The idea behind using bounding boxes is that if an object's bounding box does not collide with any other object then neither will the actual object. The same holds true for bounding objects, that is, none of the elemental objects belonging to the entity will collide if the bounding object does not cause a collision. It is important to note, however, that if the bounding object, in the case of an entity, or a bounding box in the case of an elemental object does trigger a collision, then it does not necessarily imply that the actual object is colliding with something. In this situation, the collision checking procedure must continue to the next level of spatial detail to confirm if a true collision exists. Non-moving entities (i.e., entities not attached to or part of the manipulator that is being moved) are not checked against other non-moving entities. This, further prunes the combinatorial operations.

Throughout the process of completing the collision checks, the system automatically tabulates a list of the names of all colliding object pairs. If one or more collisions were to occur, the operator then has the opportunity to display the list and ascertain the exact cause of the collision. This capability is important for manual and/or supervised error recovery. If no objects have been found to collide, the collision detector sets a global flag indicating this to the rest of the system (i.e., to the path planner and/or kinematic simulator, which may have issued a request to the collision detector).

### 8.5.3 Collision Detection Algorithm — Low-Level

The type of internal computations required to determine whether or not two objects collide depends upon the type of representations used. In general, the algorithm used to check for collisions between two CSG primitives is computationally simpler than that used to check for collisions between two general polygonal representations. To get an accurate result, however, collision checks based upon an object's polygonal representations (if they have one) are often required.

Based upon what type of geometric representations are included in the objects' database entries, two different methods for computing whether or not two objects collide are used in the above algorithm. The method used to detect a collision between two CSG primitives will be referred to as Method 1. The second method, which detects collisions between polyhedral representations, will be called Method 2.

<u>Method 1</u>

The application of Method 1 to determine object collisions is motivated by the method's computational simplicity. The actual algorithm that implements this method makes use of numerous computational "tricks", and is itself hierarchical in its approach to arriving at the final result. The ability to apply such tricks is, in part, a result of the fact that each object's CSG model includes various pieces of geometrical data, like the location and size of the sphere that bounds the CSG primitive. Thus, as an example, if the sum of the radii of the bounding spheres is less than the distance between the two sphere centers, the objects obviously do not intersect. Even simpler tests can be used in the case of boxes. If the minimum X, Y, and Z coordinates of the vertices of one object are all greater than the maximum X, Y, and Z coordinates of the vertices of the second object, then it can again be concluded that the objects do not intersect.

<u>Method 2</u>

Method 2 is more sophisticated than the first in that it actually computes the minimum distance between two convex polyhedra. To use this latter method, the objects must be expressed in terms of vertex and volumetric triangulation lists (see Section 8.4.2). The minimum distance between the two objects is equivalent to the Minkowski Set Difference and is computed using algorithms developed by Gilbert [28].

Each low-level-collision check begins by first applying Method 1, which produces a binary result indicating that the two objects either collide (YES) or they do not (NO). If the result is NO, the objects will not collide independent of whether or not either of them have explicit polyhedral representations. If the result, on the other hand, is YES, the collision checker uses Method 2 to analyze the situation taking into account the detailed object model. Note that Method 2 need not be applied if the polygonal model can provide no additional information on the object spatial occupancy, (i.e., this would be the case for a simple box shaped object). Also, as all explicitly modeled bounding objects are simple box or cylinder type CSG primitives, checks between two bounding boxes typically exercise only the faster Method 1.

## 8.6 Path Planner

Fast motion planners are required to find paths in cluttered servicing environments in order to enable safe telerobotic operations. The need to manipulate in unrehearsed environments arises due to unexpected changes in task environment, which is typical of realistic applications. A fast algorithm on a parallel processing hypercube multiprocessor (NCUBE) is implemented in the RTC subsystem to reduce path computation time to a few seconds. A multi-resolution potential-field-based path iterations implementation provides for quick convergence in most cases to a globally collision-free path without getting caught in local solutions typical of many other potential-field-based approaches. The method is based on a generalization of the 2-D motion planner ideas originally proposed in [29] and develops on some ideas presented in [30].

The path planner is designed to search for a path by minimizing a global cost function that contains three terms: the potential for colliding with an object; the potential for approaching and/or exceeding manipulator joint stops; and the desire to minimize the length of the overall path. Conceptually, the path planner operates by iteratively perturbing the location of a fixed number of points along an initial path estimate until a collision-free path that is most likely to be collision-free is found. The path planner itself, however, does not explicitly guarantee that the derived path is collision-free. Such decisions are made by the RTC subsystem's collision detector, which is normally used in conjunction with the path planner and is described in Section 8.5.

The initial path estimate used by the path planner is defined to be a linearly interpolated configuration space motion between the starting location of the arm and the designated goal location. This corresponds to the simplest means for getting from one joint configuration to another. If the initial path estimate contains a collision, the path planner then iteratively modifies this path to obtain a collision-free path.

The manner in which the path planner modifies the initial path estimate, and estimates thereafter, can be visualized by imagining that the initial and final location of the path are connected by a rubber band as illustrated for the two dimensional case in Figure 8.3.

Imagine that the rubber band representing the path has been divided into N segments. The N segments are bounded by N+1 points along the overall path; two points being the initial and final location, and the remaining N-1 points corresponding to intermediate via points. The initial path estimate corresponds to the *0*-th iteration of the path estimate.

The objective of the path planner is to shift the locations of the $N_{via}$ ($= N - 1$) path via points such that the global cost function is minimized, and thus reducing the effective repulsive "forces" acting on the path. Figure 8.3 illustrates how the path evolves from its initial estimate as the path planner searches for the path that minimizes the global cost function. At each iteration, each of the $N_{via}$ via point's locations are changed by an amount proportional to the gradient of the cost function evaluated for that path estimate.

## 8.6.1 Cost Function

The cost (or potential) function used by the path planner to obtain the repulsive forces acting on the manipulator due to various constraints is given by

$$\Xi = F_1 + F_2 + F_3 \tag{8.1}$$

where,

$$F_1 = F_1(\vec{x}_1, \vec{x}_2, \ldots \vec{x}_{N_{via}}) = \sum_{l=1}^{N_{via}} \sum_{i=1}^{N_p} \sum_{j=1}^{N_o} \left[ \prod_{k=1}^{N_{fj}} f_{j,k}(\vec{x}_i) \right]^{1/N_{fj}} \tag{8.2}$$

$$F_2 = F_2(\vec{\theta}) = \sum_{l=1}^{N_{via}} \sum_{i=1}^{N} [f_i^+(\theta_i) + f_i^-(\theta_i)] \tag{8.3}$$

$$F_3 = F_3(\vec{\theta}_0, \vec{\theta}_1, \ldots, \vec{\theta}_{N_{via}}) = \sum_{i=1}^{N_{via}} [\vec{\theta}_i - \vec{\theta}_{i-1}]^2 \tag{8.4}$$

$F_1$ represents the contribution due to obstacles in the environment, $F_2$ represents the contribution due to the existence of mechanical joint stops, and $F_3$ is a term proportional to the length of the overall path.

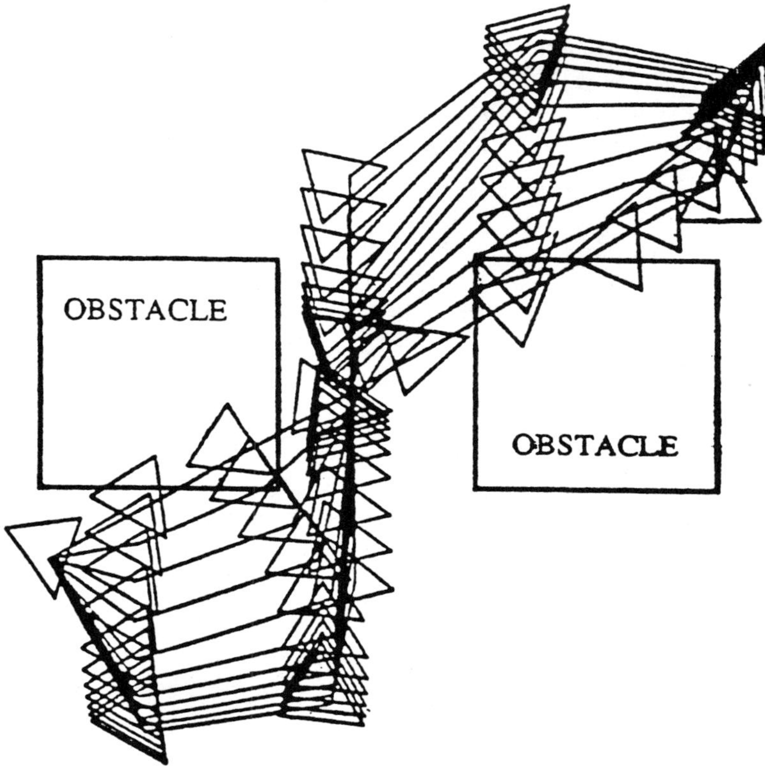


Figure 8.3: Example Path Iterations for 2-Dimensional Case

**Obstacle Cost Function:** $F_1$

The first term, $F_1$, is intended to be a summation of the costs, the gradients of which correspond to "forces" arising from each of the objects within the environment acting upon the manipulator. The path planner implicitly models the manipulator (i.e., its physical structure) by a discrete set of points. These points are chosen to represent either points on the surface of the manipulator or points contained within the physical mechanism. Using this model, the path planner represents the overall effect of the repulsive force of a single object upon the manipulator by a summation of the effect of this repulsive force upon each of the points used to model the manipulator.

The path planner models the repulsive potential of an object by combining the repulsive potential due to each face of the object. This approach takes advantage of the fact that objects are modeled as polyhedra, and these polyhedra are represented in terms of vertex and face lists. In reality, each face of the object imposes an elemental constraint upon the possible placement of a physical point on the manipulator. In the planner, this elemental constraint is represented by a potential function that decreases as you move farther away from the outward side of the face (i.e., that side that does not contain the object) and increases in the opposite direction. The gradient of this potential (or cost) function then provides the necessary repulsive force. Each face is implicitly embedded in a plane that divides the space into two halves: one that contains the object and one that does not. The space occupied by the object is equivalent to the common intersection of the half spaces associated with the object's faces.

In equation 8.2, the vector $\vec{x}_i$ is the Cartesian coordinates of point $i$ on the manipulator and $N_p$ is the number of points used to model the manipulator. The location of the point $i$, $\vec{x}_i$, is a function of the joint angles or subset thereof depending upon where along the manipulator the point is located. The scalar function $f_{j,k}(\vec{x}_i)$ represents the cost function due to face $k$ of object $j$ acting upon the manipulator at point $i$. The product term in the expression corresponds to the conjunctions of the face constraints and the power operation represents a normalizing operation.

The analytic function used to represent an elemental repulsive cost function is a sigmoid given by the expression:

$$g(x) = \frac{1}{1 + \exp(-x/T)} \tag{8.5}$$

A plot of a sigmoid function is shown in Figure 8.4.

The sigmoid function is characterized by three primary regions: in one region, the value of the function is effectively zero; in another, the value of the function is effectively 1.0, and a third monotonically increasing region that transitions between the upper and lower regions. The effective width of the transition region is parameterized by the constant $T$. The larger the value of

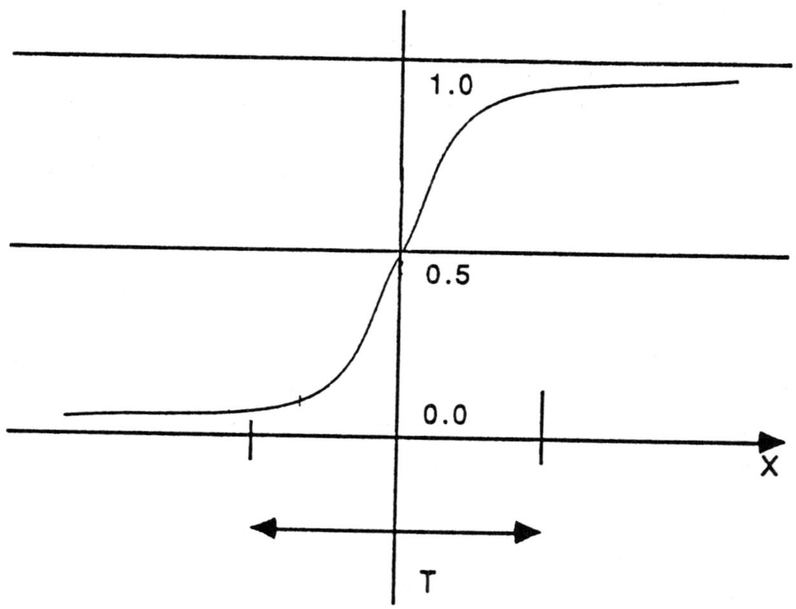


Figure 8.4: A Sigmoid Function Plotted as a Function of the Independent Scalar Variable $x$

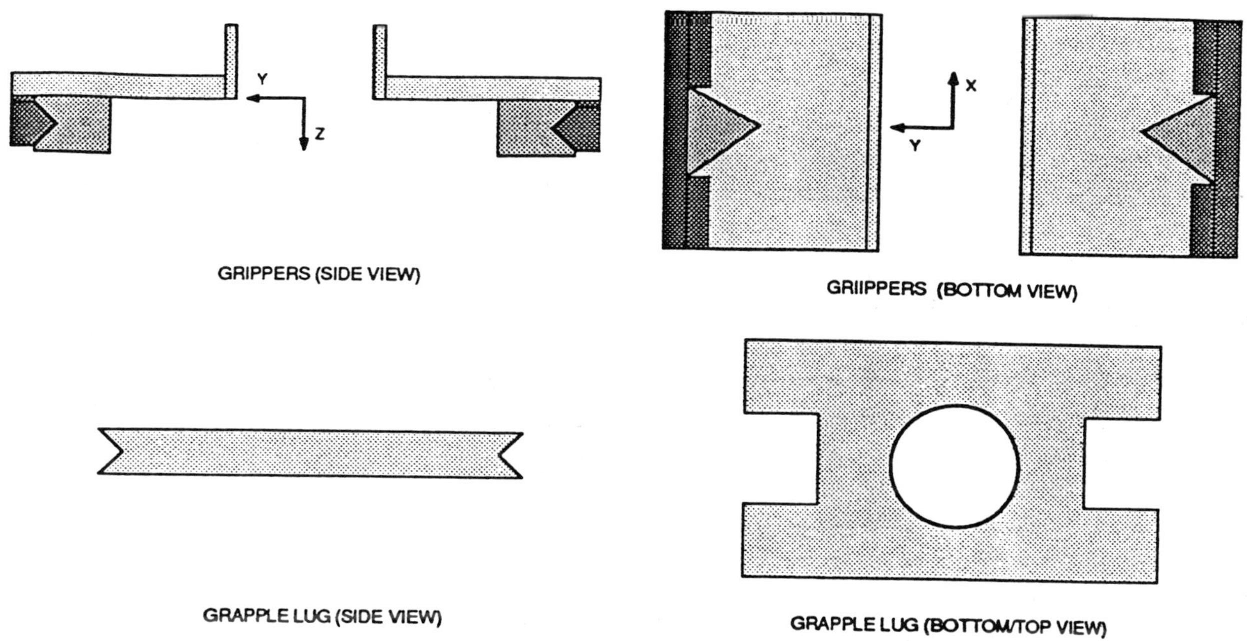


Figure 8.5: ORU Grapple Lug and Corresponding Gripper Fingers

$T$, the wider the transition region. For very small values of $T$, the sigmoid function effectively becomes a signum function. Finally, at $x = 0$, the value of the sigmoid is 0.5. The sigmoid is symmetric about both the lines $x = 0$ and $y = 0.5$.

To represent the potential function giving rise to the repulsive force generated by an object's face, the independent variable $x$ in Equation (8.5) is set equal to the perpendicular displacement between a point $\vec{p}$ in space and the plane upon which the face lies. The value of the displacement is evaluated such that it is negative for points lying on the outer side of the plane, and positive for points lying on the inner side of the plane. On the surface of the plane, and hence, on the face itself, the repulsive force is equal to 0.5 units.

The potential function contributed by an entire object is formed by the product of the potential functions generated by the object's constituent faces (see Equation (8.1)). Analytically, this product is analogous to the intersection of the face half spaces previously described. In practice, this product is normalized by the $n$th root of the number of object faces. The resulting potential function is bounded by 0 and 1, and monotonically increases towards the center of the object.

### Joint Stop Cost Function: $F_2$

The second term, $F_2$, in Equation (8.1) is a summation of the repulsive forces due to the existence of manipulator mechanical joint stops. The functions $f^+(\theta_i)$ and $f^-(\theta_i)$ are the individual potential functions associated with the positive and negative $i$th joint stops. $N$ is the number of manipulator joints. The closer a joint is to a joint stop, the larger the value of $F_2$. Hence, the minimization of $F_2$ tends to drive the path via points to configurations more towards the center of the manipulator's operating range.

The sigmoid function is also used as the analytic basis for describing the potential field giving rise to repulsive forces due to the presence of manipulator joint stops. For an $N$ degree-of-freedom manipulator, there are $2N$ joint stops corresponding to an upper and lower stop for each joint. In Equation (8.4), a sigmoid function is used to describe each joint stop repulsive force. The function $f_i^+(\theta_i)$ is the repulsive force associated with the upper joint $i$ joint stop and the function $f_i^-(\theta_i)$ is the repulsive force associated with the lower joint $i$ joint stop. These functions are themselves sigmoid functions. The sigmoid functions are biased such that when joint $i$ is at its upper stop, the function $f_i^+(\theta_i)$ evaluates to 0.5. The analogous shifting of the sigmoid is done for the lower joint stops. The summation of a joint's upper and lower joint stop repulsive force functions, produces a potential energy well describing where it is most effective to place the joint (i.e., its position) such that a joint stop will not be violated. In this context, the "safest" place to position the joint is at its center of motion. However, by appropriate selection of the sigmoid parameter $T$, a potential well that is essentially flat over the majority of the range of the joint can be easily constructed.

**Path Length Cost Function:** $F_3$

The third term, $F_3$, is a measure of the overall length of the path. It is defined as the summation of the squared distances between each successive pair of path via points. The distance measure in this case is the Euclidean norm in joint space (i.e., the vector dot product of two path via point joint angle vectors). The square of the distance is used to insure that the path via points remain spread out along the overall path as the via point locations are incrementally updated.

## 8.6.2 Convergence of the Cost Function Minimization

Convergence is one of the most important issues in the design of path planners for robotic systems. Although it has not been shown analytically, this path planner appears to exhibit exceptionally good convergence properties. These properties are due, in part, to the interpretation and use of the sigmoid parameter $T$ as a convergence control parameter. Note that as $T$ is increased, the range over which a physical constraint tends to influence the location of the path via point increases. Furthermore, the larger $T$ is, the more fuzzy the image of the objects becomes (in terms of the potential functions). The sigmoid thus acts like a low pass filter of the true physical constraints, where $T$ is inversely proportional to the filter bandwidth. This has another interesting effect. As $T$ is increased, objects near each other begin to appear as though they are one big object. As $T$ is decreased, object details become increasingly distinct. Analytically, as $T$ is increased the depth and number of valleys and local minima exhibited by Equation (8.1) tend to decrease significantly.

Based upon these observations, the following conceptual strategy was adopted in order to increase the efficiency and reliability of the path planner. The strategy involves initiating the gradient search with a relatively large value of $T$ and then incrementally reducing $T$, thus taking into account the detailed structure of the environment. The values of $T$ used and the number of search iterations performed for each value of $T$ is currently fixed and is based upon experience and experimentation. Although more complex scheduling techniques can be applied to control these parameters, the current strategy has proven to be very effective and easy to implement.

It should be noted that it is not necessary to let the search continue until the ultimate minimum of the cost function is reached. This is a reflection of the fact that there are usually many "reasonable", and perhaps often similar paths, to get from one place to another without violating the real constraints associated with the task.

The RTC subsystem path planner described in the previous sections has been implemented on a 64 node hypercube computer that is connected to the primary RTC subsystem computer. The inherently parallel form of Equation (8.1) allows the computations to be partitioned onto the various processors. Each hypercube node contains a copy of the world model and computes a subset of the required cost function gradients (i.e., forces) arising from obstacle/object interactions. These forces are propagated along the manipulator body and give rise to equivalent "torques" for the various articulated joints of the robot. This information is combined with the gradient information from the path lengths and joint stops, and is used to modify the path.

## 8.7 Task Primitives

Task level primitives enable operations such as grasping of an object. In doing so, the telerobot utilizes RTC subsystem planning, simulation, and kinematic analysis capabilities together with MCM subsystem's motion primitives. Given the complexity associated with each of the task level primitives and the functional uniqueness of each, only the "grasp" primitive will be described in detail, in this chapter. The general structure and approach taken to implement the other primitives is, however, similar. For illustrative purposes we will consider the specific case in which the task is to grasp an ORU grapple lug with the gripper previously equipped with an appropriate set of fingers (Figure 8.5).

The basic (i.e., nominal) strategy required to perform the grasp consists of the following steps:

- move the end-effector to an approach point near the object that is compatible with the desired grasp point,
- move the end-effector towards the object (i.e., its specified grasp location) until contact is made,
- compliantly level the gripper fingers (i.e., gripper palm) such that the top surface of the grapple lug and the gripper palm are in planar contact, and
- close the gripper fingers, while complying the end-effector, until a desired squeeze force is achieved.

This sequence of steps represents a straightforward approach to performing the desired grasping operation. However, the manner in which each step is performed can vary greatly depending upon the situation at hand. For instance, it depends upon how accurately the location of the grapple lug is known. If, when the approach motion completes, the fingers are too far from being centered over the grapple lug, the grasp will be bound to fail. A solution to this problem is, for example, to let the operator visually check the alignment and if necessary use teleoperation or some other means to adjust the end-effector to a new approach location. Situations such as this must be taken into account when designing and implementing task level primitives. With this introduction, the means by which the RTC subsystem implements the grasp task level primitive will now be described.

In the RTC subsystem, each task primitive has an associated state transition graph that is used to encode all of the possible methods by which the task can be performed. The state transition graph for the "grasp" primitive is illustrated in Figure 8.6.

Each node in the graph represents a different state in which the system can be in, prior to or during the grasping operation. The arcs represent the means to get from one state to another. In the real system they correspond to the name of the software routine that will perform the operations necessary to make a specific state transition (e.g., arm motions, etc.).

The nodes, or states, in the graph are defined by a state vector containing 3 elements. The first element, S1, defines the logical placement of the end-effector (i.e., whether in contact with the object, etc.) and can have one of six values. The second element, S2, of the state vector defines the logical state of the gripper fingers (i.e., correct opening, wrong opening, etc.) and can have one of four values. The third element, S3, of the state vector is a logical description of the state of the grasp itself and can have one of six values. Although the state vector can theoretically describe 96 states, not all 96 are physically realizable. Table 8.1 lists the possible values for state vector elements S1, S2, and S3 as defined by the current implementation of the grasp primitive.

The meaning of the values listed in Table 8.1 are indicated by their logical names. For example, the value "ARM_AT_APPROACH_COARSE" signifies that the manipulator's end-effector is at the desired object approach point and is not yet in contact with the object. The difference between "ARM_AT_APPROACH_COARSE" and "ARM_AT_APPROACH_FINE" is that the latter indicates the relative level of accuracy to which the end-effector has been positioned. As will be discussed later, reaching the state "ARM_AT_APPROACH_FINE" may require the use of machine vision or other techniques to achieve the positioning accuracy. The value "ARM_AT_START" simply refers to an end-effector location that is the mutual complement to the other five location states.

The current grasp primitive state transition graph contains ten arcs, each defined by its initial and terminal node. The ten node pairs and the names of the routines that perform the operations necessary to perform the defined transitions are listed in Table 8.2. Note that although there are ten arcs in the state transition graph they can be performed using only six routines.

The basic set of operations performed by the grasp primitive at its highest level are:

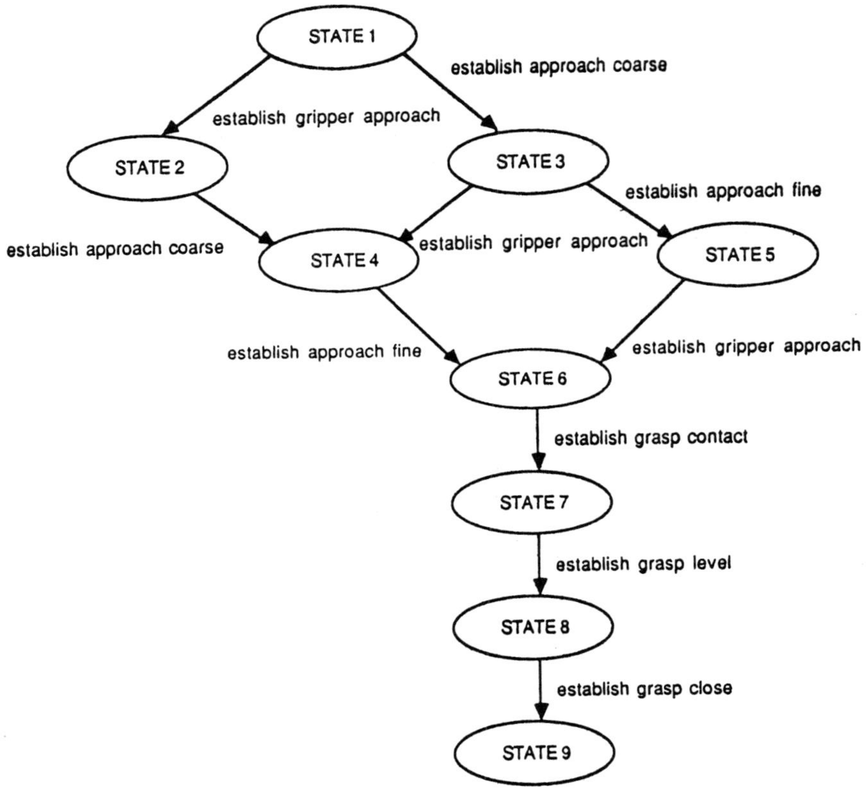


Figure 8.6: State Transition Graph for the "Grasp" Task Level Primitive

| State Vector Element | | |
|---|---|---|
| S1 | S2 | S3 |
| ARM_AT_START | GRIPPER_AT_CORRECT_OPEN | OBJECT_UNGRASPED_SECURELY_MATED |
| ARM_AT_APPROACH_COARSE | GRIPPER_AT_WRONG_OPEN | OBJECT_GRASPED_FREE |
| ARM_AT_APPROACH_FINE | GRIPPER_AT_CORRECT_CLOSE | OBJECT_GRASPED_AT_APPROACH_COARSE |
| ARM_AT_TOUCH_JUST | GRIPPER_AT_WRONG_CLOSE | OBJECT_GRASPED_AT_APPROACH_FINE |
| ARM_AT_TOUCH_LEVEL | | OBJECT_GRASPED_AT_TOUCH_MATED |
| ARM_AT_TOUCH_LEVEL_RELAXED | | OBJECT_GRASPED_AT_SECURELY_MATED |

Table 8.1: Possible State Vector Element Values for Grasp Primitive State Transition Graph

| Initial State | Final State | Routine Name |
|---|---|---|
| ARM_AT_START<br>GRIPPER_AT_WRONG_OPEN<br>OBJECT_UNGRASPED_SECURELY_MATED | ARM_AT_START<br>GRIPPER_AT_CORRECT_OPEN<br>OBJECT_UNGRASPED_SECURELY_MATED | establish_gripper_approach |
| ARM_AT_APPROACH_COARSE<br>GRIPPER_AT_WRONG_OPEN<br>OBJECT_UNGRASPED_SECURELY_MATED | ARM_AT_APPROACH_COARSE<br>GRIPPER_AT_CORRECT_OPEN<br>OBJECT_UNGRASPED_SECURELY_MATED | establish_gripper_approach |
| ARM_AT_APPROACH_FINE<br>GRIPPER_AT_WRONG_OPEN<br>OBJECT_UNGRASPED_SECURELY_MATED | ARM_AT_APPROACH_FINE<br>GRIPPER_AT_CORRECT_OPEN<br>OBJECT_UNGRASPED_SECURELY_MATED | establish_gripper_approach |
| ARM_AT_START<br>GRIPPER_AT_WRONG_OPEN<br>OBJECT_UNGRASPED_SECURELY_MATED | ARM_AT_APPROACH_COARSE<br>GRIPPER_AT_WRONG_OPEN<br>OBJECT_UNGRASPED_SECURELY_MATED | establish_gripper_coarse |
| ARM_AT_START<br>GRIPPER_AT_CORRECT_OPEN<br>OBJECT_UNGRASPED_SECURELY_MATED | ARM_AT_APPROACH_COARSE<br>GRIPPER_AT_CORRECT_OPEN<br>OBJECT_UNGRASPED_SECURELY_MATED | establish_gripper_coarse |
| ARM_AT_APPROACH_COARSE<br>GRIPPER_AT_WRONG_OPEN<br>OBJECT_UNGRASPED_SECURELY_MATED | ARM_AT_APPROACH_FINE<br>GRIPPER_AT_WRONG_OPEN<br>OBJECT_UNGRASPED_SECURELY_MATED | establish_gripper_fine |
| ARM_AT_APPROACH_COARSE<br>GRIPPER_AT_CORRECT_OPEN<br>OBJECT_UNGRASPED_SECURELY_MATED | ARM_AT_APPROACH_FINE<br>GRIPPER_AT_CORRECT_OPEN<br>OBJECT_UNGRASPED_SECURELY_MATED | establish_gripper_fine |
| ARM_AT_APPROACH_FINE<br>GRIPPER_AT_CORRECT_OPEN<br>OBJECT_UNGRASPED_SECURELY_MATED | ARM_AT_TOUCH_JUST<br>GRIPPER_AT_WRONG_CLOSE<br>OBJECT_UNGRASPED_SECURELY_MATED | establish_grasp_contact |
| ARM_AT_TOUCH_JUST<br>GRIPPER_AT_WRONG_CLOSE<br>OBJECT_UNGRASPED_SECURELY_MATED | ARM_AT_TOUCH_LEVEL<br>GRIPPER_AT_WRONG_CLOSE<br>OBJECT_UNGRASPED_SECURELY_MATED | establish_grasp_level |
| ARM_AT_TOUCH_LEVEL<br>GRIPPER_AT_WRONG_CLOSE<br>OBJECT_UNGRASPED_SECURELY_MATED | ARM_AT_TOUCH_LEVEL<br>GRIPPER_AT_CORRECT_CLOSE<br>OBJECT_UNGRASPED_SECURELY_MATED | establish_grasp_close |

Table 8.2: State Transitions and State Transition Routine Names

**Step 1:** Identifies which arm is to be used as defined in GIVENS [1]. System checks to see which arm is specified since the CAMERA arm is incapable of performing manipulation tasks.

**Step 2:** Allocates the arm by locking out other commands that attempt to make use of it. Scratch pad copies of the world model database are also assigned to support subsequent motion and trajectory planning functions.

**Step 3:** Establishes which object is to be grasped, the reference coordinate frame in which all relative motions are specified, and the end-effector coordinate frame matching method (i.e., should the physical end-effector coordinate frame be aligned with specified locations or should it point in the opposite direction?).

**Step 4:** Consults the world model database to determine appropriate approach point and grasp point. The manner in which the list of possible approach points and grasp points is pruned is determined by a set of control variables stored in GIVENS.

**Step 5:** Sets the goal state to [ARM_AT_TOUCH_LEVEL, GRIPPER_AT_CORRECT_CLOSE, OBJECT_GRASPED_SECURELY_MATED].

**Step 6:** Establishes which state the system is currently in, thus allowing the grasp procedure to enter the state transition graph at the appropriate point. (The ability to enter the state transition graph at any point is necessitated by the fact that the operator may have already performed parts of the grasping operation using teleoperation.)

**Step 7:** Compares the current state and goal state. If equal, the algorithm jumps to **Step 9**, otherwise it continues on to **Step 8**.

**Step 8:** The state transition graph is searched to determine whether or not a path leading from the current state to the goal state exists. If a path is found, the routine associated with the first transition in the path is executed. When the routine completes, the algorithm loops back to **Step 7**. If a path cannot be found an error is triggered, and the operator is notified of the failure, allowing him/her the opportunity to formulate and implement a recovery strategy (e.g., using teleoperation).

**Step 9:** All temporary databases, scratch pads, etc., are removed and all state variables updated appropriately.

The six routines that affect the transitions from one state to another in the state transition graph make direct use of the RTC kinematic simulator, path

[1] GIVENS is a data structure that holds RTC subsystem input parameters that can be set either directly by the operator or by commands sent from a higher level subsystem.

planner, collision detector, and world model database, and are responsible for issuing commands to the MCM subsystem. Each of the six routines associated with the grasp primitive are described in the following subsections.

### 8.7.1 Establish_Gripper_Approach Routine

The function of the Establish_gripper_approach routine is to open the gripper fingers to a separation that allows the subsequent grasp to be performed. The separation used constitutes a trade-off between that which would cause the end-effector to subsequently miss the object and that which would cause the fingers to interfere with other objects in the vicinity.

After simulating the finger motion and ensuring that it is collision-free, the routine formats and sends a command to the MCM subsystem requesting that it execute the Gripper Action primitive using specified parameters. Since the manipulator is not yet in contact with the environment, a non-compliant grasping mode is selected. Upon completion of the Gripper Action primitive, the routine updates the database dependent upon the report returned by the MCM subsystem and restores assorted internal motion parameters (i.e., flags) to their former values.

### 8.7.2 Establish_Approach_Coarse Routine

The "Establish_approach_coarse" routine plans and executes a motion of the specified arm from its current location to an approach point near the object to be grasped. This routine operates purely based upon the information in the world model database. Since the routine will not call upon the vision system or the operator to verify and or correct for errors in the database, it is likely that the achieved end-effector location will not coincide precisely with the desired location. Thus, the achieved position of the end-effector, measured relative to the object, may well vary from the desired separation by an appreciable amount (i.e., an amount that may require further refinement if proper contact from grasping is to be made).

The routine begins by setting up various internal variables, establishing reference coordinates, and computational modes based upon global settings previously set by the user or higher level subsystem. It also takes a complete "snapshot" of all internal variables that must either be reset upon completion of the routine or used to recover from an internal error.

An object approach point is then selected from the list of approach points associated with the object in the world model. If desired, the list may be pruned to ensure that the nominal motions from each of the approach points to the corresponding grasp points is kinematically feasible and collision-free. A heuristic is used to select an approach point for the path generation, namely the approach point that has the minimum Cartesian rotational deviation between it and the current end-effector location. The path planner, described in Section 8.6, is then invoked to plan a path to the selected approach point. The RTC subsystem's kinematic simulator and collision detector are called in conjunction with the path planner to determine whether or not the planned path is indeed feasible. Internally, the path planner, kinematic simulator, and collision detector are used iteratively to search for a feasible path using other approach/grasp points, if needed. The extent of the search is limited by a number of user specified control variables, normally set prior to running the system.

If a feasible path is found, the routine assembles all of the parameters that must be bound to the input parameters of the MCM subsystem Guarded Motion primitive. Once bound and formatted, the Guarded Motion primitive command is sent to the MCM subsystem and the approach motion is executed. Execution is monitored by the MCM subsystem which generates reports and sends them back to the RTC subsystem. On completion, the routine restores all of the internal motion flags to their former values (i.e., the complement to the snapshot procedure performed initially). The handling of execution errors and process planning errors is performed by separate error handling routines and is beyond the scope of this chapter.

### 8.7.3 Establish_Approach_Fine Routine

The routine Establish_approach_fine is designed to move the manipulator's end-effector to the desired approach location, which is specified with respect to the object using the *Local Calibration* technique described earlier in Section 8.4.5. This routine, as opposed to the Establish_approach_coarse routine, ensures that the approach location is actually achieved despite various location uncertainties. Since the application of the relative calibration procedure requires that both objects be located using the same camera view, this routine can only be used after the manipulator's end-effector has been brought into the vicinity of the object. As such, this routine, unlike the previous routine, does not require the use of the RTC path planner. Based upon the measured deviation between the end-effector and object, the system determines how far the end-effector must be moved to position it at the true approach location. In a similar fashion as the Establish-Approach routine, the kinematic simulator and collision detector are invoked for a corrective motion. The MCM subsystem is called if the proposed motion is feasible, and updates to the RTC database are performed upon receiving the reports.

### 8.7.4 Establish_Grasp_Contact Routine

The function of the Establish_Grasp_Contact Routine is to move the manipulator's gripper towards an object until the force/torque sensor indicates that contact has been made. This routine is, conceptually, the planning level complement to the Contact Motion primitive.

The first step in the routine is that of computing a nominal motion, and goal location (i.e., that location where the end-effector will attempt to move). The goal location is selected as a grasp point corresponding to the approach point achieved. The computed differential transformation is then added to the end-effector's current location, along with a fixed-overshoot transform, to provide the nominal goal location for the subsequent move-to-contact motion. The overshoot transformation is added so that the nominal goal location is slightly beyond the object, thus ensuring that contact will be made even if the object's location, as specified in the database, is slightly in error.

As before, various internal motion variables and flags are stored prior to performing a kinematic feasibility simulation along a straight line Cartesian trajectory from its current location to the specified goal location. While the simulator performs many important kinematic checks during this step, the determination of whether or not the motion requires a pose change is especially critical. This is due to the fact that the motion, if executed, will be performed by the Contact Motion primitive using a real-time Cartesian interpolation algorithm that is unable to predict whether or not a pose change will be required. Hence, if a pose change were required, the motion would fail as it approached the corresponding singularity manifold.

Once intermediate path points have been computed and kinematically checked by the simulator, they are passed to the collision detector to determine if undesired collisions will occur. Since the purpose of this routine is to plan and execute a motion to bring the end-effector into contact with the environment, it is important that the collision detector and database be configured properly prior to actually performing this collision analysis. While on the one hand, the collision detector can be expected to detect at least one collision, it shouldn't necessarily prevent the system from approving and executing the motion. A list of allowable collisions is either computed using various heuristics on object/entity information (see Section 8.5.2) or hard-wired, based upon prior experience.

As before, the command is assembled and dispatched to the MCM subsystem. Some of the information necessary to bind the parameters of the motion primitive are obtained by performing a world model query on the object itself, which usually stores a table of suitable gains, etc., needed for the grasp operation. In other cases, the GIVENS structure is accessed to obtain control parameters previously set by the operator or higher level subsystem. These are typically safety features (e.g., force thresholds, singularity thresholds, etc.).

Upon completion of the Contact Motion primitive, the MCM subsystem returns a closing report that the RTC subsystem uses to update its database. The closing report also explicitly indicates what triggered the motion to terminate (e.g., made contact, reached specified target without hitting anything, etc.). If the report indicates that contact was not successfully made, the RTC subsystem triggers an error condition, notifies the operator of the situation, and waits for the operator to intervene. In the future, automated error recovery strategies will be incorporated.

Assuming that contact has been made and that no other internal errors have been triggered, the end-effector should have reached a position at which it is just in contact with the object to be grasped. The next step in the grasp sequence would be to compliantly level the fingers on the grapple lug. However, it is conceivable that the contact configuration is not adequate and must be corrected. To account for this situation, the Establish_Grasp_Contact primitive provides the Operator with the opportunity to make slight adjustments to the location of the end-effector. The Operator can use the hand controllers to direct the system to make small vernier style perturbations to the end-effector's position and orientation.

When the operator (if he/she uses it) completes adjustments to the location of the end-effector, the Teleop primitive terminates and a closing report is again sent to the RTC subsystem. The updating of the database, based upon information in the report, signals the completion of the Establish_Grasp_Contact routine.

### 8.7.5 Establish_Grasp_Level Routine

The function of the Establish_Grasp_Level routine is to move the end-effector compliantly such that it applies a small force on the object along the approach direction (i.e., the gripper Z axis) while simultaneously complying in all directions. The resulting motion, which is completely sensor driven, tends to level the gripper palm so that the surface of the gripper palm and the contacted surface of the object become coincident. The routine and the motion it executes are designed based on the assumption that the gripper has been previously brought into contact with the object.

The Establish_Grasp_Level routine uses the Compliant Motion primitive in the MCM subsystem to actually execute the leveling operation. An interesting aspect of its use, however, lies in the fact that the primitive's "Nominal Motion" in this case is the "Null motion" (i.e., a motion to the current location). In the leveling operation, the only motion to be performed is in response to the forces and torques being felt. By configuring the primitive to apply small forces in the approach direction, it is guaranteed that the end-effector will continue to remain in contact with the object. The fact that leveling does not require a non-zero "Nominal Motion" greatly simplifies the Establish_Grasp_Level routine, since simulation and analysis are not required. It also simplifies the task of specifying primitive input parameters. For example, specification of the goal location is trivial. It is simply the current configuration of the arm as specified by the arm's state vector. Concerns are eliminated with respect to which coordinate frames motions are to be specified.

For illustrative purposes, it is worthwhile to discuss how several of the Compliant Motion primitive input parameters are set. Keep in mind that some of the input parameters are set by the Operator explicitly, via global control variables, whereas others are set internally by the Establish_Grasp_Level routine and pertain only to the upcoming instantiation of the Compliant Motion primitive.

The first of the parameters to be specified are the contact forces, which, as mentioned above, are set such that a force will be applied only along the gripper's Z axis. Forces and torques in all other directions are set to zero. Note that these contact forces and torques are those which will be actively generated by the manipulator. The magnitude of force to be applied is determined by the operator or the higher level subsystem as it is obtained from a set of control variables which are set ahead of time. If the associated control variable has not been explicitly set, the default value of 20.0 Newtons is used.

The next most critical parameters to be set are the Compliant Motion primitive spring constants. The spring constants, in this case, are all set to zero so that the system does not continually try to drive the end-effector back to the Nominal trajectory, in this case the initial location of the end-effector.

There are several parameters that must be specified to indicate the desired compliant behavior. For starters, the six mode flags, indicating which Cartesian degrees-of-freedom to affect active compliance, are all set to 1 (i.e., true). Similarly, the six mode flags, indicating in which Cartesian degrees-of-freedom to apply position (0) or force (1) control, are all set to 1. The coordinate frame in which compliant motions are made, the COMPLY frame, and the coordinate frame in which force and torque sensor readings are interpreted, the SENSE frame, are both specified to be coincident with the gripper's finger coordinate frame. The origin of this coordinate frame is located at the center of the surface of the gripper's palm. Its axes are aligned according to conventional practice. Finally, the force control gains are set externally by the operator or higher level subsystem. Typical values used are 0.00003 mm/Newton for the three translational degrees-of-freedom and 0.02 degrees/Newton-mm in the three rotational degrees-of-freedom.

The last input parameters set are those that specify the conditions upon which the primitive will terminate. Under normal conditions, the primitive is designed to terminate after executing for 4.0 seconds. To guard against off nominal behavior, the primitive's input parameters are also set to terminate the motion if the deviation in the position and/or orientation of the end-effector exceeds specified thresholds. For example, the contact motion performed prior to the Establish_Grasp_Level routine may have contacted the object significantly off to one side. Hence, compliantly leveling the gripper may cause the gripper to "roll" off the object to one side. The input parameters used in this case cause the MCM subsystem to monitor such behavior and take appropriate action prior to damaging anything. Although the operator will usually have visually verified the previous contact configuration, there are potential situations in which he/she cannot directly view the grasp, and hence, automatic monitoring is critical.

Having set the Compliant Motion primitive parameters, the Establish_Grasp_Level routine issues a compliant motion primitive command to the MCM subsystem. Upon completion of the primitive, the MCM subsystem returns a closing report indicating, among other things, the reason for termination. If the motion was terminated for any reason other than time, the RTC subsystem will trigger an error, notify the operator of the situation, and wait for the operator to intervene. If the primitive terminates on the end-time condition, the RTC subsystem updates the database with the measured state of the manipulator. This completes the Establish_Grasp_Level routine.

### 8.7.6 Establish_Grasp_Close Routine

The function of the Establish_Grasp_Close routine is to coordinate the motions involved in compliantly grasping an object. In this context, compliant grasping refers to the task of closing the gripper fingers while simultaneously complying the position and orientation of the end-effector. The latter is required to ensure that 1) binding does not occur, 2) excessive forces and torques are not induced upon the object, and 3) the gripper fingers "seat" properly on the object. This routine uses the Gripper Action primitive to execute the desired gripper motions.

Typically, this routine would be executed only after the gripper has been brought into contact with the object and has been leveled. The routine does, however, contain a number of features that allow it to function relatively well in the absence, or partial absence, of these conditions. These additional capabilities are often required in situations where the operator has used teleoperation to prepare for the actual grasp as opposed to relying upon the completely automated grasp capability. The operator, for example, may not have leveled the gripper onto the object completely.

Like the Establish_Grasp_Level routine, this routine does not need to explicitly simulate and analyze the motions to be performed. The motions, for the most part, will be governed by external events (i.e., forces and torques) that cannot be modeled or predicted. Furthermore, there is no need to check for collisions since, given the context of the situation, the only objects that could collide are those that should collide.

In terms of compliant features, this routine uses essentially the same parameter values and settings as used by the Compliant Motion primitive in the Establish_Grasp_Level routine. The reader is referred to the previous section for details.

The remaining parameters to be set pertain to the gripping action. In particular, the desired grip force and finger speed are set according to values specified either by the operator or a higher level subsystem. The desired finger separation is determined on-line by considering the orientation of the end-effector relative to the object, the geometry of the object, and an overshoot parameter. The overshoot parameter is a relatively small distance that is subtracted from the ideal grasp separation and ensures that the gripper fingers will reach the desired grip force. While the grippers can measure the squeeze force being applied, they are unable to servo to a particular force. Hence, the fingers are driven in position control mode to a separation slightly less than necessary and the force sensor is used to terminate the motion when a particular grip force threshold is reached.

To ensure that the compliant motion of the end-effector can respond appropriately to the forces and torques induced during the grasp, the finger speed is set relatively low. For rigid objects, the grip force is usually specified to be at or near the gripper's maximum rated squeeze force of 40 lbs.

After evaluating and setting the parameters described above, the Establish_Grasp_Close routine issues a Gripper Action primitive command to the MCM subsystem. Upon completing the primitive, the MCM subsystem returns a closing report to the RTC subsystem. From this, the RTC subsystem determines whether or not the primitive completed successfully. If an off nominal result is observed, the RTC subsystem triggers an error, notifies the operator, and waits for the operator to intervene.

Even if the primitive itself terminates properly, the grasp may not have occurred exactly as desired. For example, in grasping geometrically complex objects, like the grapple lug, it is quite possible that the gripper fingers made contact with a face adjacent to the optimal contact face. Such grasps may, from a dynamic standpoint, be unstable and hence must be identified. At present the means for identifying such situations lies in a comparison between the measured finger separation and the desired finger separation. If these are within a specified tolerance, the system assumes that the grasp is successful. If this test fails, the RTC subsystem triggers an error, notifies the operator, and waits for the operator to intervene.

The final operation performed by the Establish_Grasp_Close routine is that of modifying the database spanning tree to account for the fact that the object(s) grasped is (are) now physically attached to the gripper. This is extremely important since all subsequent arm motions must take into account the fact that the object will move with it. The task of modifying the database is non-trivial and will not be discussed in detail in this chapter. The basic principle, however, involves a recursive search through the database to determine if the object grasped is part of a larger entity and, if so, what objects are contained within it. Once this is done, the eldest relative of the grasped object, presumably the prime node of the entity can be disconnected from its current parent object and reattached to the gripper. This operation, although conceptually simple, involves a fair amount of computations since all of the kinematic transformations associated with the grasped object must be updated appropriately. Completion of this task signals the completion of the Establish_Grasp_Close routine.

## 8.8 Conclusions

We have presented a description of the capabilities of the JPL telerobot testbed and identified its utilization for automated assembly. It is hoped that the discussion and the associated references will prove valuable to others in the telerobotics community.

**Acknowledgments**

The research described in this document was performed at the Jet Propulsion Laboratory, California Institute of Technology, under contract with the National Aeronautics and Space Administration. The JPL testbed implementation integrates the contributions of a number of people and wherever possible these contributions are acknowledged by means of appropriate references to their work throughout the chapter.

## 8.9 References

[1] P. Schenker, "NASA research and development for space telerobotics," *IEEE Transactions on Aerospace and Electronic Systems*, vol. 24, pp. 523–534, Sept 1988.

[2] J. Matijevic, W. Zimmerman, and S. Dolinsky, "The NASA/OAST telerobot testbed architecture," in *Proceedings of NASA Conference on Space Telerobotics*, (Pasadena, California), Jan. 31-Feb. 2, 1989.

[3] A. Lokshin, J. Beahan, and J. Balaram, "Intelligent pipelined control architecture for remote robotic applications," in *1990 ASME Annual Winter Meeting*, (Dallas, TX), November 1990.

[4] P. Backes and K. Tso, "UMI: An interactive supervisory and shared control system for telerobotics," in *Proceedings of the 1990 IEEE International Conference on Robotics and Automation*, (Cincinnati, Ohio), pp. 1096–1101, May 13-18, 1990.

[5] J. Matijevic, "Telerobot project: System/subsystem interface specifications." Jet Propulsion Laboratory. Sec 347. November 1988. (internal document).

[6] S. Hayati, T. Lee, K.Tso, P. Backes, and J. Lloyd, "A testbed for a unified teleoperated-autonomous dual-arm robotic system," in *Proceedings of the 1990 IEEE International Conference on Robotics and Automation*, (Cincinnati, Ohio), pp. 1090–1095, May 13-18, 1990.

[7] A. Bejczy and J. K. Salisbury, "Controlling Remote Manipulators through Kinesthetic Coupling," *Computers in Mechanical Engineering*, vol. 2, no. 1, pp. 48–60, 1983.

[8] S. Hayati, K. Tso, and T. Lee, "Dual arm coordination and control," *Journal of Robotics and Autonomous Systems*, vol. 4, no. 5, pp. 333–344, 1989.

[9] K. Kreutz, M. Milman, and A. Lokshin, "Dynamics, coordination, and load balancing of multiple rigid robot arms moving a commonly held rigid object." Engineering Memorandum 347-89-268, Jet Propulsion Laboratory (internal document).

[10] H. V. and R. Paul, "Robot manipulator control under UNIX RCCL," *International Journal of Robotics Research*, vol. 5, pp. 94–111, Winter 1987.

[11] J. Lee, S. Hayati, *et al.*, "Implementation of RCCL, a robot control C library, on a MicroVax II," in *Proceedings of the SPIE Conference on Advances in Intelligent Robotics Systems, Vol. 726*, (Cambridge, MA.), pp. 26–31, Oct. 1986.

[12] J. Lloyd, M. Parker, and R. McClain, "Extending the RCCL programming environment to multiple robots and processors," in *Proceedings of the 1988 IEEE International Conference on Robotics and Automation*, (Philadelphia, PA), pp. 465–469, April 1988.

[13] B. Wilcox, K. Tso, T. Litwin, S. Hayati, and B. Bon, "Autonomous sensor-based dual-arm satellite grappling," in *Proceedings of NASA Conference on Space Telerobotics*, vol. 3, (Pasadena, California), pp. 307–316, Jan. 31-Feb. 2, 1989.

[14] A. Bejczy and Z. Szakaly, "Universal Computer Control System (UCCS) for Space Telerobots," in *Proceedings of the 1987 IEEE International Conference on Robotics and Automation*, (Raleigh, NC), pp. 318–324, Mar. 30-Apr. 3, 1987.

[15] D. Gennery, T. Litwin, B. Wilcox, and B. Bon, "Sensing and perception research for space telerobotics at JPL," in *1987 IEEE International Conference on Robotics and Automation*, (Raleigh, NC), pp. 311–317, March 31-April 3, 1987.

[16] B. K. T. Wilcox, "Real-time model-based vision system for object acquisition and tracking," in *International Conference on Optical and Digital Pattern Recognition*, vol. 754, (Los Angeles, CA), SPIE, January 15, 1987.

[17] J. Balaram *et al.*, "A run-time control architecture for the JPL telerobot," in *Proceedings of Workshop on Space Telerobotics* (G. Rodriguez, ed.), vol. 1, pp. 221–222, July 1987.

[18] J. Balaram and G. Rodriguez, "An ADA Run-time Control Architecture for Telerobots," in *Proc. Intelligent Robots and Computer Vision* (D. Casasent, ed.), vol. 726, (Cambridge, Massachusetts), pp. 462–469, SPIE, Oct 1986.

[19] R. Desai *et al.*, "GARE: Geometric Analysis and Reasoning Engine," in *3rd International Workshop on ICAD Systems*, (Texel, Netherlands), 1989.

[20] R. Doshi, *Task Planning for 'Simple' Assembly.* PhD thesis, Department of Computer Science, University of Minnesota, October 1990.

[21] E. Kan, J. Tower, G. Hunka, and G. VanSant, "The JPL telerobot operator control station: Part I - hardware," in *Proceedings of NASA Conference on Space Telerobotics*, vol. 2, (Pasadena, California), pp. 51–62, Jan. 31-Feb. 2, 1989.

[22] E. Kan, B. Landall, S. Oxenberg, and C. Morimoto, "The JPL telerobot operator control station: Part II - software," in *Proceedings of NASA Conference on Space Telerobotics*, vol. 2, (Pasadena, California), pp. 63–73, Jan. 31-Feb. 2, 1989.

[23] R. Lam *et al.*, "Diagnosing faults in autonomous robot plan execution," Tech. Rep. JPL Publication 88-14, Jet Propulsion Laboratory, California Institute of Technology, Pasadena, California, 1988.

[24] P. Backes, "Generalized compliant motion task description and execution within a complete telerobotic system," in *Proceedings of the 1990 IEEE International Conference on Systems Engineering*, (Pittsburgh, PA), August 1990.

[25] A. Requicha, "Representation for rigid solids: Theory, methods, and system," *ACM Computing Surveys*, vol. 12, pp. 437–464, Dec. 1980.

[26] B. Bon, B. Wilcox, T. Litwin, and D. Gennery, "Operator-coached machine vision for space telerobotics," in *Proceedings of the Symposium on Advances in Intelligent Systems, Conference on Cooperative Intelligent Robotics in Space*, vol. 1387, (Boston, MA), pp. 337–342, SPIE, Nov 1990.

[27] S. Hayati and J. B. Balaram, "Supervisory telerobotics testbed for unstructured environments," *submitted to Journal of Robotic Systems*, 1992. (special issue on Robots in Unstructured and Radioactive Environments).

[28] E. Gilbert, D. Johnson, and S. Keerthi, "A fast procedure for computing the distance between complex objects in three-dimensional space," *IEEE Journal of Robotics and Automation*, pp. 193–203, April 1988.

[29] J. Park and S. Lee, "Connectionist model path planning based on artificial potential fields." unpublished.

[30] J. Balaram, "Probabilistic methods for robot motion determination," in *Proceedings of the Symposium on Advances in Intelligent Robotics System*, (Cambridge, MA), SPIE, Nov 1988.

# Index


Assembly, 60, 302
    by astronauts, 83, 112, 113
    path planning, 155
    space-based, 112
    tests, 80–90
Assembly sequence planning, 4, 73, 96, 117, 143
    AND/OR graph, 149, 150
    face connection, 149
    geometric feasibility, 145, 146
    line connection, 149
    point connection, 147
    space-truss domain, 145
    structural feasibility, 145–147
    task planner, 142
    tetrahedral truss, 104–109

Collision avoidance, 9, 316–318
Convex Hull, 146
Coordination, 28, 31, 315
Coordinators, 28–31

Design for assembly, 115
    robotic assembly, 116, 136
Disassembly, 68, 145
    algorithm DISASSEMBLE, 150
Discrete event dynamic systems
    concurrency, 7
    conflict, 7
    deadlock, 7
    performance analysis, 8
Dispatchers, 28–31

Error recovery, 8, 12, 57, 67, 89
Extra vehicular activity, 40, 111–113

Fiducial marks, 91, 128, 131

Geometric modeling, 157–159, 311, 312
Grasping, 326–332
Grippers, 16, 242, 303
    approach routine, 332, 333

Hierarchy, 28–31, 69–76, 88, 300
    disassembly strategy, 150
    task representation, 142

Integration, 6–8
Intelligent machine, 8, 9, 302
Internal vehicular activity, 40

Laser scanner, 211–223
    calibration, 211, 213–221
    intrinsic and extrinsic parameters, 211
    mathematical model, 212
    pincushion distortion, 213, 219

Manipulator Control
    adaptive, 10

- compliant motion, 308
- contact motion, 308
- dual arm, 2, 11
- flexibility, 10, 138, 308
- force, 10, 13, 308
- force reflection, 297
- gravity compensation, 6
- guarded motion, 306
- gyroscopic forces, 257
- hybrid force control, 13
- impedance force control, 13
- master/slave, 27
- position accommodation, 13, 308
- remote, 11
- self-motion, 27
- sensor-based, 298, 335
- supervised autonomy, 11, 40, 67, 69, 300
- teleoperation, 11, 298

Mobile Robots, 10, 19, 20

Neural Networks, 8

Nodes, 49
- bolted/flex drive, 117, 121, 125, 132
- collet/flex drive, 117, 120, 124, 132
- hammer-head, 122, 125, 132, 136
- test descriptions, 126–131, 135
- threaded collar, 119, 124, 132

Non-structural volume element, 147, 148

Non-structural volume element node, 147, 148

Non-structural volume element strut, 147, 148

Obstacle avoidance, 163

Path planning, 2, 9, 64, 65, 96, 97, 259
- A* algorithm, 171
- assembly path planning, 155
- computational complexity, 177
- potential fields, 156, 159, 160, 169, 318
- singularities, 180

Petri nets, 7, 9, 12, 29–31
- ordinary, 29
- software, 18
- transducers, 29–31

Quaternions, 195

Robotic transporter, 15, 16

Sensing
- crossfire, 242
- force/torque, 2, 17, 66, 69, 91, 137, 242, 247, 303
- infrared, 19
- light beam, 247
- position, 243, 248
- ultrasonic, 19

Shuttle Remote Manipulator System, 271

Simulation, 89, 268–276
- cooperating manipulators, 257, 276–282
- graphics, 99
- manipulator with gyroscopic forces, 257

Spatial operator algebra, 259

Structural volume element, 147, 148

Struts
- identification convention, 60
- insertion, 65
- robot acquisition, 61
- storage, 51–56

Task planning, 2, 9, 142, 300, 305
Telerobotics, 39, 113, 298
Testbeds
  CIRSSE, 2–4, 13–20, 234–238
  Jet Propulsion Laboratory, 298–302
  Johnson Space Center, 113–115, 129
  Langley Research Center, 43–60
Trajectory generator, 23
Truss joint (see nodes)
Truss structures
  assembly, 2, 4, 60, 101–104, 242
  linear truss beam, 97

Vision, 90–95, 126–130, 304
  active techniques, 185
  camera calibration, 186, 187, 235, 313–315
  camera calibration errors, 207–210
  extrinsic camera parameters, 188
  image (point) correspondence problem, 186
  intrinsic camera parameters, 188, 189
  laser pointer, 186
  lighting in space, 5, 6
  midpoint of the common normal, 226
  mobile cameras, 236
  passive techniques, 185, 186
  point estimation methods, 223, 232
  pose, 185, 211
  singular value decomposition, 191
  stereo cameras, 2, 186
  structured lighting, 186, 211, 235
  three-dimensional, 2, 185
  Tsai's calibration method, 187, 188–195
  velocity estimation, 13
  visual servoing, 13

Zero-gravity, 6